TODAY'S TECHNICIAN ™

Classroom Manual for

Basic Automotive Service and Systems

Third Edition

TODAY'S TECHNICIAN™

Classroom Manual for
Basic Automotive Service and Systems
Third Edition

Clifton E. Owen

Griffin Technical College
Griffin, Georgia

THOMSON

DELMAR LEARNING™   Australia Brazil Canada Mexico Singapore Spain United Kingdom United States

THOMSON

DELMAR LEARNING

Today's Technician: Basic Automotive Service and Systems, Third Edition Classroom Manual

Clifton E. Owen

Vice President, Technology and Trades ABU:
David Garza

Director of Learning Solutions:
Sandy Clark

Senior Acquisitions Editor:
David Boelio

Product Manager:
Matthew Thouin

Marketing Director:
Deborah Yarnell

Marketing Coordinator:
Mark Pierro

Director of Production:
Patty Stephan

Production Manager:
Andrew Crouth

Content Project Manager:
Barbara L. Diaz

Content Project Manager:
Cheri Plasse

Editorial Assistant:
Andrea Domkowski

Library of Congress Cataloging-in-Publication Data:
Owen, Clifton E.
  Manual for basic automotive service and systems / Clifton E. Owen. — 3rd ed.
    p. cm.
  Includes index.
  ISBN 1-4180-2128-8
  1. Automobiles—Maintenance and repair.
I. Title. II. Title: Basic automotive service and systems.
  TL152.095 2006
  629.28'72—dc22
                                    2006023157

Card Number:
ISBN: 978-1-4180-2128-3
ISBN: 1-4180-2128-8

NOTICE TO THE READER

Publisher does not warrant or guarantee any of the products described herein or perform any independent analysis in connection with any of the product information contained herein. Publisher does not assume, and expressly disclaims, any obligation to obtain and include information other than that provided to it by the manufacturer.

The reader is expressly warned to consider and adopt all safety precautions that might be indicated by the activities herein and to avoid all potential hazards. By following the instructions contained herein, the reader willingly assumes all risks in connection with such instructions.

The publisher makes no representation or warranties of any kind, including but not limited to, the warranties of fitness for particular purpose or merchantability, nor are any such representations implied with respect to the material set forth herein, and the publisher takes no responsibility with respect to such material. The publisher shall not be liable for any special, consequential, or exemplary damages resulting, in whole or part, from the readers' use of, or reliance upon, this material.

CONTENTS

PREFACE

Thanks to the support the *Today's Technician™ Series* has received from those who teach automotive technology, Thomson Delmar Learning is able to live up to its promise to provide regular revisions of the series. We have listened and responded to our critics and our fans and present this new, updated and revised third edition. By revising our series regularly, we can and will respond to changes in the industry, changes in technology, changes in the certification process, and the ever-changing needs of those who teach automotive technology.

The *Today's Technician™ Series* features textbooks that cover all mechanical and electrical systems of automobiles and light trucks (the Heavy-Duty Trucks portion of the series does the same for heavy-duty vehicles). The individual titles correspond to the main areas of ASE (National Institute for Automotive Service Excellence) certification. Additional titles include remedial skills and theories common to all of the certification areas and advanced or specific subject areas that reflect the latest technological trends. Each text is divided into two volumes: a Classroom Manual and a Shop Manual.

Unlike yesterday's mechanic, the technician of today and the future must know the underlying theory of all automotive systems and be able to service and maintain those systems. Dividing the material into two volumes provides the reader with the information needed to begin a successful career as an automotive technician without interrupting the learning process by mixing cognitive and performance learning objectives into one volume.

The design of the *Today's Technician™ Series* is based on features that are known to promote improved student learning. The design was further enhanced by a careful study of survey results, in which the respondents were asked to evaluate particular features. Some of these features can be found in other textbooks, while others are unique to this series.

Each Classroom Manual contains the principles of operation for each system and subsystem. The Classroom Manual also contains discussions on design variations of key components used by the different vehicle manufacturers. This volume is organized to build upon basic facts and theories. Its primary objective is to allow the reader to gain an understanding of how each system and subsystem operates. This understanding is necessary to diagnose the complex automobiles of today and tomorrow. Although the basics contained in the Classroom Manual provide the knowledge needed for diagnostics, diagnostic procedures appear only in the Shop Manual. An understanding of the basics is also a requirement for competence in the skill areas covered in the Shop Manual.

A spiral-bound Shop Manual covers the "how-tos." This volume includes step-by-step instructions for diagnostic and repair procedures. Photo Sequences are used to illustrate some of the common service procedures. Other common procedures are listed and accompanied with line drawings and photos that allow the reader to visualize and conceptualize the finest details of the procedure. This volume also contains the reasons for performing the procedures, and notes when particular services are appropriate.

The two volumes are designed to be used together and are arranged in corresponding chapters. Not only are the chapters in the volumes linked together, the contents of the chapters are also linked. This linking of content is evidenced by marginal callouts that refer the reader to the chapter and page on which the same topic is addressed in the other volume. This feature is valuable to instructors. Without this feature, users of other two-volume textbooks must search the index or table of contents to locate supporting information in the other volume. This is not only cumbersome but creates additional work for an instructor when planning the presentation of material and when making reading assignments. Page references show students exactly where to look for supporting information.

Both volumes contain clear and thoughtfully selected illustrations, many of which are original drawings or photos specially prepared for inclusion in this series. This means that the art is a vital part of each textbook and not merely inserted to increase the numbers of illustrations.

The page layout is designed to include information that would otherwise break up the flow of information presented to the reader. The main body of the text includes all of the "need-to-know" information and illustrations. In the wide side margins of each page are many of the special features of the series. These include items that are truly "nice-to-know" information such as: simple examples of concepts just introduced in the text, explanations or definitions of terms that will not be defined in the glossary, examples of common trade jargon used to describe a part or operation, and exceptions to the norm explained in the text. This type of information is placed in the margin, out of the normal flow of information. Many textbooks insert this type of information in the main body of the text; this tends to interrupt the thought process and cannot be pedagogically justified. By placing this information off to the side of the main text, the reader can select when to refer to it.

Highlights of this Edition—Classroom Manual

The entire text was updated with information on new tools, equipment, and vehicle systems in use today and possibly in the near future. It has also been rearranged so that the chapters follow a sequence that is common to most automotive training curricula. The text was updated throughout to include information on alternative vehicle power systems. Some of these new topics include electric and hybrid-power plants and the use of new alternative fuels. The chapter on tools has been expanded to offer photos of tools instead of line art. The discussion on precision measuring now includes a more detailed explanation of how to use metric measuring tools to match the industry's trend toward the use of metric measurements in vehicle specifications. Included in the chapter is a section on engine horsepower and torque and their relationship to each other and vehicle operation.

Each of the succeeding chapters provides more updated information on specific vehicle systems, starting with the engine. The engine chapter includes updated information on computers, sensors, actuators, and emission controls, while the driveline chapter is expanded to cover the operation of constant variable transmissions. The last chapter includes updated information on auxiliary systems and adjustable brake pedals. Within this chapter is a more detailed explanation, supported by updated figures on climate control systems.

Highlights of this Edition—Shop Manual

Along with the Classroom Manual, the Shop Manual was revised and updated in both text and figures to emphasize the changing technology in tools, equipment, and procedures. The chapters were rearranged to match the Classroom Manual's layout and more closely match a typical automotive school's sequence of instruction. Included in the first several chapters is more information on safety and informational sources within Canadian and United States environmental and safety agencies. Included in the safety chapter is an overview of emergency first aid dos and don'ts. Tools usage is covered in more detail, with an emphasis on tool and equipment inspection and proper use. Service information is discussed to aid in the collection of diagnostic and repair procedures. Service repair orders and the use of the vehicle identification number speak to the importance each plays in vehicle service.

Precision measurements are more closely examined to explain how and when a precise measurement should be performed. Less precise but equally necessary measures are also discussed to show why and when they are needed and how they affect the diagnostic procedure.

The discussion of visual inspection addresses using the five human senses to begin the diagnosis and theories of operation to find the root cause of a vehicle problem. Updated explanations of all major systems of a typical vehicle are included throughout the text. New and revised photo sequences and figures illustrate and support the written text. Job Sheets have been revised or replaced with new ones to promote a better understanding of the written information and the application of that information. The last chapter, Auxiliary System Service, guides the reader through procedures from working around air bags to testing relays. This chapter provides an excellent means to close the text by introducing the testing and diagnosing of electrical systems, one of the largest systems on a vehicle.

CLASSROOM MANUAL

Cognitive Objectives

These objectives define the contents of the chapter and define what the student should have learned upon completion of the chapter. *Each topic is divided into small units to promote easier understanding and learning.*

Marginal Notes

New terms are pulled out and defined. Common trade jargon also appears in the margin and gives some of the common terms used for components. This allows the reader to speak and understand the language of the trade, especially when conversing with an experienced technician.

A Bit of History

This feature gives the student a sense of the evolution of the automobile. This feature not only contains nice-to-know information, but also should spark some interest in the subject matter.

An **automatic transmission** changes forward gears by using valves and hydraulic pressure. A **manual transmission** requires the driver to select a gear using a shift lever.

33

Cross-References to the Shop Manual

Reference to the appropriate page in the Shop Manual is given whenever necessary. Although the chapters of the two manuals are synchronized, material covered in other chapters of the Shop Manual may be fundamental to the topic discussed in the Classroom Manual.

Shop Manual
page 31

Author's Note

This feature includes simple explanations, stories, or examples of complex topics. These are included to help students understand difficult concepts.

Summaries

Each chapter concludes with summary statements that contain the important topics of the chapter. These are designed to help the reader review the contents.

Terms to Know List

A list of new terms appears next to the Summary. Definitions for these terms can be found in the Glossary at the end of the manual.

---

of a well-trained, motivated technician will likely increase. This further emphasizes the technician's golden rule: Fix it right the first time.

Summary

❏ The electrical system has become the single most important system regarding vehicle operation.

❏ Starting and charging systems provide the means to start the engine and power the electrical system.

❏ The power system is the component that changes heat from burning or reactive chemicals to mechanical energy.

❏ Power systems are constantly in a state of change regarding fuels and operations.

❏ Alternative fuels may provide a means to reduce the use of hydrocarbon fuels while lowering harmful emissions.

❏ The powertrain transmits the power-system output to the drive wheels.

❏ The transmission or transaxle provides a means to harness the engine's output to the load based on vehicle load and environment.

❏ Steering allows the driver to direct the vehicle in the desired direction.

❏ Suspension helps protect vehicle components and aids in a better ride.

❏ A brake system provides the driver a way to stop and control the vehicle.

❏ The frame supports the body and all of its mounted components.

❏ The body protects the passengers and vehicle components from outside hazards.

❏ The climate control system, though not needed to operate the vehicle, reduces driver and passenger fatigue and increases alertness.

❏ Accessories include many features or subsystems that assist the driver or passenger and provide a better interior environment.

Terms to Know
Automatic (transmission)
Biodiesel
Combustion chamber
Corporate Average Fuel Economy (CAFE)
Fuel cell
Hybrid power
Manual (transmission)
Powertrain Control Module (PCM)
Refrigerant
Thermodynamics
Transmission Control Module (TCM)

Review Questions

Short Answer Essay

1. Discuss how the vehicle electrical system has been expanded in use over the last twenty years.

2. Explain how the starter circuit turns the crankshaft during engine startup.

3. Discuss the use of alternative fuels in today's and future vehicles.

4. Explain the purpose of the powertrain.

5. Explain the general usage of a computer or computers in today's vehicles.

6. List five of the major computer modules used on light vehicles.

7. List and explain the purpose of the power train's maj[...]

8. Explain the primary operational difference(s) between [...] automatic transmission.

Cautions and Warnings

Throughout the text, warnings are given to alert the reader to potentially hazardous materials or unsafe conditions. Cautions are given to advise the student of things that can go wrong if instructions are not followed or if a nonacceptable part or tool is used.

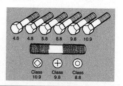

Figure 5-18 Grade marking system for metric bolts, hex head cap screws, and studs.

Inch System		Metric System	
Grade	Identification	Class	Identification
Hex nut grade 5	3 dots	Hex nut property class 9	Arabic 9
Hex nut grade 8	6 dots	Hex nut property class 10	Arabic 10
Increasing dots represent increasing strength.		Can also have blue finish or paint dab on hex flat. Increasing numbers represent increasing strength.	

Figure 5-19 Grade marking system for U.S. (English) and metric nuts.

the part or the fastener and cause the parts to fail. Overtightening causes fasteners to stretch. These no longer have the strength they had before the overtightening.

To avoid these problems, fasteners must be tightened with a torque wrench. There are torque specification charts for each important fastener in the service manual for the car you are working on. Always follow these specifications.

Pitch Gauges

CAUTION: Never use a regular impact wrench to tighten a fastener. The impact wrench can quickly exceed the required torque and damage the fastener or part.

There are two different thread systems in use, each with fine and coarse threads, which can cause a great deal of confusion. Metric threads cannot be used with English threads. Fine threads cannot be used with coarse threads. These different types of threads can make it difficult for the technician to tell one thread from another.

> When threads are damaged by using the wrong sizes or starting a fastener incorrectly with a wrench, we call this "cross threading" or "stripping."

120

Review Questions

Short-answer essays, fill-in-the-blanks, and multiple-choice-type questions follow each chapter. These questions are designed to accurately assess the student's competence in the stated objectives at the beginning of the chapter.

❏ ATRA provides a means for automatic transmission tech[...] recognition.

❏ MACS is designed for air conditioning technicians.

❏ AERA and SAE provide technical and educational suppo[...] technicians.

❏ Manufacturer and vendor training is essential for technicians wishing to keep up to date on various aspects of vehicle repair.

Review Questions

Short Answer Essay

1. Describe the role of the independent shop within the automotive repair industry.

2. Explain how sales and service work together to gain customers.

3. Explain how certifying agencies can assist in promoting customer trust and validating technician technical knowledge.

4. Describe the purpose of NATEF.

5. Discuss how customer education can help an automotive repair business.

6. List the eight automotive areas tested semiannually by ASE.

7. List the steps needed for a technician to be certified as a Master Automotive Technician.

8. Discuss *your* definition of business ethics.

9. Explain why employees share in the legal responsibilities of the employer.

10. List the level of management typically found in an automotive shop.

Fill-in-the-Blanks

1. A dealership is an automotive business operating under _____ _____ in conjunction with an automotive manufacturer.

2. If standards or profits are not met, the manufacturer may withdraw the _____ _____ of the dealership.

3. NATEF conducts certification of _____ _____.

4. The logo for NATEF is a(n) _____ with the _____ name on it.

5. ASE certified technicians wear a(n) _____ ASE patch.

6. Vehicle manufacturer training programs can be certified by _____.

7. Air conditioning technicians can be certified through _____.

8. ATRA offers certification to _____ _____ technicians.

9. Air conditioning technicians must be _____ certified by either MACS or ASE.

10. A(n) _____ is also known as a supplier.

31

SHOP MANUAL

To stress the importance of safe work habits, the Shop Manual dedicates one full chapter to safety. Other important features of this manual include:

Performance-Based Objectives

These objectives define the contents of the chapter and what the student should have learned upon completion of the chapter.

Author's Note

This feature includes simple explanations, stories, or examples of complex topics. These are included to help students understand difficult concepts.

Photo Sequences

Many procedures are illustrated in detailed Photo Sequences. These photographs show the students what to expect when they perform particular procedures. They also familiarize students with a system or type of equipment that the school may not have.

Basic Tools Lists

Each chapter begins with a list of the basic tools needed to perform the tasks included in the chapter.

Special Tools Lists

Whenever a special tool is required to complete a task, it is listed in the margin next to the procedure.

Photo Sequence 3
Typical Procedure for Connecting and Using an Impact Wrench to Remove Lug Nuts

P3-1 Raise the car on a hoist or jack and jack safety stands.

P3-2 Remove the wheel cover from the wheel.

P3-3 Connect the air impact wrench to the air hose quick-disconnect coupling.

P3-4 Install the correct size impact socket on the wrench drive.

P3-5 Switch the impact wrench to the reverse (counterclockwise) drive direction.

P3-6 Place the wrench and socket over the lug nut and pull the trigger to remove the lug nut.

P3-7 Remove each of the other lug nuts and remove the wheel.

P3-8 Install the wheel back on the car and start each lug nut by hand.

P3-9 Switch the impact wrench to the forward (clockwise) direction.

Service Tips

Whenever a short-cut or special procedure is appropriate, it is described in the text. Generally, these tips are things commonly done by experienced technicians.

Warnings and Cautions

Throughout the text, cautions are given to alert the reader to potentially hazardous materials or unsafe conditions. Warnings are also given to advise the student of things that can go wrong if instructions are not followed or if a nonacceptable part or tool is used.

References to the Classroom Manual

Reference to the appropriate page in the Classroom Manual is given whenever necessary. Although the chapters of the two manuals are synchronized, material covered in other chapters of the Classroom Manual may be fundamental to the topic discussed in the Shop Manual.

Marginal Notes

New terms are pulled out and defined. Common trade jargon also appears in the margins and gives some of the common terms used for components. This allows the reader to speak and understand the language of the trade, especially when conversing with an experienced technician.

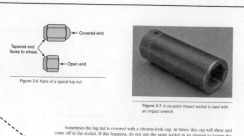

Figure 3-6 Parts of a typical lug nut.

Figure 3-7 A six-point impact socket is used with an impact wrench.

Sometimes the lug nut is covered with a chrome-look cap. At times, this cap will shear and come off in the socket. If this happens, do not use the same socket in an attempt to loosen the nut. It will be slightly too big and may round the nut corners. Use a slightly smaller (13/16 in place of a 7/8 for example) socket to remove the nut. Removing a lug nut with rounded corners sometimes becomes an almost impossible task without damaging the wheel.

SERVICE TIP: You should make match marks on the wheel rim and one of the studs with chalk or grease pencil. Always replace the wheel in the same position in case it has been balanced on the car.

WARNING: Using a loose-fitting impact socket on a wheel lug nut can cause the corners of the hex to be rounded off and not fit a wrench properly.

CAUTION: Always use impact sockets with an impact wrench. Standard sockets can fracture and explode when used on an impact wrench. Always wear eye protection when using an impact wrench.

You are ready to set up and connect the air-operated impact wrench (Figure 3-8). The wrench has a socket drive on the end. Install the impact socket you plan to use on the wrench drive.

The wrench will need to be connected to the shop air source to get its power. The shop air hose is connected...

A **female coupling** houses the locks that hold the male coupling in place after hookup. This coupling also has the release mechanism.

A **male coupling** slides into a mated female unit similar to the manner in which a ratchet's drive stud fits into a socket. The male is fitted with grooves and ridges for locking.

40

Figure 2-3 The accountant acts as the service department's cashier in many shops.

technician to earn her money while the hourly person performs the low-profit quick-turnaround jobs. This results in an efficient use of personnel while providing excellent training for the new tech. In many cases when customers are waiting for the "30-minute guarantee or money back lube job," every technician in the shop is available to perform such jobs.

The accountant may also act as the cashier for the service department (Figure 2-3). The customer will pick up his or her vehicle keys and pay the repair cost at the cashier's window. However, the accountant does much more. The information on the customer, the vehicle, the repair accomplished, and so on is usually collected and entered into a computerized database. The information is used to remind customers of scheduled maintenance and may be used to resolve a customer's complaint. In larger shops, there may be several individuals working in a service department's accounting office.

Parts

The parts manager usually reports to the service manager. The parts department is responsible for ensuring that sufficient routine maintenance items are present (Figure 2-4). Also, parts personnel must order, deliver, pick up, and enter parts on the repair order. The department operates a pickup window for walk-in customers. A running inventory is kept at all times to prevent shortages of routine items. In addition, dealerships are linked by telephone and satellite communications to other dealerships. In this manner, it may be possible to get an out-of-stock part from a dealer on the other side of town instead of waiting several days to get one from the manufacturer's warehouse. The parts are charged against the repair order or a separate invoice, which is sent to the accountant for payment collection.

Classroom Manual page 25

Classroom Manual page 27

...Pay

...le repairs. ...shops will ...cost over

17

The fire will probably be a Class B type, which involves flammables such as fuel, oil, and solvents. However, you must understand the type of fire you are combating and whether or not the fire extinguishers in your shop are operable.

The fire extinguisher normally found in an automotive shop is either a Class B or a multipurpose extinguisher that can be used on different types of fires. Look for a "B" enclosed in a square on the fire extinguisher (Figure 1-9). It may have more than one symbol indicating that it is a multipurpose extinguisher. Obviously, inspection of a fire extinguisher cannot be done as you try to stop the fire. Fire extinguishers must be inspected periodically, preferably by the same person. There are usually several fire extinguisher points throughout the shop. Look at the **tag** and gauge of each to make sure the extinguisher is charged and ready to work. The safety clip or wire must be in place and there must be no damage to the fire extinguisher. The inspector initials and dates the tag if the extinguisher is ready for use. Study the different types of extinguishers available, learn their locations, and the proper use of each type during the inspection.

The use of a Class B fire extinguisher is discussed because this is the most common type of fire in an automotive shop. Class B fires must have their oxygen supply cut off before they go out. Water or other spreading material is not applicable on Class B fires because it spreads the fuel. When a fire is discovered, sound the alarm and locate a Class B fire extinguisher. Position yourself about 8 feet from the base of the fire, **upwind** if possible. Remove the safely clip, aim the nozzle at the base of the fire, and pull the trigger. Continue using the extinguisher until the fire is out or the extinguisher is empty. If necessary, spray the fire area to cool any hot spots that may re-ignite the fire. If you or any of your co-workers feel the fire is too large for your shop's fire extinguishers, leave the area and notify the local fire department.

Photo Sequence 1 explains the use of a dry chemical fire extinguisher.

First Aid

This section is not intended to make the reader a certified medical first responder. First aid is defined as actions to protect the victim and reduce the injury until trained medical personnel are on the scene. Remember that we are not discussing extreme cases, but common and usually non-life-

The extinguisher's **tag** is usually a small card tied directly to the extinguisher. The inspector may enter his or her initials, date, and status of the extinguisher.

Upwind means the wind is blowing on the person's back. If the wind is in the firefighter's face (downwind), the flame and the firefighting material can be blown toward the fighter.

Figure 1-9 The fire extinguisher highlighted is a multipurpose fire extinguisher for use against Class A, B, and C fires.

10

Case Studies

Case Studies concentrate on the ability to properly diagnose the systems. Each chapter ends with a Case Study in which a vehicle has a problem, and the logic used by a technician to solve the problem is explained.

ASE-Style Review Questions

Each chapter contains ASE-Style Review Questions that reflect the performance objectives listed at the beginning of the chapter. These questions can be used to review the chapter as well as to prepare for the ASE certification exam.

Job Sheets

Located at the end of each chapter, the Job Sheets provide a format for students to perform procedures covered in the chapter. A reference to the ASE Task addressed by the procedure is referenced on the Job Sheet.

ASE-Style Challenge Questions

Each technical chapter ends with five ASE challenge questions. These are not more review questions, rather they test the students' ability to apply general knowledge to the contents of the chapter.

Terms to Know List

Terms in this list can be found in the Glossary at the end of the manual.

Diagnostic Charts

Some chapters include detailed diagnostic charts that list common problems and most probable causes. They also list a page reference in the Classroom Manual for better understanding of the system's operation and a page reference in the Shop Manual for details on the procedure necessary for correcting the problem.

CASE STUDY

Older technicians have been heard to say "I have an educated arm, I do not need a torque wrench." They think they can feel the correct tightness for the fasteners they are tightening and do not need a torque wrench.

A group of technicians got a chance to prove this at a recent automotive tools trade show. A torque wrench manufacturer had an engine on a stand with the cylinder head bolts ready for tightening. Technicians were offered a chance to tighten a head bolt to the specified 85 ft.-lbs. without a torque wrench. If they could get to within 10 ft.-lbs. of the specification they would win a prize. Technician after technician took a turn. After each tightening, the torque on the head bolt was measured. The trade show lasted 3 days and not one prize was given. So much for the educated arm theory!

Terms to Know

Condition seizing	Nut splitter	Quick-disconnect coupling
Female coupling	Penetrating oil	Spring-loaded switch
Lug nuts	Pre-loaded	Race
Male coupling		

ASE-Style Review Questions

1. The use of an air impact wrench is being discussed.
   Technician A says the air impact wrench works well for tightening wheel lug nuts.
   Technician B says the air impact wrench should be used primarily for disassembly work.
   Who is correct?
   A. A only   C. Both A and B
   B. B only   D. Neither A nor B

2. The use of an air impact wrench is being discussed.
   Technician A says ... be used on the ...
   Technician B says ...
   Who is correct?
   A. A only
   B. B only

3. The use of an air ...
   Technician A says ...
   reverse switch be...
   Technician B say...
   when using an ai...
   Who is correct?
   A. A only
   B. B only

4. The use of a torque wrench to tighten lug nuts is being discussed.
   Technician A says torque specifications are found in the shop service manual.
   Technician B says the rule is to get the lug nuts as tight as possible.
   Who is correct?
   A. A only   C. Both A and B
   B. B only   D. Neither A nor B

Table 3-1 Guidelines for Service

Raise a vehicle with a lift.

Steps		Classroom Manual	Shop Manual
	1. Position the vehicle between the posts.	xx	xx
	2. Position the lift arms and pads under the vehicle.	xx	xx
	3. Raise the vehicle until its wheels clear the floor. Check for stability.	xx	xx
	4. Lift the vehicle to work height and set the locks.	xx	xx

Table 3-2 Guidelines for Service

Lift a vehicle with a jack.

Steps		Classroom Manual	Shop Manual
	1. Position the vehicle, place in park, and block wheels.	xx	xx
	2. Place jack lift pad under the vehicle's lift point.	xx	xx
	3. Unlock the handle and knob.	xx	xx
	4. Operate the handle to raise the jack.	xx	xx
	5. Place jack stands under the vehicle.	xx	xx
	6. Lower the vehicle to the jack stand.	xx	xx

	Classroom Manual	Shop Manual
...move the lug nuts on a wheel.		
...pport on safety stands.	xx	xx
...heel position.	—	xx
...correct size socket.	xx	xx
...nd.	xx	xx
...pin lug nuts into	xx	xx

Job Sheet 1

Name _____ Date _____

Filling out a Work Order

NATEF Correlation

This job sheet addresses the following NATEF task:
A.1. Complete work order to include customer information, vehicle identifying information, customer concern, related service history, cause, and correction.

Objective

Upon completion and review of this job sheet, you should be able to prepare a service work order based on customer input, vehicle information, and service history.

Tools and Materials Needed

An assigned vehicle or the vehicle of your choice
Service work order or computer-based shop management package
Parts and labor guide

Work Order Source

Describe the system used to complete the work order. If a paper repair order is being used, describe the source.

Procedures

1. Prepare the shop management software for entering a new work order or obtain a blank paper work order.
2. Enter customer information, including name, address, and phone numbers on the work order.
3. Locate and record the vehicle's VIN.
4. Enter the necessary vehicle information, including year, make, model, engine type and size, transmission type, license number, and odometer reading.
5. Does the VIN verify that the information about the vehicle is correct? _____
6. Normally, you would interview the customer to identify his or her concerns. However, to complete this job sheet, assume the only concern is that the valve (cam) cover is leaking oil. This concern should be added to the work order.
7. The history of service to the vehicle can often help diagnose problems as well as indicate possible premature part failure. Gathering this information from the customer can provide some of this information. For this job sheet assume the vehicle has not had a similar problem and was not recently involved in a collision. Service history is further obtained by searching files based on customer name, VIN, and license number. Check the files for any related service work.

---

7. The use of a gear puller is being discussed.
   Technician A says gear pullers can be used to pull parts of a shaft.
   Technician B says gear pullers can be used to pull parts out of a hole.
   Who is correct?
   A. A only   C. Both A and B
   B. B only   D. Neither A nor B

8. The removal of a broken stud is being discussed.
   Technician A says first try to remove the broken stud with a screw extractor.
   Technician B says first try to remove the broken stud with a chisel.
   Who is correct?
   A. A only   C. Both A and B
   B. B only   D. Neither A nor B

9. The removal of a rounded-off nut is being discussed.
   Technician A says it can be removed with a flat chisel.
   Technician B says it can be removed with a nut splitter.
   Who is correct?
   A. A only   C. Both A and B
   B. B only   D. Neither A nor B

10. A nut is rusted onto a stud.
    Technician A says soak it in antiseize compound.
    Technician B says soak it with penetrating oil.
    Who is correct?
    A. A only   C. Both A and B
    B. B only   D. Neither A nor B

ASE-Style Challenge Questions

1. Technician A says the lug nut should be tightened in a circular sequence since the wheel's lug nut area is rounded.
   Technician B says by tightening the lug nuts in a star sequence will cause uneven stress on the rim.
   Who is correct?
   A. A only   C. Both A and B
   B. B only   D. Neither A nor B

2. Technician A says to use chrome sockets when tightening chrome lug nuts because the chrome indicates a match of fastener to wrench.
   Technician B says if the chrome covering comes off the lug nut, then a smaller size socket may be needed.
   Who is correct?
   A. A only   C. Both A and B
   B. B only   D. Neither A nor B

3. Using gear pullers are being discussed.
   Technician A says the puller jaws should not pull on the outer circumference of the gears.
   Technician B says the center bolt of the puller should be aligned with or on the bearing journal.
   Who is correct?
   A. A only   C. Both A and B
   B. B only   D. Neither A nor B

4. Technician A says to remove any broken screw drill completely through the screw so the extractor can get a full-length grip.
   Technician B says a chisel may be used to remove a broken screw?
   Who is correct?
   A. A only   C. Both A and B
   B. B only   D. Neither A nor B

5. Technician A says the use of antiseize may reduce the chance of a fastener sticking or corroding to its mating threads.
   Technician B says a nut splitter can be used to remove a stud from its thread bore.
   Who is correct?
   A. A only   C. Both A and B
   B. B only   D. Neither A nor B

55

REVIEWERS

The author and publisher would like to extend special thanks to the following instructors for reviewing this material:

Elias Alba
Eastfield College
Dallas, TX

Steve Bertram
Palomar College
San Marcos, CA

Ron Chappell
Santa Fe Community College
Gainesville, FL

Tom Fitch
Monroe Community College
Rochester, NY

Robert Klauer
Metro Tech High School
Phoenix, AZ

Eddie Shumate
Forsyth Technical Community College
Winston-Salem, NC

Automotive Systems and Operations

Upon completion and review of this chapter, you should be able to:

❏ List the major systems of a vehicle and explain their purpose.

❏ List and explain the major subsystems of the electrical system.

❏ Discuss the need for and use of computerized systems on a vehicle.

❏ List the different types of power systems for a vehicle and discuss their purpose.

❏ List and describe the purpose of the powertrain and its major components.

❏ Discuss the purpose and list the major components of the steering and suspension system.

❏ List the major components of the brake system and describe their general operation within the system.

❏ Discuss the purpose of the frame and body.

❏ Discuss the general operation of climate control systems and their major components.

Introduction

The systems discussed in this chapter are the basic operating systems of the vehicle. They enable the operator to regulate the power produced by the engine, select a driving gear, and steer and brake the vehicle while experiencing a smooth, comfortable ride. Though the computer system is actually part of the electrical system and is not required to make a vehicle functional, federal standards for fuel mileage and exhaust emissions make it almost impossible for the manufacturer to meet those standards through strict mechanical means. The climate control system also is not needed for vehicle operation, but it is included as standard equipment on almost every vehicle sold today.

Electrical Systems

At one time the only purpose of the electrical system was to start and operate the engine. Soon after the advent of electrical starters and generators, interior and exterior lights were added to the vehicle. Every system on today's vehicle involves the use of electricity and electronics, including tire pressure and adjustment of the steering column and foot pedals to match the physique of the driver. Several general subsystems are found within the overall electrical system.

Starting and Charging

Prior to the use of electric starter motors, the operator had to stand in front of the vehicle or wherever the engine was mounted and hand crank the engine. Assuming the hand crank was not kicked backwards, causing injury, this was a seemly reliable method to turn the crankshaft. It was used on farm tractors as late as the mid-1950s. However, if the spark or fuel was not correct, the hand crank could be kicked backwards, injuring the operator; or, even after many attempts, the engine might still refuse to start. In the late 1920s electrical starters began to appear on vehicles. Electric-current-producing generators were added at the same time as the direct-current storage battery.

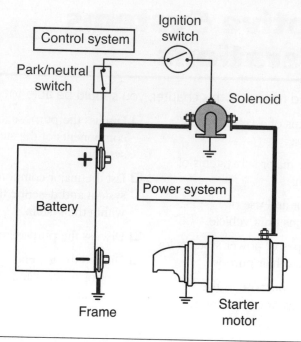

Figure 1-1 A typical starter motor circuit. The control circuit is highlighted.

The starter system is comprised of two subsystems: the power system and the control system (Figure 1-1). The starter motor relay or solenoid is wired directly to the battery's positive terminal. The drive end of the starter motor has a gear that can engage another gear to turn the crankshaft. The control circuit generally consists of the battery, ignition switch (or button), and the relay/ solenoid. When power is supplied through the ignition switch to the relay/solenoid, large contacts within the relay connect the starter motor to the battery. At the same time magnetic forces within the starter motor wiring or windings force the motor's drive gear into contact with the gear that drives the crankshaft (Figure 1-2). Hence the crankshaft moves the pistons through their cycles, drawing air and fuel into the cylinders, igniting the mixture, and causing the engine to run under its own power. By the mid 1970s, a switch called the park/neutral or clutch switch had been installed into the starter motor control circuit to prevent engine startup if the transmission was in any drive gear or if the clutch was engaged (pedal up).

The charging system uses a belt-driven alternating-current (AC) generator to produce the electrical current to operate the vehicle and all of its electrical devices (Figure 1-3). A rectifier with one-way electrical devices located near the generator output converts AC to direct current (DC). The battery is shared between the starting and charging systems and is used to store DC for use when the generator is not operating, such as during engine startup. DC is required because alternating current cannot be stored.

AUTHOR'S NOTE: Up until about the mid-1990s the generator was officially referred to as an alternator due to the former use of the DC generator. During the 1990s, vehicle manufacturers and government agencies adopted definitions that clarified many automotive terms. At that time the term *alternator* was defined as the alternating-current generator and was quickly shortened to *generator*. However, the term *alternator* is still widely accepted in shops, by customers, and by parts vendors.

Lighting, engine controls, and other subsystems receive their electrical power from the alternator based on need. In other words, if the lights are off then power is not used but is available. These electrical subsystems are included within other vehicle systems such as the engine. They will be discussed in later chapters that pertain to particular systems.

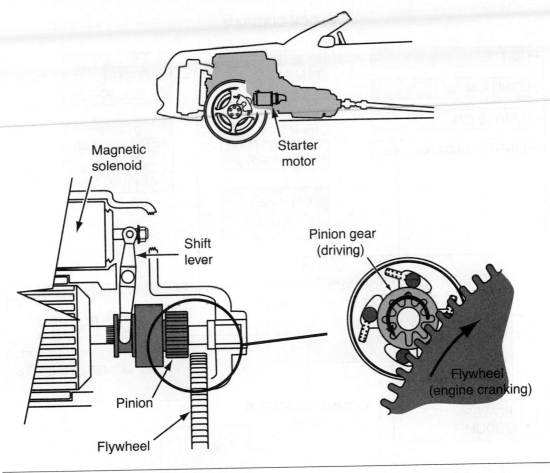

Figure 1-2 Typical location of a starter motor.

Figure 1-3 A typical alternating-current generator.

Computer Systems

The computer system on a vehicle can be thought of as the vehicle's "intranet." Up to forty different computer modules may be wired into one communications bus (Figure 1-4). Data is sent and received by each computer module and acted upon per its individual programming. The

BLOCK DIAGRAM

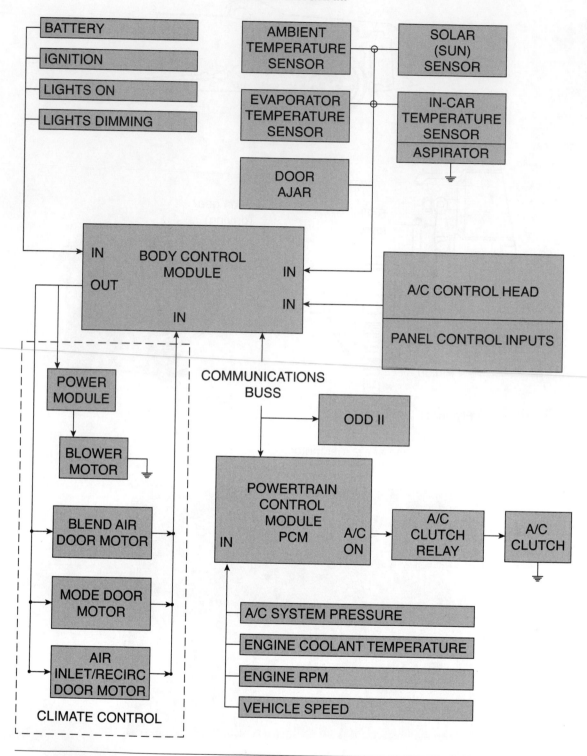

Figure 1-4 Typical computer system layout, mid-1990s. The newest versions are based on this early generation of electronics.

largest single computer module is the **powertrain control module (PCM).** This module not only controls traffic on the vehicle's "intranet," but also controls the operation of various electrical actuators on the engine, sends commands to other modules, requests and processes data from sensors on the engine and other components, and makes some embedded (programmed) decisions based on that data. The PCM may control the transmission directly or provide data or commands to the **transmission control module (TCM),** which controls shift solenoids and processes information (signals) from transmission sensors.

The original reason for vehicle on-board computers was to assist manufacturers' ability to meet federal **Corporate Average Fuel Economy (CAFE)** and federal exhaust emission requirements. This just was not possible with strictly mechanical means like the carburetor and mechanical/electrical ignition systems. Electronics provided the manufacturers with a means to measure exhaust emissions, make necessary and immediate changes in the air–fuel mixture, and fire the spark plugs quickly and more accurately.

As the computer systems improved, it became more efficient to either measure or control other mechanical systems. The next two systems to be computerized, after the engine, were the automatic transmission and the antilock braking system (ABS). Since the mid-1990s, almost every mechanical device on the vehicle is measured, monitored, or controlled by a computer. This has greatly increased fuel mileage, reduced harmful emissions, and provided a smoother ride and happier passengers.

The major modules in the vehicle are the PCM, TCM, body control module (BCM), air-bag control module, and climate control module. Other modules may control antitheft devices, door locks, power seats and windows, suspension and steering, and many passenger comfort features such as radios, DVD players, and on-board navigation.

Power Systems

Most current vehicles use one type of basic power system: internal combustion, reciprocating piston-driven engine fueled either with gasoline or diesel. Since the early 1980s, research has been undertaken with the intent of replacing this type of power system, especially in light vehicles. Research continues to explore the use of more "exotic" means of generating vehicle power.

Internal Combustion Engines

The internal combustion engine uses expanding gases encased in a sealed chamber to apply force to a piston. This is **thermodynamics** (see Chapter 7), the use of heat to produce mechanical motion or energy (Figure 1-5). In a gasoline-fueled engine, air and fuel are mixed prior to entering the **combustion chamber,** either at the intake valve or in the throttle body. The air–fuel mixture is compressed by a piston to about one-eighth or less of its original volume and is ignited by a timed spark (Figure 1-6). In a diesel-fueled engine, the air is first compressed, then the diesel fuel is injected into the chamber under extremely high pressure. The diesel fuel is ignited by the superheated air.

AUTHOR'S NOTE: When gas (vapor) is compressed its temperature rises drastically. On a shop's air compressor the steel line from the compressor to the storage tank reaches temperatures high enough to sear the skin if touched. On the working outlet of the air hose from the tank, the air is cool and could carry a lot of moisture. This is caused by the sudden drop in pressure between the tank and outlet end of the air hose. Moisture in the air condenses under these conditions. See Chapter 7 for more information on pressure and temperature.

Corporate Average Fuel Economy (CAFE) is a federal requirement concerning fuel mileage or usage. It is an average calculated from a manufacturer's vehicles with the worst (lowest mileage) versus the best (highest) mileage. The average must meet the federal standard of miles per gallon: 27.5 mpg on cars, and 20.7 mpg on light trucks.

The **combustion chamber** is the area of the cylinder where the air–fuel mixture is compressed and ignited.

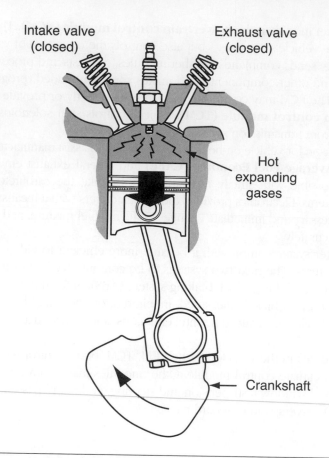

Intake valve
(closed)

Exhaust valve
(closed)

Hot
expanding
gases

Crankshaft

Figure 1-5 The expanding gases caused by combustion force the piston downward. Note that both valves are closed.

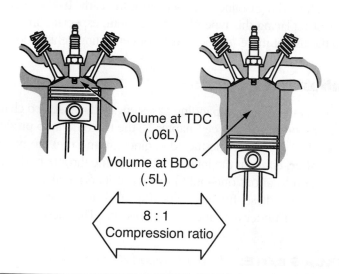

Volume at TDC
(.06L)

Volume at BDC
(.5L)

8 : 1
Compression ratio

Figure 1-6 Compression ratio: 1½ L of air/fuel is compressed into a .06-L combustion chamber, giving a compression ratio of approximately 8:1.

The piston being propelled by the expanding gases is connected to a crankshaft (Figure 1-7). The crankshaft is shaped to change the linear motion (back and forth) to a rotary motion. The crankshaft is connected to a powertrain system that delivers the engine power to the drive wheels. This seems to be a straightforward operation, but in order to be effective certain issues must be addressed throughout the entire operation.

First the air and fuel must be in the right proportion to each other and must be delivered during particular phases of piston travel (Figure 1-8). The best all-around air–fuel mixture is 14.7 parts of air to 1 part of fuel. This is accomplished by nature in a carburetor system, which mixes the air and fuel based on atmospheric pressure and engine vacuum. Atmospheric pressure is usually considered to be 14.7 pounds per square inch and vacuum is any pressure less than atmospheric. This was a fair system for many years; however, it left a lot to be desired when considering the accuracy of the air–fuel mixture. Electronic-controlled fuel injection is now the norm for delivering fuel to the combustion chamber (Figure 1-9) and is much more effective in controlling the air–fuel mixture.

The entry of the air–fuel mixture into the combustion chamber is controlled by timed mechanical valves (Figure 1-10). The exhaust of the spent combustion gases is also controlled by mechanical valves. New engine designs may use electrically controlled and actuated valves, thus eliminating the need for many mechanical parts and providing better engine efficiency. Chapter 8 features a detailed explanation of the configurations of engines and their operations.

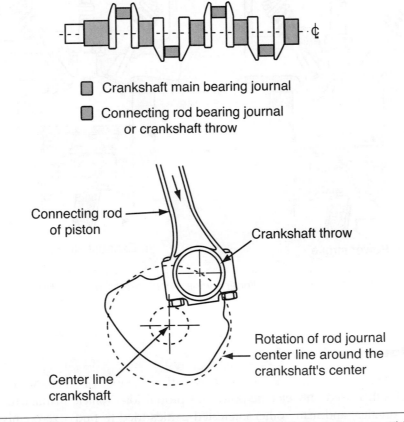

Figure 1-7 As the throw rotates, it turns the crankshaft on its main bearings. This changes the linear motion of the piston into the rotary motion of the crankshaft. Shown is a piston midway through its downward movement.

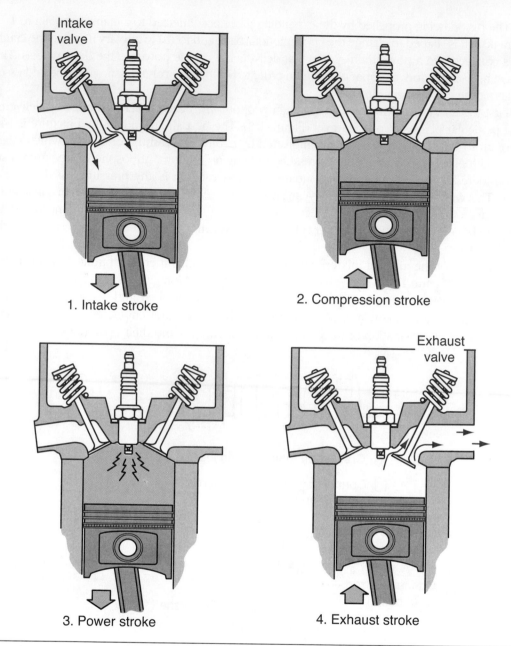

1. Intake stroke

2. Compression stroke

3. Power stroke

4. Exhaust stroke

Figure 1-8 Typical cycle of a four-cylinder engine. Note position of the valves.

Electric-Powered Vehicles

During the late twentieth century and into the twenty-first, millions of dollars were spent on developing vehicles that used only electric power for propulsion. Vehicle manufacturers, governments, electric utilities, and many other interested parties tried to find a zero-emission vehicle power source. The original intent was a vehicle operating on a battery (power pack) with a 300,000-mile life, each recharge lasting about 300 miles, and capable of being quickly recharged from 80 percent discharge 1,000 times. The idea was a throwaway 300,000-mile vehicle with many recyclable components. Many facets of electric power and its incorporated controls were researched, developed, and proven, and now are in use on various vehicle applications today. The single biggest problem was the batteries comprising the power pack. Though the batteries improved greatly, it became clear after years of research that the type of battery needed for this

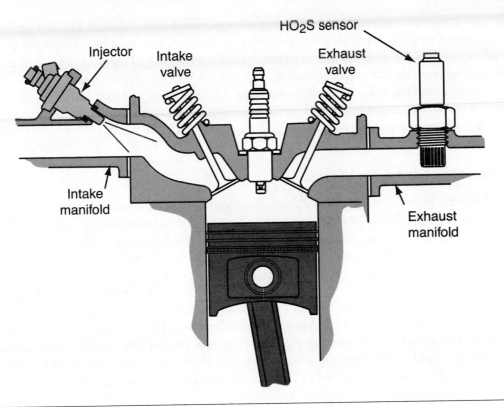

Figure 1-9 The injector opens for a specific period of time measured in microseconds. Pressurized fuel is allowed into the air stream and into the cylinder.

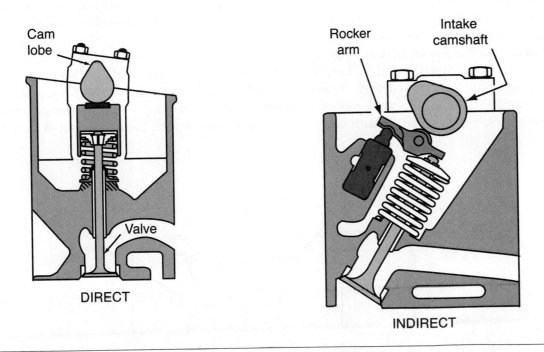

Figure 1-10 The overhead camshaft acts either directly on a valve or moves a cam follower that moves the valve.

application could not be made to withstand 1,000 quick recharges, or even just 100 recharges. Fast recharging tends to decrease the durability of the battery, and most people did not want to wait 10 to 16 hours for their vehicle to recharge after a few hundred miles of driving. The extra electrical load from air conditioning, radios, and other power accessories also shortens the life of the recharge and the miles driven on a recharge. In 2004, most research and development of a true electric-powered vehicle ceased. However, the proven control packets with their various sensors and actuators made the transition to present-day vehicles. The reader may desire to research additional data on "drive-by-wire" vehicles.

Hybrid-Powered Vehicles

Hybrid power, in the case of motor vehicles, is a combination of an internal combustion engine coupled with a large electric-current generator and an electric power pack (Figure 1-11). The internal combustion engine is used to drive the generator, which recharges the power pack and, when commanded, can assist in powering the vehicle. The power pack delivers electric current to one or more electric drive motors, either attached to a power splitter or directly to the drive wheels. On some hybrid-powered vehicles, the engine runs at a constant speed all the time, whereas on others the engine only operates when it is needed to recharge or provide drive power. As of 2004, the most prominent hybrid-powered vehicles offered for sale were the Honda Insight and Toyota Prius. Both use four-cylinder gasoline engines. During 2005, other vehicle manufacturers offered hybrid vehicles for sale while Honda and Toyota expanded and upgraded their hybrid offerings. This trend is expected to increase dramatically over the next several years. Many of the controls developed and perfected during electric vehicle research are now used on hybrid-power vehicles. While hybrid-powered vehicles will continue to grow in popularity, new vehicle power systems are being developed.

Fuel Cell Power

Fuel cell technology is the utilization of electricity created by hydrogen (H) that has been run through a catalyst that strips away the hydrogen electron (Figure 1-12). The electron is directed to an electric device such as a motor and then routed to a second catalyst. The hydrogen ion formed after the loss of its electron goes through the second catalyst where it is combined with an oxygen molecule (O_2) and the freed hydrogen electron to form water (H_2O). This is the only emission by-product of a fuel cell. Although this technology is in the first stages of research and development, it seems to work well. The system, when fully developed, will deliver sufficient power to satisfy the vehicle owner and passenger while reducing the harmful emissions to near

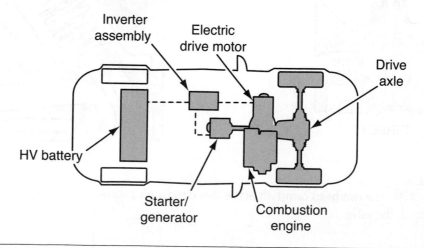

Figure 1-11 Simplified representation of a hybrid powertrain.

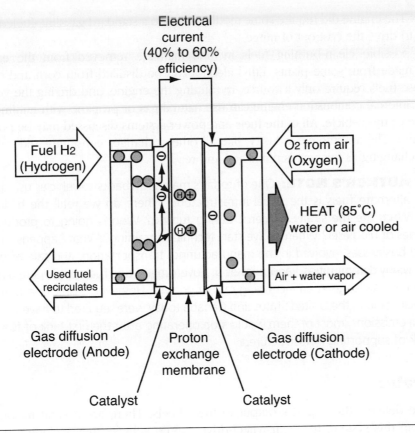

Electrical current (40% to 60% efficiency)

Fuel H₂ (Hydrogen)

O₂ from air (Oxygen)

HEAT (85°C) water or air cooled

Used fuel recirculates

Air + water vapor

Gas diffusion electrode (Anode)

Proton exchange membrane

Gas diffusion electrode (Cathode)

Catalyst

Catalyst

Figure 1-12 General layout of a fuel cell and its operation.

zero. A full market offering of fuel-cell vehicles may be well into the future, but there is a good possibility that the technician graduating from high school in 2006 may be repairing a fuel-cell-powered vehicle before retiring from their automotive career.

Alternative Fuels

Every type of engine requires some type of fuel. As of 2005, hydrocarbon fuels are being blended in different chemical compounds to lower harmful exhaust emissions. A component of retailed hydrocarbon fuels—gasoline and diesel—is sulfur. Current fuel blends have reduced much of the sulfur and other compounds that create harmful emission gases during combustion.

A "new" type of diesel is becoming popular in some areas of the country. **Biodiesel,** generally speaking, is the use of vegetable oils (grease) left over after cooking some foods or treating some biological material. The oils are readily available and only require minor adjustments to the engine's fuel-delivery system. The oils are mixed with diesel to form biodiesel fuel. It is not quite that simple of an operation, of course, but it can be done cheaply and the exhaust emissions are almost completely harmless. Some individual drivers today use the grease or waste from restaurants for biodiesel fuel.

Hydrogen

Some vehicle manufacturers are developing internal combustion engines that use hydrogen for fuel. Hydrogen can be manufactured by removing it from other natural elements, and, when ignited, it burns clean with very few harmful emissions. BMW has produced a race vehicle fueled by hydrogen in a slightly modified internal combustion engine that produces excellent speed and great torque. The engine did require some modifications to a standard gasoline-fueled engine, but not enough to drive the cost out of range.

Other possible clean-burning fuels include methane retrieved from the earth, rotting garbage, or made from some plants. Ethyl alcohol can be distilled from corn and other grains. Some of these fuels require only a source, retrofitting the engine, and driving the vehicle. Vehicles with an internal combustion engine can use natural gas or propane with minimum changes to the engine or the vehicle. All of the fuels and power systems discussed may be possibilities in the near future. The automotive industry, like all others, changes constantly and vehicle power systems are changing as this text is published and read.

> **AUTHOR'S NOTE:** One of today's main drawbacks to vehicles using fuel cells or alternate fuels is the social infrastructure. Where do we refill the hydrogen tank? What are the safety concerns during fueling? Who is going to produce the large quantities of biodiesel? When do we start technician training? What happens to all the people and businesses involved in the manufacturing, transportation, and sale of gasoline and diesel? Many communities, various levels of government, and fuel suppliers/retailers are eliminating this challenge at an increasing rate of growth. However, if only one-tenth of the total vehicles sold in the United States and Canada today were required to have absolutely zero harmful emissions, most of them would stop operating after the first tank of fuel because of the lack of supporting infrastructure.

Powertrains

The driveline delivers the engine's output to the wheels. There are several major, expensive components in this system. In a rear-wheel-drive (RWD) vehicle, they include a transmission, differential, axles, drive shafts, and all of the mountings and housings. In the front-wheel drive (FWD), in the same order, the components are the transaxle, final drive, drive axles, mountings, and the housing. There is no drive shaft in a FWD. Most of the driveline is located in the engine compartment or under the vehicle. This system is covered in detail in Chapter 9, "Powertrains."

An **automatic transmission** changes forward gears by using valves and hydraulic pressure. A **manual transmission** requires the driver to select a gear using a shift lever.

The transmission/transaxle provides a means of selecting a gear size that smoothly marries engine output to vehicle load (Figure 1-13). It may be an **automatic** or a **manual,** where the driver selects forward gears through lever and linkages. The differential/final drive allows the drive wheels to turn at different speeds when the vehicle is turning a corner or curve (Figure 1-14). Drive shafts provide the link between the RWD transmission and differential. Drive axles deliver the power from the transaxle final drive to the wheels on a front-wheel-drive vehicle.

Some vehicles have four- or all-wheel drive. Four-wheel drives allow the driver to use the front wheels to pull the vehicle when the rear wheels lose traction. All-wheel systems drive all four wheels at all times.

One type of automatic transmission that is being installed in a few 2005 models is the constant variable ratio (CVR). It was first introduced by SAAB in production vehicles of the late 1990s. It has no gears per se like most of the current transmissions. Instead it has two opposing pulleys whose sides can be pushed together or moved apart. As the sides of one spread open, the other pulley's sides are pushed together. This, in effect, causes the two pulleys to change their diameters. The drive belt between the two pulleys moves up and down within the pulleys as their sides move. In low range or reverse the drive pulley is closed, making a large pulley. The driven pulley is fully open, creating a small pulley. Torque is now sufficient to accelerate the vehicle. As vehicle speed increases, the drive pulley sides open (diameter gets smaller) and the driven pulley sides close (diameter gets larger), creating more speed and less torque. The applied drive ratio can be any ratio between the two extremes. See Chapter 9 for more information on this and other powertrain components.

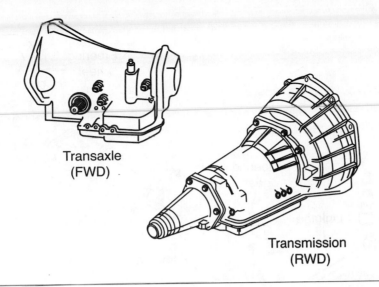

Transaxle
(FWD)

Transmission
(RWD)

Figure 1-13 Typical transaxle (FWD) at top and a transmission (RWD) at bottom.

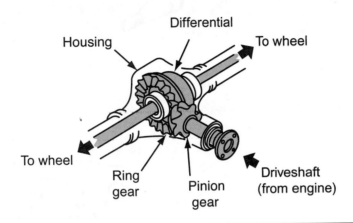

Differential

Housing

To wheel

To wheel

Ring
gear

Pinion
gear

Driveshaft
(from engine)

Figure 1-14 Power flow through a RWD differential.

Steering and Suspension Systems

Steering, as the name implies, allows the driver to steer the vehicle in the direction desired. The steering system provides a system of linkages that allows the driver to turn the front wheels (Figure 1-15). The steering wheel is connected through a shaft to the steering gear unit. The gear multiplies the input force to make steering easier.

Most systems have power-assisted steering. Valves in the gear unit direct pressurized, hydraulic fluid against a piston. This reduces driver effort. The major components are the steering column with the steering wheel, linkage, steering knuckle, and gearing. Newer automobiles have rack-and-pinion systems where the gear unit has been replaced (Figure 1-16). The rack system takes less room and connects the steering gear assembly directly to the wheels.

The suspension supports the vehicle and helps provide a smoother ride. Major components include springs and shock absorbers. Generally, each wheel has some type of spring and shock absorber. Any vehicle component mounted above the springs is supported by the springs and is referred to as sprung weight. When a wheel strikes a bump or pothole, the springs allow the wheel assembly to move away from the bump, thereby preventing some of the impact being delivered

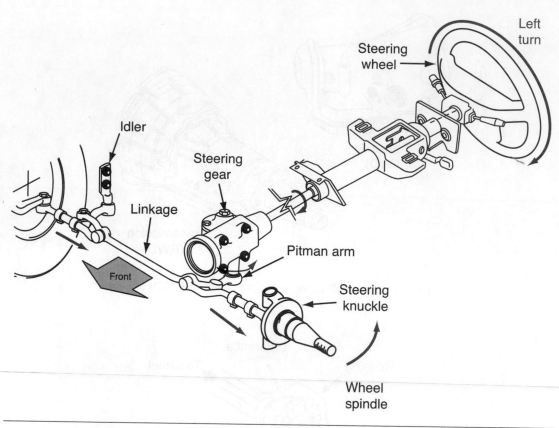

Figure 1-15 Movement of steering linkage during a left turn.

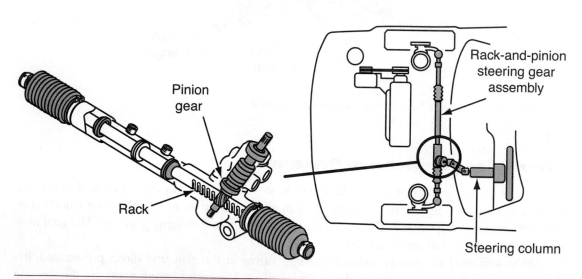

Figure 1-16 A rack-and-pinion assembly provides a direct link between the steering wheel and wheels through the pinion gear and rack.

to the passenger compartment. Springs have a tendency to keep bounding and rebounding even after the bump has been passed. Shock absorbers are designed to slow and stop the movement of the spring (Figure 1-17). They also help prevent some of the impact being transmitted to the passengers. Shock absorbers do not support the weight of the vehicle. Suspension and steering are covered in Chapter 10, "Suspension and Steering."

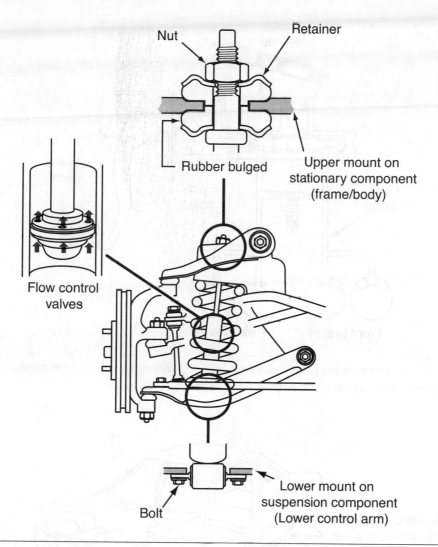

Nut

Retainer

Rubber bulged

Upper mount on stationary component (frame/body)

Flow control valves

Lower mount on suspension component (Lower control arm)

Bolt

Figure 1-17 The control valves restrict the amount of fluid flow from the top chamber to the bottom chamber. This slows piston movement, which helps control spring movement and absorbs some of the road shock.

Brake Systems

There has to be a means for a controlled stop of the vehicle once it is moving. The brake system connects the driver to the wheel through linkage and hydraulic circuits (Figure 1-18). The brake system on a vehicle is hydraulically activated by a mechanical linkage in the passenger compartment. The applied forced is transmitted via hydraulics to the brakes at the front and rear axles. Mechanical devices apply the force to the braking components at the wheels.

Many new systems use an antilock system to prevent brake lockup during emergency stops. Antilock systems rely upon electronic devices and computers to control braking. Brakes are discussed in detail in Chapter 11, "Brakes."

Frame and Body

Much as bones support the body, the frame provides support for all components of the vehicle. There are various frame designs, but newer systems use a *unibody design* where the body itself provides some of the support (Figure 1-19). This makes the vehicle lighter and more fuel efficient.

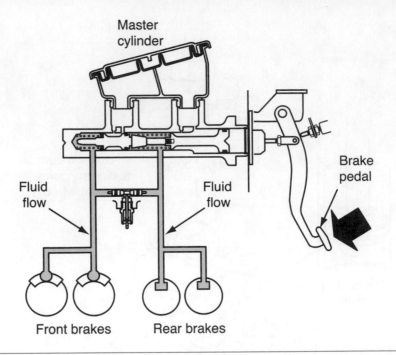

Figure 1-18 Force on the pedal moves the master cylinders forward, forcing pressurized fluid to generate braking force at the wheels on this split system.

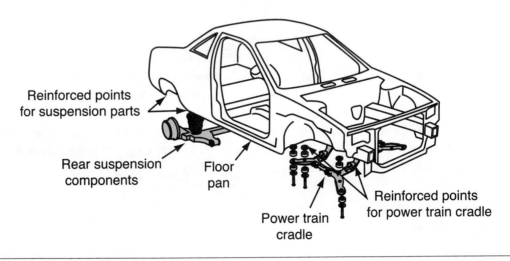

Figure 1-19 A unibody covers the vehicle components and has reinforced mounting points for the major underbody components.

The frame has attachment points for suspension and steering components, bumpers, and the body. A *full frame* has rigid parallel channels with crossmembers, which help keep the parallel channels in place and provide mounting points for items like the engine and transmission (Figure 1-20). Full frames are heavy, but can support heavy loads. Most passenger cars now use the unibody, while trucks still use the full frame. Some rear-wheel-drive cars still use a version of the full frame.

Unibody construction relies on the body to support the passengers and interior components. However, the unibody by itself cannot support the torque actions and movement of the powertrain. The steering and suspension systems also stress a unibody. To counter this problem, subframes are used to support the powertrain and at least some portion of the steering and suspension systems (Figure 1-21). The subframes are constructed to support various loads without stressing the rest of the vehicle. Some vehicles only have a subframe at the front of the vehicle supporting the engine, transaxle, steering, and the front suspension. The rear portion has strengthened areas for attaching the rear wheels and suspension. On vehicles with two true subframes, the body becomes the connection between the two units. This reduces weight greatly, but a small "parking lot" accident can twist the body and subframes out of alignment on older designs.

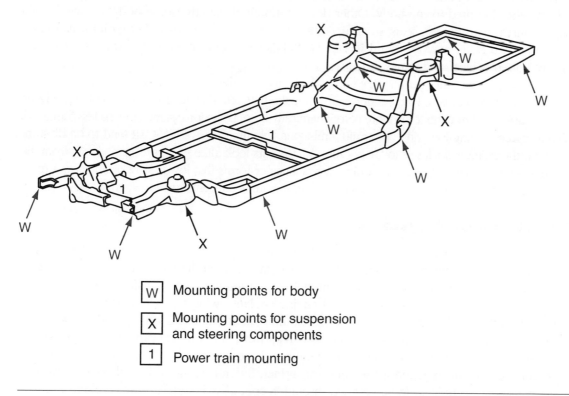

W	Mounting points for body
X	Mounting points for suspension and steering components
1	Power train mounting

Figure 1-20 A full frame or ladder design provides mounting points and support for the vehicle's major components, including the body.

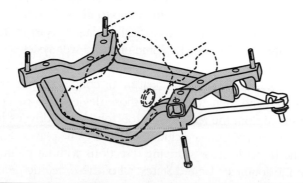

Figure 1-21 A typical subframe for supporting the engine, transaxle, and some suspension/steering components.

The *body* is the shell that surrounds most of the vehicle's components and passenger compartment. The body also provides a mount for several automotive subsystems. Almost all lights, interior and exterior, are mounted to body panels. The doors, hood, and trunk lid become supporting members of the body when closed. Door posts or pillars provide latching and locking devices for the door and help support the roof. Front and rear windshields and other glass components are mounted to the body. Other devices mounted directly to the body are windshield wipers, instrument panels, seats, steering columns, and rearview mirrors. Further information on the frame and body may be found in Delmar Publisher's publication *Motor Auto Body Repair, 3E,* by Robert Scharff and James E. Duffy.

The body forms the structure that moves the air aside as the vehicle is moving. Because of CAFE and emissions concerns, it became necessary to make that action as smooth as possible. The body was reshaped and molded to be more aerodynamically fluid. Less air resistance means less fuel used, with a corresponding lowering of harmful emissions. Old-time fans of NASCAR tend to be disgruntled with the racing vehicles of today because "all the cars look alike." The problem is that, from the manufacturers' viewpoint, there are only a few shapes or forms that lower air resistance on a moving vehicle. So, most of today's vehicles bear at least a surface resemblance to each other.

Climate Control Systems

Climate control includes the heating and air conditioning systems. In many models, the two systems have been combined into a single system that may be controlled manually or automatically. Either system can keep the passengers comfortable by using liquids to transfer heat.

The heater system blows air over hot engine coolant owing through a small radiator-type device. This warms the air and the passenger compartment. Air conditioning works in a similar manner, but uses a special liquid referred to as **refrigerant,** or freon, and varying pressures within the system. Both systems can affect the operation of the engine and both should be serviced or supervised by an experienced technician. In fact, federal, state, and local laws require an air conditioning technician to be certified through a special written test. Climate control is covered in Chapter 12, "Auxiliary Systems and Accessories."

Accessories

Accessories include not just the radio and windshield wipers of the old days. Today's vehicle accessories range from standard features such as the radio to global positioning systems. The seats, the pedals, and the steering wheel can be automatically adjusted just by touching a button, before you even get physically near the vehicle. Some of the newer systems don't even require the driver to press a button. Electronic sensors and controls will unlock the door, deactivate the antitheft protection, start the engine, turn on the heat or air conditioning, move the steering wheel and seat to allow easy entry, and then adjust all of those devices to a particular driver. This is all done based on a radio signal from the remote control "fob" in the driver's hand or pocket as she approaches the vehicle. Other accessories include DVD and game players for the passengers. Navigation features allow the driver to access driving directions without pulling over for assistance. Like air conditioning in the southeast United States, all of these "exotic" features will be standard on future vehicles. Each requires some degree of technical expertise for diagnosing and repairs. Though this means better pay and employment for the automotive technician, the downside is the increased cost of parts and labor. The cost of the parts may eventually be lower, but the labor cost of a well-trained, motivated technician will likely increase. This further emphasizes the technician's golden rule: Fix it right the first time.

Summary

- ❏ The electrical system has become the single most important system regarding vehicle operation.
- ❏ Starting and charging systems provide the means to start the engine and power the electrical system.
- ❏ The power system is the component that changes heat from burning or reactive chemicals to mechanical energy.
- ❏ Power systems are constantly in a state of change regarding fuels and operations.
- ❏ Alternative fuels may provide a means to reduce the use of hydrocarbon fuels while lowering harmful emissions.
- ❏ The powertrain transmits the power-system output to the drive wheels.
- ❏ The transmission or transaxle provides a means to harness the engine's output to the load based on vehicle load and environment.
- ❏ Steering allows the driver to direct the vehicle in the desired direction.
- ❏ Suspension helps protect vehicle components and aids in a better ride.
- ❏ A brake system provides the driver a way to stop and control the vehicle.
- ❏ The frame supports the body and all of its mounted components.
- ❏ The body protects the passengers and vehicle components from outside hazards.
- ❏ The climate control system, though not needed to operate the vehicle, reduces driver and passenger fatigue and increases alertness.
- ❏ Accessories include many features or subsystems that assist the driver or passenger and provide a better interior environment.

Terms to Know

Automatic (transmission)
Biodiesel
Combustion chamber
Corporate Average Fuel Economy (CAFE)
Fuel cell
Hybrid power
Manual (transmission)
Powertrain Control Module (PCM)
Refrigerant
Thermodynamics
Transmission Control Module (TCM)

Review Questions

Short Answer Essay

1. Discuss how the vehicle electrical system has been expanded in use over the last twenty years.
2. Explain how the starter circuit turns the crankshaft during engine startup.
3. Discuss the use of alternative fuels in today's and future vehicles.
4. Explain the purpose of the powertrain.
5. Explain the general usage of a computer or computers in today's vehicles.
6. List five of the major computer modules used on light vehicles.
7. List and explain the purpose of the power train's major components.
8. Explain the primary operational difference(s) between a manual transmission and an automatic transmission.

9. Discuss the operation of a brake system. '

10. Explain the overall operation of an internal combustion engine.

Fill-in-the-Blanks

1. The largest single vehicle computer is the _____ _____ _____ .

2. The charging system uses a belt-driven _____ _____ _____ to produce electrical energy.

3. CAFE is a federal requirement on _____ _____ or usage.

4. _____ is the use of heat to produce mechanical motion or energy.

5. The _____ is shaped to change linear motion into rotary motion.

6. The entry of the air–fuel mixture is controlled by timed, mechanical _____ .

7. Vegetable oils can be mixed with diesel to form _____ .

8. BMW has produced a race vehicle fueled by _____ fuel.

9. Newer vehicles have a _____ _____ _____ steering system.

10. Antilock brake systems are used to _____ brake lockup during emergency stops.

Multiple Choice

1. The split brake system has _____ pistons in the master cylinder.
   A. two   C. three
   B. one   D. four

2. The vehicle body
   A. protects the vehicle components.
   B. provides protection for the passenger.
   C. provides mounting points for many vehicle components.
   D. All of the above

3. CAFE helped cause the development of the
   A. unibody.   C. aerodynamically shaped body.
   B. full-frame support.   D. None of the above

4. The vehicle heater
   A. blows air over the radiator for heating.
   B. may be paired with the air conditioning system.
   C. is always a separate system.
   D. draws electric power from the battery during engine operation.

5. The following subsystems are part of the electric system, EXCEPT:
   A. ABS.   C. starter.
   B. ignition.   D. manual transmission.

6. When the ignition key is in the "start" position, control circuit current is routed through all of the following EXCEPT:
   A. clutch switch.   C. starter motor.
   B. P/N switch.   D. starter relay.

7. The traffic on the vehicle's communication bus is controlled by the
   A. PCM.   C. intranet server.
   B. BCM.   D. Both A and B

8. Spent air–fuel mixture is routed through/around the
   A. intake valve.
   C. compression valve.
   B. exhaust valve.
   D. power valve.

9. The best air–fuel mixture ratio for reducing emissions and improving performance and fuel mileage is
   A. 14 to 7.1.
   C. 14.7 to 1.
   B. 1 to 14.7.
   D. 8 to 1.

10. A typical hybrid-powered vehicle operates with a(an)
   A. internal combustion engine.
   C. power pack.
   B. power splitter.
   D. All of the above

The Automotive Business

Upon completion and review of this chapter, you should be able to:

❑ Discuss the different types of businesses supporting the automotive industry.

❑ Describe the main differences between dealerships, independents, franchises, and service stations.

❑ Discuss how departments can increase business.

❑ Discuss the general duties and responsibilities of the service manager.

❑ Discuss the legal and ethical responsibilities of a business and its employees.

❑ Discuss the role of accrediting agencies within the automotive industry.

Introduction

Employees must work effectively within the organization so they can show a pro table return for their **investment** within the company. This includes knowing the company's organization and the general duties of each department, working with the various departments, and protecting the investment through safe, profitable, and ethical work practices.

Investment is the amount of time, money, and other resources used to establish a business.

Shop Manual
pages 15–19

Global Automotive Business

Automotive service centers used to be fairly small operations. Usually there was a person who owned and managed the business while performing repairs and customer service at the same time. Even manufacturer repair service centers were small-scale compared to today's multimillion-dollar shops. Small operations still exist today, but the majority of automotive repair businesses must meet the demand for highly trained technicians, expensive diagnostic equipment, customer waiting rooms, extensive parts departments, and the facilities and staff to support them. The various levels of business interlock throughout the industry (Figure 2-1).

A modern-day automotive repair business includes every type of employee position, from custodian to general manager. There are accountants, clerks, salespeople, service writers, technicians, department and service managers, vehicle preps, and a general manager to oversee the entire operation. In any case, the person who performs the work that directly impacts profit-making is the technician.

Local Automotive Repair Business

Local car repair shops are a great community asset in terms of both taxes collected and employment opportunities offered. They are probably one of the largest single-type employers in the area. Individual manufacturing plants may employ several hundred people, but all of those employees and their families have vehicles that must be maintained and repaired. That requires many repair shops, sales, and parts vendors with their numerous trained technicians and supporting staff, ranging from the cleanup crews to accountants to attorneys and the waste/recycling companies. Many people who work in the automotive business do not realize their

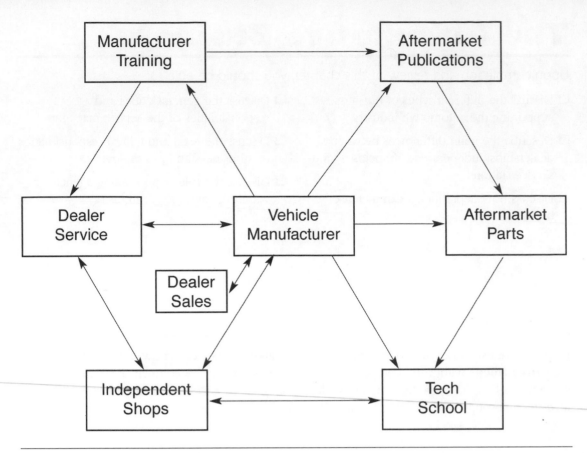

Figure 2-1 Relationship of automotive business.

importance to the community at large. Not only do they support the automotive side, they pay taxes, property and personal, that help support community activities. The automotive employer and its employees provide a great amount of money and labor to support local charities, churches, and schools—not to mention the money they spend locally for their personal or professional needs. The individual technician plays just as important a role in his community as he does within the automotive industry itself.

Dealership Operations

Shop Manual
pages 15–19

Franchises are businesses locally owned and operated under the policies of a larger company. For example, McDonald's is a franchise operation.

Normally, dealerships are automotive businesses operating under independent ownership in conjunction with a vehicle manufacturer. Operations of this type usually sell and service only certain makes of vehicles. The owners conduct day-to-day business, either directly or through a **general manager**. The business must conform to certain standards set forth by the vehicle manufacturer (Figure 2-2). These standards may include policies on new car sales, accounting and financing procedures, service operations, and the layout of the facilities. In some instances, the manufacturer may withdraw the **franchise** license if standards and profit margins are not met.

Sales and service departments working together can create a customer relationship that draws repeat business and favorable word-of-mouth advertising. The technician is required to have extensive knowledge of the makes and models sold at the dealership, and is required to perform warranty repairs.

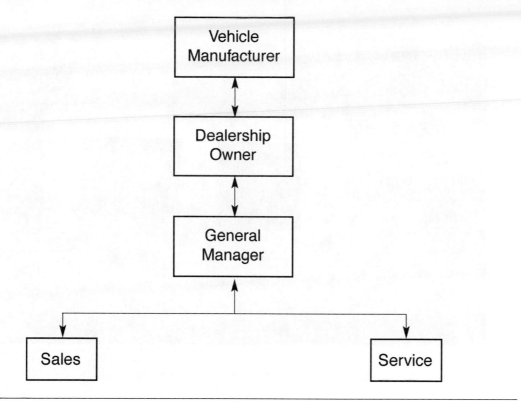

Figure 2-2 Relationship between dealership and manufacturer.

Independent Repair Facilities

Independent repair shops usually concentrate on vehicle repairs and service only (Figure 2-3). They may stock some of the most common repair parts, but the majority of needed parts come from aftermarket **vendors** or the local dealership. Independents usually take in any vehicle that needs repair. This places a heavy technical load on the technicians. They must know how to repair almost any vehicle and all automotive systems. Independent shop owners usually have to pay for their technicians' update training in addition to salaries and travel expenses. Most independents make use of classes offered by local parts vendors or the local technical schools.

Other Automotive Repair Businesses

Franchise automotive businesses are independently owned and operated. They are licensed under the business umbrella of a national or regional chain (Figure 2-4). The owner must follow the policies of the company issuing the franchise license. Precision Tune and Quik Lube are two examples. These types of businesses are set up to perform only certain types of repairs or maintenance. Businesses such as Precision Tune make engine-performance repairs or maintenance while Quik Lube types usually perform oil and lubrication work. Since their inception, each business has changed its operations to include work not directly related to its original base business.

Local service stations are another part of the repair business. Usually, they handle brakes, tires, minor tune ups, and other light repairs. Other local businesses repair only certain systems of the automobiles. Generally, they specialize in brakes, tires, exhaust, and related repairs. However, most specialized shops, including service stations, are realizing that they cannot survive today without accepting repair work outside of their original business intentions.

Each of these automotive businesses make up the automotive repair facilities in this country. To support them, there are numerous parts makers and suppliers, hazardous waste disposal

Shop Manual
pages 15–19

Vendors are businesses that sell parts or service to other businesses and the public. For example, NAPA is an automotive vendor.

Figure 2-3 Independent shops repair almost any type and make of vehicle.

companies, and many other local, national, and international firms that contribute to keeping a vehicle on the road.

The trend in dealership operation is the megadealership. As mentioned earlier, dealerships have to meet certain corporate standards set by the vehicle manufacturer but are locally owned. In recent years, Mr. Roger Penske (see Figure 2-4) and others have formed corpora-

Figure 2-4 Franchise businesses usually service one system of the vehicle, but may perform related system repairs.

tions that own multidealerships throughout the world under one company umbrella. As of early 2006, AutoNation is the largest single megadealership company. United Auto Group, with Mr. Penske as the chair and chief executive officer, is the second largest. Each of these corporations is publicly traded, meaning an individual can buy shares in the business through the New York Stock Exchange or, as it is commonly known, Wall Street. However, the general operation of the local dealership is the same as before; just the owners and some business policies have changed.

Technician Training and Certification

The technician's training usually involves some type of certification. Certification is usually done through technical training by some type of educational agency and/or certification by an outside source, such as those discussed next. But technical training is not all that is required. The technician not only has technical responsibilities, but also legal and ethical responsibilities he must be aware of and must practice.

Legal Responsibilities

Not only are the technicians expected to be technically proficient, they must understand the legal side of automotive repair. The customer is paying for professional service and has a right to expect it. Poor repair work, damage to the vehicle, and poor attitude toward the customer may lead to legal proceedings against the technician and the shop.

In addition to legal responsibilities to the customer, federal, state, and local laws and regulations require business owners and employees to help protect the community and environment from harm. Automotive technicians share this legal responsibility. Proper handling of hazardous materials and waste is of prime concern to the automotive industry. Not only can a poorly tuned engine emit hazardous materials and gases, the by-products of poor repair work also may damage the environment. The business owner is ultimately responsible for the actions of the firm's employees, but the technician plays a large part in fulfilling the legal responsibilities of the company.

Business Ethics

Most people hold the automotive industry in high esteem. Two segments of the industry, sales and service, are the exceptions. Much of this distrust stems from the high cost of vehicles and their repairs. Further supporting this distrust is the fact that most people do not understand the workings of an automobile, and they have no reference with which to compare the costs of buying and maintaining a car to that of other large purchases. Educating customers is one way to resolve the problem. Very few automobile owners understand that their vehicle's computer is as powerful as the personal computer they have at home or that the technicians are constantly attending some type of training. A business will survive much better in today's highly competitive automotive arena if it can educate its customers to some extent. A customer may pay well over $1,000 for a personal computer (PC) and balk at paying $600 for a vehicle computer. A technician or service writer can win over the customer by showing the capabilities of that $600 computer versus the PC.

Automotive businesses expect to be treated fairly by their suppliers and must treat their own customers in the same manner—this is the essence of **business ethics.** By treating each customer as a valued individual, charging a fair price for service, and admitting and correcting mistakes, the shop and all its employees can expect to be in business for a long time.

Business ethics are hard to define. While one person's idea may be different from another's, the managers set the rules. Treating customers, managers, and employees fairly is the basic rule.

Figure 2-5 Master technicians must have 2 years of experience and pass all eight exams.

National Institute for Automotive Service Excellence (ASE)

Shop Manual
pages 17–19

ASE is the logo for the National Institute for Automotive Service Excellence.

ASE tests technicians for their technical knowledge. Technicians are authorized to wear the shoulder patch only if they have successfully completed a written test and have 2 years' working experience. They must recertify every 5 years to maintain their standing. The Master Automotive Technician patch is given only if a technician passes all eight automotive tests and meets the experience requirement (Figure 2-5). Certifications are also offered in heavy truck, alternative fuels, parts, machining, service writing, and school buses. The automotive tests given in May and November each year are engine rebuilding, engine performance, brakes, steering and suspension, heating and air conditioning, electrical, automatic transmission, and manual drivelines.

The shop itself can be certified if it meets ASE standards for shop operation and employee technical certification. The shop can then display the blue ASE sign and use the ASE symbol in its advertising. The ASE sign and shoulder patch can be a valid advertising tool. ASE also publicizes the technician's role within the automotive business and to the public at large. Learn more at ASE's Web site: http://asecert.org.

National Automotive Technicians' Education Foundation (NATEF)

Shop Manual
pages 17–19

The **NATEF** key is the logo for a certification administrated by the National Automotive Technicians' Education Foundation.

NATEF, a division of ASE, certifies automotive training programs. The programs may be offered at public and private high schools or post-secondary schools and manufacturer schools. Students who graduate from a NATEF certified program have an advantage over other persons applying for an automotive technician position.

To meet NATEF standards for certification, the training program must be capable of teaching standard tasks published by NATEF. The program must have all of the tools required by NATEF and have ASE-certified instructors in each area requesting certification. The program is certified for 5 years with an update at the halfway point. Once certified, the school can use the NATEF key logo in its advertising and catalog (Figure 2-6). Generally, high schools will be certified in only four areas: brakes, electrical/electronic systems, engine performance, and suspension. Learn more at NATEF's Web site: http://www.natef.org.

AUTOMOTIVE TECHNOLOGY

Figure 2-6 The right to use the NATEF key logo in advertising is awarded after successfully completing the NATEF training program procedures.

Automotive Youth Educational Systems (AYES)

AYES is an education program directed toward high school students. It is supported by almost all vehicle manufacturers and other automotive-based organizations. This program provides some funding, equipment, and expert support to the school and offers internships and employment, usually at local dealerships. The student receives excellent training in basic automotive repair. This training will aid the graduate in gaining employment and/or entering post-secondary or manufacturer programs for further training.

Automatic Transmission Rebuilders Association (ATRA)

ATRA is an association of transmission rebuilders. It offers a certification to any technician who desires, but it is directed toward technicians who only do automatic transmission repairs. The test is given annually at selected sites and is brand and model specific in its questions. The test is designed for technicians who have experience in transmission diagnosis and repairs. ATRA publishes *Gears* magazine, which may be found at its Web site: http://www.Atra-gears.com.

ATRA and **MACS** are two examples of automotive certification organizations. There are others, but the most nationally recognized are ASE and NATEF.

Mobile Air Conditioning Society (MACS)

MACS is similar to ATRA in its testing. It is designed for automotive air conditioning technicians, but, like ASE, is offered to any technician. An EPA requirement for all air conditioning technicians is to be certified by an approved agency. This test will be discussed further in Chapter 12, "Auxiliary Systems and Climate Control." Visit the MACS's Web site for further information: http://www.macsw.org.

Automotive Engine Rebuilders Association (AERA)

Shop Manual
pages 17–19

AERA primarily provides technical, product, and education support to technicians and shops specializing in the rebuilding of automotive engines. It also offers financial grants to assist in education expenses. It charges a minimum fee for membership, as do the other organizations of this type. Learn more at AERA's Web site: http://www.aera.org.

Society of Automotive Engineers (SAE)

SAE provides much technical support to governments, manufacturers, all levels of repair shops, and individual technicians. SAE, like other organizations, charges an annual fee for this expertise. At the technician level, information is available on repairs, future products and systems, and the overall goals for emission control, fuel economy, vehicle performance, and the technical aspects of the automotive industry. Further information on SAE can be found at its Web site: http://www.sae.org.

Manufacturer and Vendor Training

Manufacturer training is conducted by schools set up and supported by a vehicle manufacturer. They range from entry-level training usually conducted at a post-secondary school to update training and certification of dealership technicians conducted through on-line classes, computer-based training (CBT), or at a manufactured-funded site. Other technicians may attend manufacturer update training after paying a fee and if space is available. The manufacturer may elect to certify dealer technicians in various areas of the vehicle(s) and in most cases supports or accepts the certification from the independent agencies just discussed.

Training conducted or sponsored by the local parts vendor is open to all technicians. Classes may require a fee payment. Vendor training is conducted for two reasons. First, updates on diagnostics and repairs are presented. This information is usually very pertinent, timely, and presented in a professional manner. Of course, they are also there to sell their products. The class presentation includes methods of using their product to perform quick and accurate diagnosis and repair. Though the underlying cause is to sell the vendor's products, most of these classes provide much technical information helpful to the technician. Most technicians understand this and do not mind listening to a little sales pitch as long as the overall information increases their income. In fact, technicians may attend to see the new diagnostic tools that will help solve problems that might be present in new-model vehicles. Specialty Equipment Market Association (SEMA) is one such aftermarket vendor offering automotive products, training, and possible financial aid assistance to technicians. Visit its Web site at: http://www.sema.org. Other organizations, such as the Automatic Transmission Rebuilders Association (ATRA) and the Automotive Engine Rebuilders Association (AERA), offer training in their respective areas. Their Web sites, respectively, are http://www.arta.com and http://www.aera.org. Many other groups also offer on-line information and training. The site http://www.howstuffworks.com provides much basic information with links to other automotive sites as well.

Terms to Know

AERA
ASE
ATRA
Business ethics
Franchise
General manager
Investment

Summary

❑ Management is responsible for all business operations.

❑ All employees share the legal responsibilities of the business.

❑ Education of the customer is another way to gain customer trust.

❑ ASE provides a nationwide certification program for technicians.

❑ NATEF certifies automotive training programs.

❏ ATRA provides a means for automatic transmission technicians to gain specific skill recognition.

❏ MACS is designed for air conditioning technicians.

❏ AERA and SAE provide technical and educational support for various establishments and technicians.

❏ Manufacturer and vendor training is essential for technicians wishing to keep up to date on various aspects of vehicle repair.

Terms to Know
(*continued*)
MACS
NATEF
Vendors

Review Questions

Short Answer Essay

1. Describe the role of the independent shop within the automotive repair industry.

2. Explain how sales and service work together to gain customers.

3. Explain how certifying agencies can assist in promoting customer trust and validating technician technical knowledge.

4. Describe the purpose of NATEF.

5. Discuss how customer education can help an automotive repair business.

6. List the eight automotive areas tested semiannually by ASE.

7. List the steps needed for a technician to be certified as a Master Automotive Technician.

8. Discuss *your* definition of business ethics.

9. Explain why employees share in the legal responsibilities of the employer.

10. List the level of management typically found in an automotive shop.

Fill-in-the-Blanks

1. A dealership is an automotive business operating under _____ _____ in conjunction with an automotive manufacturer.

2. If standards or profits are not met, the manufacturer may withdraw the _____ _____ of the dealership.

3. NATEF conducts certification of _____ _____.

4. The logo for NATEF is a(n) _____ with the _____ name on it.

5. ASE certified technicians wear a(n) _____ ASE patch.

6. Vehicle manufacturer training programs can be certified by _____.

7. Air conditioning technicians can be certified through _____.

8. ATRA offers certification to _____ _____ technicians.

9. Air conditioning technicians must be _____ certified by either MACS or ASE.

10. A(n) _____ is also known as a supplier.

ASE-Style Review Questions

1. Business operations are being discussed.
   Technician A says an independent repair facility stocks most of the commonly used repair parts that may be kept on hand at the shop.
   Technician B says most franchise shops are usually owned by a national business chain and operated by local managers.
   Who is correct?
   - **A.** A only
   - **B.** B only
   - **C.** Both A and B
   - **D.** Neither A nor B

2. Business ethics are being discussed. *Technician A* says good business should be applied to all customers.
   Technician B says educating the customer could be a part of the shop's business ethics.
   Who is correct?
   - **A.** A only
   - **B.** B only
   - **C.** Both A and B
   - **D.** Neither A nor B

3. Legal responsibilities of the business are being discussed.
   Technician A says each employee shares in legal responsibilities.
   Technician B says the owner bears ultimate responsibility.
   Who is correct?
   - **A.** A only
   - **B.** B only
   - **C.** Both A and B
   - **D.** Neither A nor B

4. Program certification is being discussed.
   Technician A says ATRA certifies air conditioning programs.
   Technician B says MACS certifies air conditioning technicians.
   Who is correct?
   - **A.** A only
   - **B.** B only
   - **C.** Both A and B
   - **D.** Neither A nor B

5. *Technician A* says that to achieve master automotive certification, nine tests must be successfully completed, including the Advanced Engine Performance Test.
   Technician B says only eight tests must be taken and only 2 years of experience are needed for master certification.
   Who is correct?
   - **A.** A only
   - **B.** B only
   - **C.** Both A and B
   - **D.** Neither A nor B

6. *Technician A* says business ethics means treating all customers honestly and fair.
   Technician B says business ethics means treating vendors correctly.
   Who is correct?
   - **A.** A only
   - **B.** B only
   - **C.** Both A and B
   - **D.** Neither A nor B

7. The service manager's duties and responsibilities are being discussed.
   Technician A says the training of a technician is his or her responsibility.
   Technician B says he or she usually has authority over the parts manager.
   Who is correct?
   - **A.** A only
   - **B.** B only
   - **C.** Both A and B
   - **D.** Neither A nor B

8. Independent repair shops are being discussed.
   Technician A says this type of shop is needed to help reduce customer costs.
   Technician B says independents may send work to the local dealership.
   Who is correct?
   - **A.** A only
   - **B.** B only
   - **C.** Both A and B
   - **D.** Neither A nor B

9. Technician certification and training is being discussed.
   Technician A says NATEF ensures certain tasks are taught.
   Technician B says NATEF ensures that the technician is certified.
   Who is correct?
   - **A.** A only
   - **B.** B only
   - **C.** Both A and B
   - **D.** Neither A nor B

10. Air conditioning certification is being discussed.
    Technician A says an air conditioning technician can be EPA certified by ASE only.
    Technician B says ATRA deals with air conditioning technicians.
    Who is correct?
    - **A.** A only
    - **B.** B only
    - **C.** Both A and B
    - **D.** Neither A nor B

Automotive Tools and Equipment

Upon completion and review of this chapter, you should be able to:

❏ Identify and describe the safe use of the common wrenches used in the automotive shop.

❏ Identify and describe the correct use of the power wrenches used in the automotive shop.

❏ Identify and explain the safe use of the metalworking tools commonly used in the automotive shop.

❏ Identify and explain the safe use of the common threading tools used in the automotive shop.

❏ Describe the major types of lifts.

❏ Describe test equipment.

❏ Describe the purpose of computerized test equipment.

Introduction

Knowing how to use and care for tools properly is one of the most important parts of being an automotive technician. There is a correct tool for each and every automotive repair job. It is also important to use each tool in the correct way. Technicians use a number of these tools on every repair job. This chapter discusses the most common tools. Remember, a successful technician must know the name of each tool, what it does, and what tool works best for each automotive repair job.

Most of the tools used by a technician are called hand tools. They get their name from the fact that they are operated by the technician's own muscle power. Power tools, on the other hand, get their power from electricity, compressed air, or hydraulics.

Common Hand Tools

Wrenches

Many automotive parts and components are held together or fastened with bolts and nuts. Wrenches, which come in many different sizes, are designed to tighten or loosen these bolts and nuts. Bolts and nuts also come in different sizes.

The size of a wrench is determined by the size of the nut or bolt head it fits on. The wrench shown in Figure 3-1 has the number 14 stamped on it. This means the opening of the wrench

Shop Manual
pages 31–32

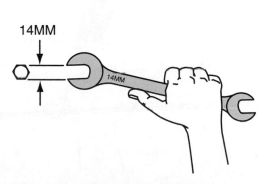

14MM

Figure 3-1 The size stamped on a wrench shows what bolt head or nut it fits.

measures 14 millimeters across the opening and it will fit on a bolt head or nut that measures 14 millimeters across.

Wrench sizes are given either in metric or U.S. (English) system units (see Chapter 6, "Mechanical Measurements and Measuring Devices"). Metric wrench sizes are given in millimeters, for example, 7, 8, 9, 10, 11, 12, and 13 millimeters. Sizes for U.S. (English) system wrenches are given in fractions of an inch, for example, $5/16$, $3/8$, $7/16$, $1/2$, and $9/16$ inches. Because both U.S. (English) and metric system bolts and nuts are common, the technician will need both metric and U.S. (English) wrench sets.

Open-End Wrenches

An open-end wrench is one common type of wrench (Figure 3-2). These wrenches have an opening at the end that is placed on the bolt or nut. The opening is usually at an angle to the handle. Often one of the gripping surfaces or jaws is thicker than the other. This makes turning in a tight space easier. A hexagonal nut can be turned continuously just by turning the wrench over between each swing (Figure 3-3). Open-end wrenches are made in many different sizes and shapes. Most open-end wrenches have two open ends of different sizes.

When using open-end wrenches, always try to pull the load as shown in Figure 3-4. When a wrench is pulled, it is usually away from some obstruction. When a wrench is pushed, it is to-

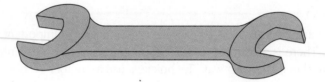

Figure 3-2 An open-end wrench.

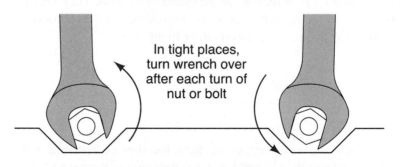

In tight places, turn wrench over after each turn of nut or bolt

Figure 3-3 An open-end wrench is flipped over to turn nuts in tight places.

Apply force in direction indicated

Figure 3-4 Pull on an open-end wrench when tightening.

ward an obstruction. A hand injury can result if the wrench slips off the nut or bolt head. If an open-end wrench must be pushed toward an obstruction, do not wrap fingers around the handle. Use the heel of the hand as shown in Figure 3-5. Be prepared with good footing and a clean floor in case something breaks.

Box-End Wrenches

Box-end wrenches like the one shown in Figure 3-6 are designed to fit around a bolt or nut. They usually have twelve points or corners to grip the nut or bolt head. They allow the technician to apply a lot of force with less chance of the wrench slipping off the nut or bolt. Like other wrenches, they come in many sizes. The angle of the head is offset to the handle to give turning room in a tight space. Two common offsets are 15 and 45 degrees, as shown in Figure 3-7.

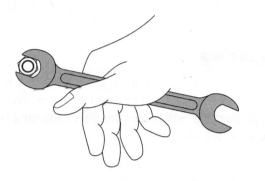

Figure 3-5 Correct way to push on an open-end wrench.

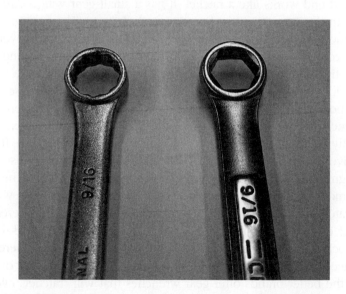

Figure 3-6 A twelve-point box-end wrench (left) and a six-point box-end wrench (right). A six-point box-end wrench usually provides a better fit and reduced chances for fastener damage.

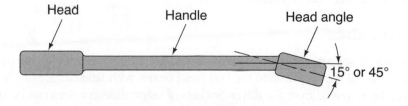

Figure 3-7 The box-end of the wrench may be offset to the handle 15 or 45 degrees.

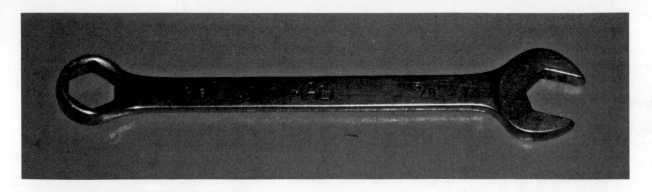

Figure 3-8 A combination wrench.

Combination Wrenches

Combination wrenches like the one shown in Figure 3-8 are very useful. They are called combination wrenches because they have one box-end and one open-end. The open-end side is used where space is limited. The box-end side is used for final tightening or to begin loosening. They are usually the same size at both ends.

Gear Wrenches

Gear wrenches are a version of a combination wrench, open at one end and boxed at the other. However, the boxed end works like a ratchet. It has a small gear with a directional lock in the body surrounding the point section that fits over the bolt head. The body is relatively thin, like a typical combination wrench, and will fit into close areas. Once fitted over the bolt and the ratchet in the correct position, the gear wrench works like a ratchet driving a socket wrench. This allows quick tightening or loosening of fasteners in close areas. The technician must be aware of and consider two possible problems with a gear wrench. Because of the small ratcheting gear it may not be possible to apply great torque to loosen or tighten a fastener without damaging the wrench. Usually using the open-end or a regular box-end wrench to break the fastener loose or apply the final torque takes care of this problem. In this case, the better, and usually more expensive, brand will be the better buy. A second possibility also directly relates to the price and design of the wrench. In extremely close areas, the ratchet in the wrench may have too much spin or movement to help very much. The number of degrees the wrench has to move before the lock "ratchets" is different on different brands and even on different models. For instance, if the outer end of the wrench has to move 2 or 3 inches (about 8–10 degrees of rotation) before the lock shifts, then the wrench will only be useful when there is at least 2 to 3 inches of clearance (Figure 3-9). If not, then the wrench becomes an overpriced box-end wrench. Some of the better brands offer gear wrenches that will "ratchet" every 3 degrees. On a wrench 8 inches long, this is about 1 inch of travel. In other words, you can fit into really confined areas. But this also means the gear teeth and ratchet will be very fine, which could mean a weaker gear that in turn leads back to the first problem. Tool brand should be considered closely when purchasing gear wrenches.

Socket Wrenches

Socket wrenches and attachments are often grouped together and called "sockets." They fit all the way around a bolt or nut and can be detached from a driver.

Socket wrenches, or sockets, have the wrench part and the driver part made in two different pieces. The wrench part fits all the way around the bolt head or nut with little danger of slipping off. The wrench part can be removed from the driver. Sockets of many different sizes can be used with one handle. Sockets usually come in sets and are made in all the English and metric sizes (Figure 3-10).

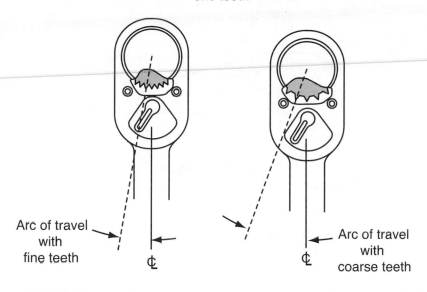

Gear ratcheted
one tooth

Arc of travel
with
fine teeth ℄

Arc of travel
with
coarse teeth ℄

Figure 3-9 Gear wrenches with fine ratchet teeth allow for shorter wrench movement and thus can be used in more constricted spaces.

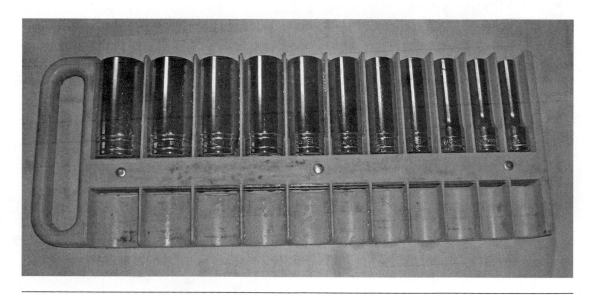

Figure 3-10 Typical deep-well socket set.

Sockets are attached to a square drive lug on the handle by a square hole at one end of the socket (Figure 3-11). These drive holes and lugs are also made in different sizes. For small bolts and nuts, such as automotive trim parts, socket sets with a $1/4$-inch square drive are useful. For general purpose work, a $3/8$-inch drive set is popular. Heavier work requires a $1/2$-inch drive socket set. Even larger sizes, $3/4$ inch and 1 inch, are made for driving very large socket wrenches.

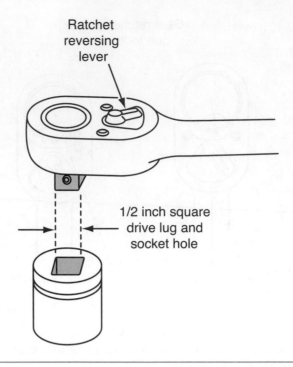

Ratchet reversing lever

1/2 inch square drive lug and socket hole

Figure 3-11 The square drive lug on the ratchet handle fits the square hole in the socket.

Socket wrenches come in two basic lengths: the standard or common length, and long or deep sockets. The latter is longer so that it fits over a long bolt or stud. A standard and deep socket are shown in Figure 3-12. Deep sockets come in all the same common wrench sizes as the standard size sockets.

Sockets are also available with different numbers of corners, or points, inside the socket to grip the bolt or nut. The three common numbers of points are six, eight, and twelve (Figure 3-13). The twelve-point socket is easiest to slip over a bolt or nut because of its many corners. The six-point socket is the hardest to slip over the bolt or nut. The fewer the points or corners, the stronger the socket. The six-point socket is stronger than the twelve-point version and less likely to slip and round the corners of a fastener.

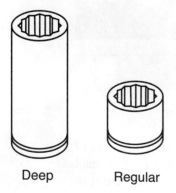

Deep   Regular

Figure 3-12 Deep- and standard-length sockets.

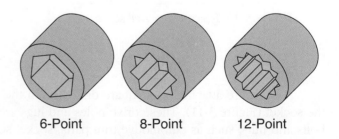

6-Point   8-Point   12-Point

Figure 3-13 Sockets are made with different numbers of points.

Socket Drivers, Handles, and Attachments

Socket wrenches require handles or drivers in order to be used. A large number of handles and attachments are available to drive socket wrenches. The most commonly used type of driver is called a *ratchet handle* (Figure 3-14). It has a square drive that fits into the square hole in the socket wrench. The socket is then placed over a bolt or nut. The bolt or nut is tightened or loosened by rotating the socket handle. A freewheeling or ratchet mechanism inside the ratchet handle allows it to drive the nut in one direction and to move freely in the other direction without driving the nut. This permits fast work in a small space because the socket does not have to be removed from the nut each time it is turned. A lever on the ratchet handle allows the mechanic to choose which direction the ratchet will drive and which direction it will turn free (Figure 3-15).

The *speed handle* (Figure 3-16) is another popular socket driver. The socket wrench is installed on the end of the speed handle. The technician pushes on the end of the handle to hold the socket firmly on the nut or bolt. At the same time, the technician turns the crank-shaped handle in a direction to tighten or loosen. The combination of a swivel handle and crank allows very quick driving of a socket. The speed handle is used when a large number of bolts or nuts must be removed or replaced. Technicians often use this handle when removing an automatic transmission fluid pan. A *breaker bar* (Figure 3-17) is another common socket driving tool. The breaker bar is used to break loose bolts or nuts that are very tight. The long handle allows the technician to get a lot of force into the turning. Its drive end has a hinge that will permit driving at different angles.

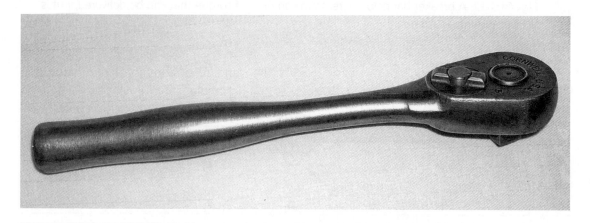

Figure 3-14 A ³/₈-inch drive ratchet.

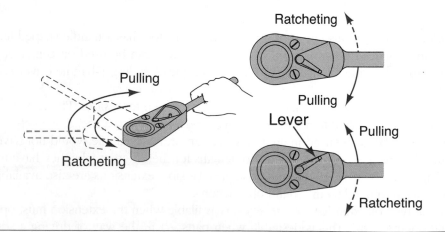

Figure 3-15 Switching the lever allows the ratchet to pull in either direction.

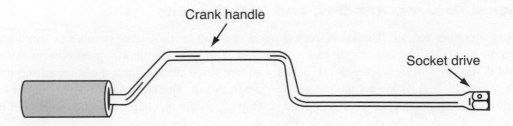

Figure 3-16 A speed handle allows for faster movement of a fastener, but reduces the amount of torque that can be delivered.

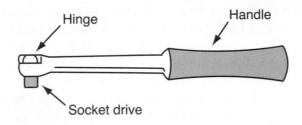

Figure 3-17 A breaker bar helps increase the amount of torque that can be delivered, but is slow at turning a fastener.

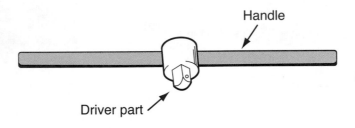

Figure 3-18 A sliding T-handle can be used in very tight spaces, but usually cannot provide much torque and is very slow.

A *sliding T-handle* is also used to drive sockets. This tool has a handle shaped like the letter T. The handle slides through the driver part. The handle can be used on center or it can be moved outward to provide a longer lever for more torque. A sliding T-handle driver is shown in Figure 3-18.

Socket wrenches are often attached to the drivers through *extensions* (Figure 3-19). One end of the extension is connected to a handle or driver and the other to a socket wrench. The extension allows the socket to be used in an area where an ordinary handle would not have enough room to turn. Extensions come in a variety of lengths for the different jobs they have to do. The common lengths are 3, 6, 12, 18, 24, and 36 inches. Flexible extensions are also available that allow the technician to bend them around obstructions.

A special type of extension, or *universal,* is available when the extension must operate at a sharp angle (Figure 3-20). This is desirable when parts get in the way of driving a socket. It is

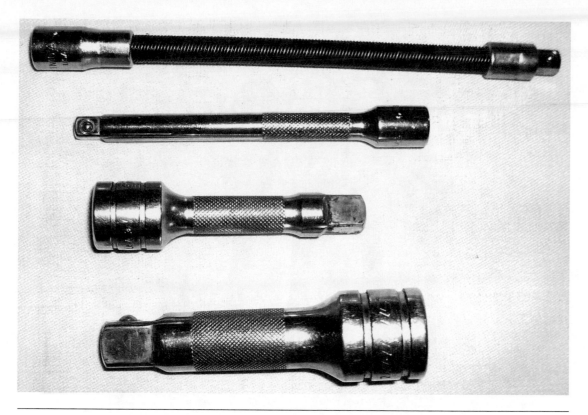

Figure 3-19 Extensions allow the wrench to reach "buried" fasteners and can be used with any matching driver. However, the longer the extension the more flex, which may reduce the torque being delivered.

Figure 3-20 A universal allows for torque to be delivered at an angle to the socket and fastener. Torque delivered becomes less as the operating angle of the universal increases.

made in two parts, which allow it to bend at near 90 degrees. Some universals are part of the socket wrench itself. They are usually known as universal sockets. The primary advantage of a universal socket is the shorter overall length that allows the socket to fit into close areas.

A socket, extension, and handle can be combined into one tool called a *nut driver* (Figure 3-21). The handle is often made like a screwdriver handle. Nut drivers are usually made in small sizes for very small nut and bolt heads.

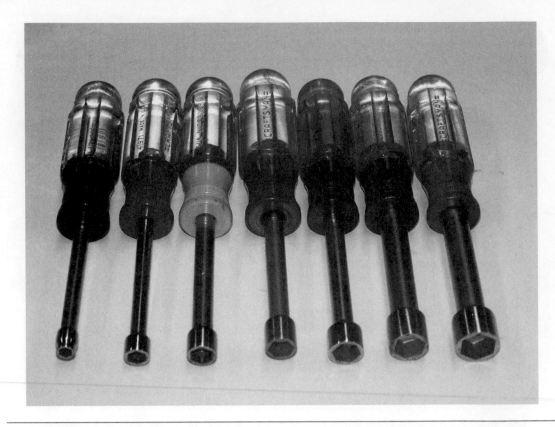

Figure 3-21 Nut drivers are usually used at low torque and on smaller fasteners. They are used for dash/instrument panel work.

Adjustable Wrenches

The wrench shown in Figure 3-22 is called an **adjustable wrench** because it adjusts to fit bolts and nuts of different sizes. There are adjustable wrenches from about 4 inches to about 20 inches long. The longer the wrench, the larger the opening will adjust. For example, a 6-inch adjustable opens $^3/_4$ inch wide and the 12-inch opens $1^5/_{16}$ inches.

The adjustable wrench is not as strong as a box-end or open-end wrench. The adjustable jaw can break if overloaded. The technician must be careful to adjust the jaws to fit snugly against the flats of the nut or bolt head. When tightening, the wrench must be placed on the bolt or nut so that stress falls on the stationary jaw. If the wrench is used incorrectly, the adjustable jaw can break and cause injury. The right and the wrong way to use an adjustable end wrench are shown in Figure 3-23.

Allen Wrenches

Some automotive components are fastened with hollow head Allen screws. These screws require special **Allen wrenches** (Figure 3-24). The hexagonal, or six-sided, Allen wrench fits in the hexagonal head of the Allen head screw. The fit is tight and prevents the wrench from slipping out of the screw head. Allen wrenches are available in sets. They are sized according to the size of the Allen screw in which they fit. They are made in U.S. (English) system sizes such as $^3/_{32}$ and $^1/_8$. They are also made in metric sizes such as 6 millimeter, 8 millimeter, and 10 millimeter. Some Allen wrenches are made to fit on socket drivers, as shown in Figure 3-25.

Figure 3-22 The "fit everything" wrench. The adjustable wrench provides a range of openings to fit many fasteners. It is not as strong as fixed-size wrenches.

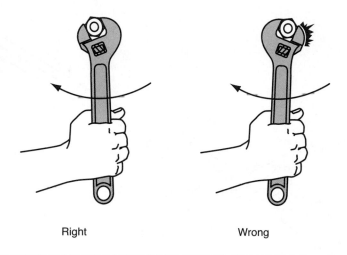

Right   Wrong

Figure 3-23 Using the adjustable wrench incorrectly can result in injury, damage to the wrench, and possible damage to the fastener or component.

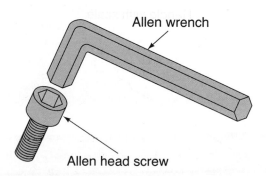

Allen wrench

Allen head screw

Figure 3-24 An Allen head screw and Allen head wrench set.

Figure 3-25 This type of Allen wrench is driven by most sizes of drivers, but is normally $1/4$-inch or $3/8$-inch size in automotive work.

Torque Wrenches

When some automotive parts are reassembled after repair, the bolts and nuts must be tightened a certain amount. A special socket handle called a **torque wrench,** which measures torque, is used for this purpose. Torque is the turning or twisting force. A common measurement of torque is foot-pounds (ft.-lb.). This means a force, measured in pounds, is acting through a distance of 1 foot. For example, imagine a wrench that is 1 foot long. Then push on the end of the wrench with 50 pounds of force. We have just applied 50 foot-pounds of torque.

There are many types of torque wrenches. One popular type (Figure 3-26) uses a beam and pointer assembly. During tightening, the beam on the wrench bends as the resistance to turning increases. The torque is shown on a scale near the handle. Another type of torque wrench (Figure 3-27) has a ratchet drive head and a breakover hinge. Its adjustable handle and scale allows the mechanic to adjust the wrench to a certain torque setting. When the torque setting is reached, the breakover hinge swivels to signal the technician to stop turning.

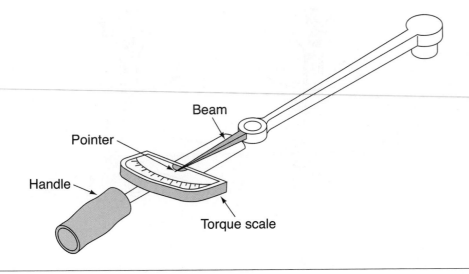

Figure 3-26 A beam-type torque wrench.

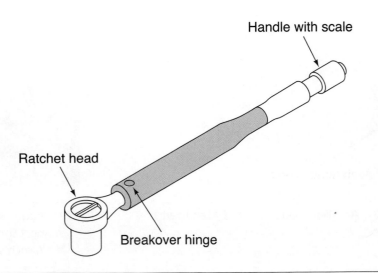

Figure 3-27 A torque wrench with a breakover hinge.

Another type has a dial on the handle (Figure 3-28). This dial has a needle that points to the amount of torque the wrench is delivering. Many tool suppliers offer electronic torque wrenches with a digital display.

Several different torque measurement systems are in use. Specifications are given in metric or U.S. (English) units. Typical U.S. (English) units are inch-pounds and foot-pounds. There are 12 inch-pounds in 1 foot-pound. Metric system torque specifications are most often given in Newton-meters.

Special-Purpose Wrenches

Most automotive service jobs are done with regular open-end, box-end, combination, and socket wrenches. Some jobs, however, require special wrenches. These are usually variations of the common wrenches that have special shapes to get around obstructions. For example, a *brake bleeder wrench* (Figure 3-29) is a box-end wrench that is bent to fit on the small bleeder valves on brake wheel cylinders. Another special wrench is the *flare nut* or *fuel line fitting wrench* (Figure 3-30). This one is a combination of an open-end and a box-end wrench. It is used on fuel line fittings.

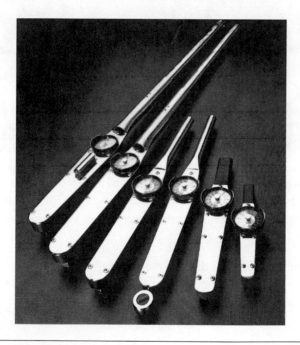

Figure 3-28 A dial readout-type torque wrench.

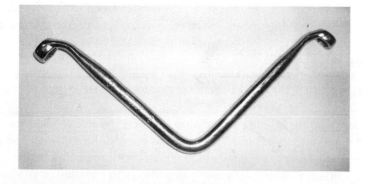

Figure 3-29 A bleeder wrench should never be used for anything except bleeding the brake system. It can bend or break under heavy torque.

Figure 3-30 Always use a line or flare-nut wrench to loosen line and hose fittings. When possible, also use it to tighten the same fittings.

An open-end wrench might round off the soft metal fittings. A box-end wrench would be impossible to slip over a fuel line. There are, of course, many other special-purpose wrenches. Most technicians collect these wrenches for special jobs.

Pliers

Pliers act as an extension of the technician's fingers. They allow the technician to grip parts with great force. In addition, some pliers are made for cutting things like wire or cotter keys. Good technicians know which pliers to use for every job. They never use pliers when some other tool will do the job better; for example, pliers are never used to loosen or tighten a nut.

Combination Pliers

Combination pliers (Figure 3-31), one of the most commonly used types, have a slip joint where the two jaws are attached. This slip joint can be set for either of two jaw openings: one for holding small objects and one for holding larger objects. These pliers are used for pulling out pins, bending wire, and removing cotter pins. They come in many different sizes. The pliers most commonly used in automotive work is the 6-inch combination. Some combination pliers have a side cutter to allow the cutting of wire and cotter pins. The better grades of combination pliers are made of drop-forged steel and can take a great deal of hard usage.

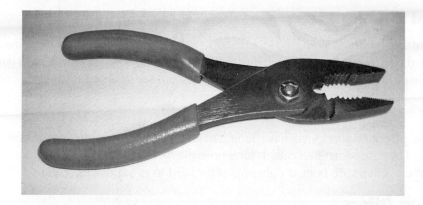

Figure 3-31 The rule for pliers' use: They are to hold, not turn, fasteners or components.

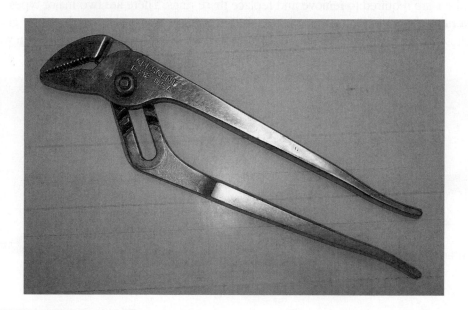

Figure 3-32 Channel-lock pliers.

Channel-Lock Pliers

Channel-lock pliers (Figure 3-32) are used to grip large objects. They get their name from the channels or ribs that allow the jaws to be set at many different openings. The design of the channels permit quick, nonslip adjustments with practically parallel jaws. These pliers are slim enough, with long enough handles, so that the 45-degree jaws will reach and firmly grip objects that would be out of reach of ordinary pliers. The jaw teeth are sharp and deep to take a firm grip on pipes and hoses. The interlocking design of the rib joints works well to prevent slipping under load.

Channel-lock is a trade name that has been accepted as a type of pliers. These types of pliers are actually rib joint pliers.

A BIT OF HISTORY

Channel-lock pliers are often called "water pump pliers." This is the only tool with which a technician can remove the large nut from the water pump in cars made from the 1920s to the 1940s.

Diagonal Cutting Pliers

Diagonal cutting pliers (Figure 3-33) are used to cut electrical wire and cotter pins. Their jaws have hardened cutting edges. Because they are made for cutting wire, they have hard cutting edges. Like other pliers, they are made in many different sizes and are grouped by their overall length.

Diagonal cutting pliers have cutting edges on the jaw for cutting pins or wire. These pliers are often called "dikes," which is short for "diagonals."

Needle Nose Pliers

Needle nose pliers (Figure 3-34) have long, slender, tapering jaws that are useful in gripping small objects. They are sometimes called "long-nose pliers." Some needle nose pliers are equipped with side cutters; others are bent at right angles for hard-to-get-at places (Figure 3-35).

Snap Ring Pliers

Many automotive components are held together with rings that snap into place under tension. Special pliers are required to remove and replace these rings. There are two major types of **snap ring pliers** as shown in Figure 3-36. *Inside snap ring pliers* have jaws that close and grip when the handles are closed like an ordinary pair of pliers. *Outside pliers* have jaws that open when the

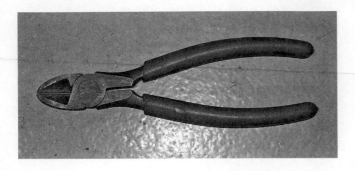

Figure 3-33 Diagonal side cutters or dykes are used to cut wires, not cable or metal.

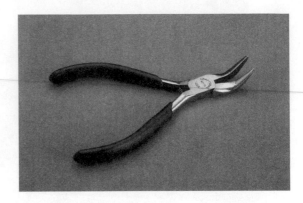

Figure 3-35 Bent needle nose pliers.

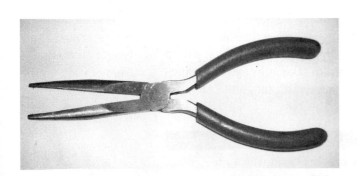

Figure 3-34 Needle nose pliers are good at picking up or holding when the component cannot be reached by other means. The longer the needle length the less torque can be applied. Excessive torque will warp the ends.

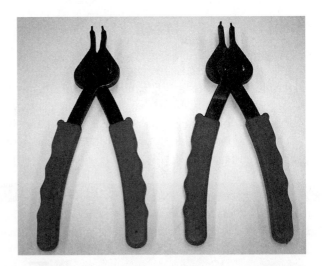

Figure 3-36 Snap ring pliers are used to remove/install snap rings. They are commonly used within transmissions and some air conditioning compressors.

handles are drawn together. The jaws of each are designed to safely engage the powerful snap rings to be removed or installed.

Vise Grips

Vise grips (Figure 3-37) are compound-lever pliers that can be locked to the part so the technician can remove her hand. The adjustable lower jaw is retracted by an overcenter spring when the locking handle is released. Adjustment of the jaw opening size is made by turning a screw at the end of the primary handle. Vise grips do not take the place of wrenches. They are, however, often used to hold nuts that have had their corners rounded off by the misuse of wrenches.

> **AUTHOR'S NOTE:** "Vise Grips," like "Channel-Lock," started out as the brand name of a certain type of pliers. The brand name stuck and is generally recognized as a specific type of pliers. Vise grips pliers are actually locking pliers. In instances like this and others, the brand name has become more familiar to people than the proper technical name.

Special-Purpose Pliers

There are many special-purpose pliers designed for certain repair jobs. The pliers shown in Figure 3-38 are used to remove and replace the spring tension clamps on some coolant hoses. The special groove in the jaws grip the ends of the hose clamps.

The pliers shown in Figure 3-39 are *battery pliers*. The jaws on these pliers are made to grip terminal nuts on battery connections. The tool shown in Figure 3-40 is used to remove hub-caps and grease caps. The round pad on the one jaw is used as a hammer to install grease caps.

The *crimping tool* shown in Figure 3-41 is used for electrical work. It has cutting edges to cut wire. There are special cutting edges for stripping the insulation off of wire. The jaws of the tool are designed to squeeze or crimp solderless terminals and connectors on the ends of wire.

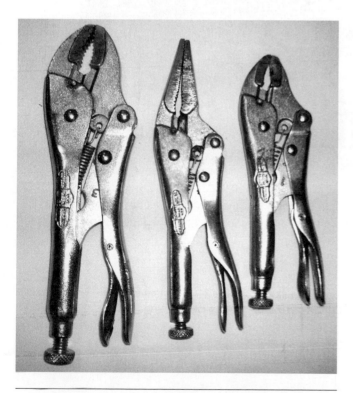

Figure 3-37 Like all pliers, vise grips are used to hold. They can provide a tight and locked fit for better holding.

Figure 3-38 This type of pliers is restricted in its use. Note the grooves used to capture and hold the ends of a hose clamp.

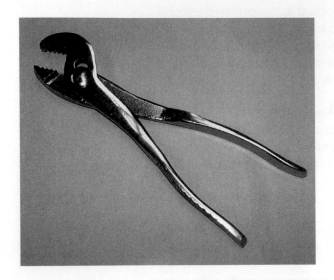

Figure 3-39 Battery terminal pliers.

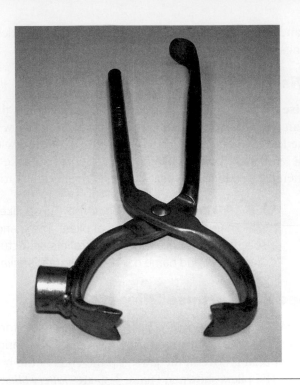

Figure 3-40 Grease cap pliers.

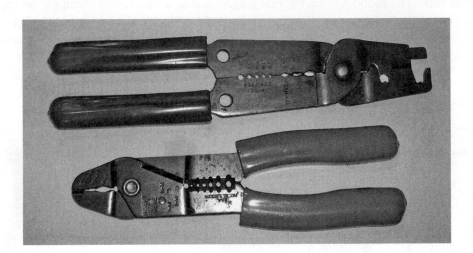

Figure 3-41 Crimpers are used to crimp electrical terminals onto an electrical conductor or wire. The top one can be used to crimp spark plug connectors onto spark plug wires on older vehicles.

Screwdrivers

Many automotive components are held together with screws. A screwdriver is used to turn or drive a screw. There are many different types of screws and screw heads. Consequently, there are many different types of screwdrivers to drive them.

Standard Screwdrivers

A standard **screwdriver** is the most common type. It is used to drive screws with a straight slot in the top. The main parts of the standard screwdriver are shown in Figure 3-42. The technician grips and turns the *handle*. The steel part extending beyond the handle is the *shank*. The part that

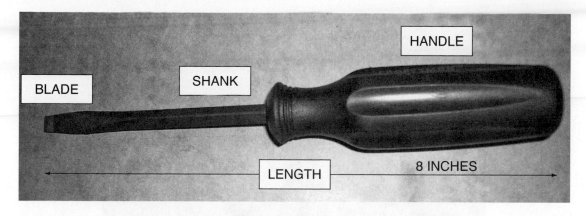

Figure 3-42 A typical 8-inch flat-tip screwdriver with a $\frac{1}{4}$-inch blade.

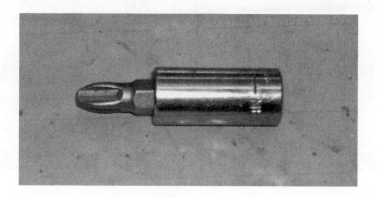

Figure 3-43 A screwdriver socket is almost always driven by a $\frac{1}{4}$-inch ratchet in the automotive business. Larger sizes are available.

fits into the screw head is the *blade,* which is often called a *bit.* The blade, or bit, is flat and ground at a right angle to the shank. The handle may be made from wood or plastic. The shank is designed to withstand a great deal of twisting force. The length of these screwdrivers is determined by their overall length. The larger the screwdriver, the larger its blade.

Screwdriver bits are available as socket drivers (Figure 3-43). These can be driven with the same drivers and attachments as socket wrenches.

Recessed-Head Screwdrivers

There are a number of screws with recessed-type heads. **Recessed-head screwdrivers** are required to fit these screws. The most common example is the *Phillips screwdriver* (Figure 3-44). This screwdriver is made to fit Phillips-head screws. Phillips-head screws are used primarily to hold moldings and other trim usually on the body or interior of the car. The heads of these screws have two slots that cross at the center and do not extend to the edges of the heads. This design has the advantage that the screwdriver head will not slip out of the slots and scratch the finish. The blades are sized on a numbering system from 0 to 6, with 0 being the smallest and 6 the largest.

The *Reed and Prince* (Figure 3-45) is another common recessed-head screwdriver. Reed and Prince cross-slot screws are similar to, and often confused with, Phillips-head screws. But the Reed and Prince slots are deeper and the walls separating the slots are tapered (Figure 3-46). The technician must be careful not to mix up the two types of screws. Using the incorrect screwdriver can result in damaged screw heads.

Recessed-head screwdrivers are made to fit recessed-head screws like Phillips, clutch, Reed and Prince, and torx.

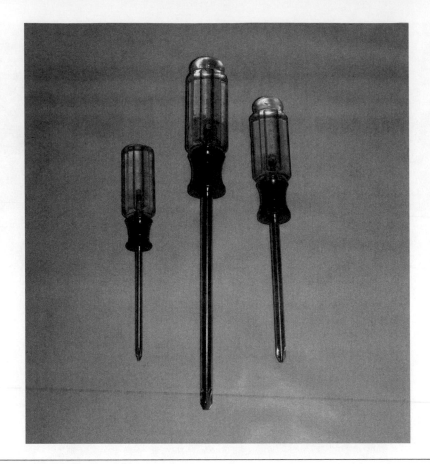

Figure 3-44 Typical Phillips-head screwdrivers. Note the different tip sizes along with the thinner shank and grip.

Figure 3-45 Reed and Prince screwdriver. This screwdriver usually doesn't work well with Phillips-head screws.

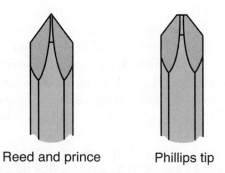

Reed and prince   Phillips tip

Figure 3-46 The Reed and Prince bit is longer than the Phillips.

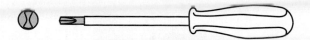

Figure 3-47 Clutch screwdriver. This screwdriver is not commonly used on vehicles.

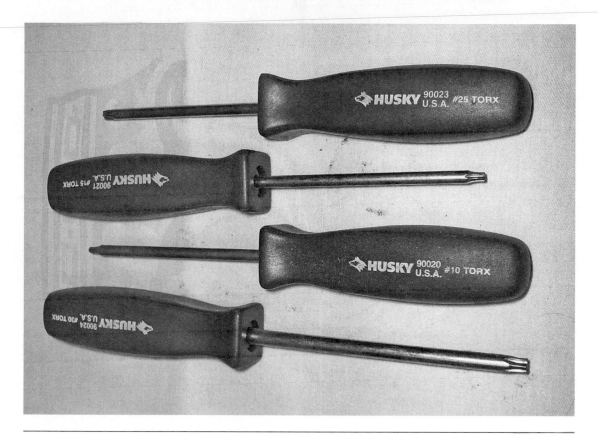

Figure 3-48 Torx-type fasteners are becoming more common on vehicle low-torque fasteners. They may be like the type shown or in a socket type driven by a ratchet.

Clutch screwdrivers are made for turning clutch-bit screws (Figure 3-47). These are also known as figure-eight screws. The screw head they fit looks like the number eight. Like the Phillips and Reed and Prince screws, clutch-type screw heads are designed to keep the blade from slipping off the screw head.

Another common recessed tip screwdriver, called a *torx-type screwdriver,* is shown in Figure 3-48. It has a six-prong tip and is often used to fasten automatic transmission parts.

There are many other types of recessed-head screws. Each requires a different type of screwdriver. Screwdrivers are also available that have interchangeable bits on the end of the shanks. Bits are available for each of these recessed-head screws.

Offset Screwdrivers

Offset screwdrivers (Figure 3-49) are designed for working in a tight space. The blades at opposite ends are at right angles to each other. The technician can change ends of the offset screwdriver after each swing and continue to turn the screw (Figure 3-50). These screwdrivers come in all the common blade or bit sizes.

Hammers

Hammers are used in many different trades. Every hammer has two basic parts: a head and a handle. The handles are made from wood or plastic. The heads are made from different materials for

Shop Manual
pages 47–52

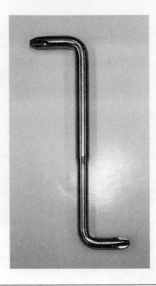

Figure 3-49 An offset screwdriver is handy in some dash/instrument work, but may be awkward to use in tight places.

Figure 3-50 The tips of an offset may be reversed or right angled to each other.

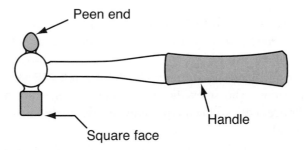

Figure 3-51 A technician's favorite tool, a ball-peen hammer. It is used to remove/install pin fasteners, shape metal, remove/install/position some components, and drive metal cutters like the chisel.

different jobs. The automotive technician will need to know how to use several different types of hammers. A good technician knows when and how to use the correct type of hammer.

Ball-Peen Hammers

The **ball-peen hammer** (Figure 3-51) is one of the most common hammers used by automotive technicians. These hammers have one face on the head that is square to the end of the handle. The opposite end is rounded and is called the ball peen. The square end is used for general hammering, such as driving a pin punch or chisel. The rounded ball peen is used to form or peen over the end of a rivet, as shown in Figure 3-52.

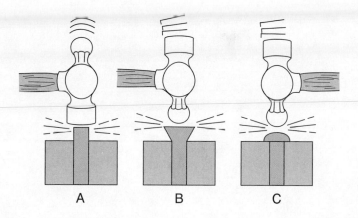

Figure 3-52 Using a ball-peen hammer to set a rivet.

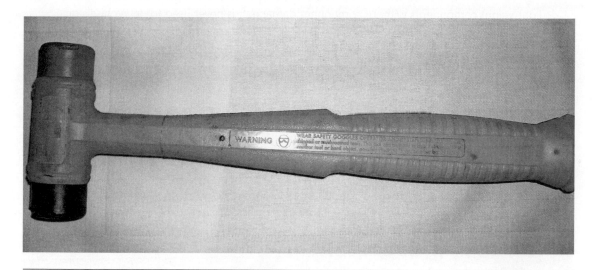

Figure 3-53 A plastic-head hammer can be used to position, remove, or install some lightweight components without damage to the components. It is never used to drive chisels or punches.

The hardened head of a ball-peen hammer should not be used to hammer on automotive parts. Because the head is harder than the part, the part will be damaged. Ball-peen hammers of different sizes are listed according to the weight of their head. Small ones weigh as little as 4 ounces; big ones weigh over 32 ounces.

Plastic Tip Hammers

Plastic tip hammers (Figure 3-53) have a head made from plastic. These heads are much softer than the ball-peen, steel-head hammers. They are used to align or adjust parts that might be damaged by a steel-head hammer. Plastic hammers will not damage metal parts because the plastic is softer than metal. Plastic can also absorb the force of the impacts instead of transmitting them to the metal parts. The technician should always remember to use a hammer that is softer than the material being hammered on.

Brass Hammers

Hammers with brass heads (Figure 3-54) are softer than ball-peen hammers. **Brass hammers** are much heavier than plastic ones. They can be used when more force is required. Brass hammers

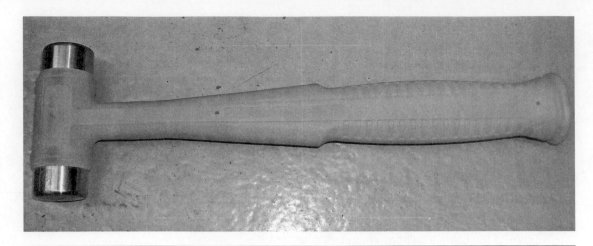

Figure 3-54 A brass-headed hammer is used to shape, install/remove, and position some lightweight components without component damage. It is never used to drive chisels or punches.

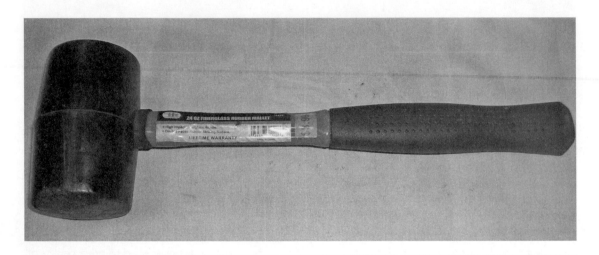

Figure 3-55 A rubber mallet is almost always used to shape metals. It is never used to drive chisels or punches.

are often used for driving pilot pins and other materials that will be harmed or distorted by steel hammers.

Rubber Mallets

Rubber mallets (Figure 3-55) are softer than brass or plastic. These are used to seat hub caps, wheel covers, and other trim parts. They are soft enough so that they will not damage the finish of parts like wheel covers.

Punches and Files

Punches are used with ball-peen hammers to drive pins or to make the center of a part to be drilled. The most common types are the *pin punch, starter punch, center punch,* and *cotter pin remover.*

Pin and Starter Punches

Some automotive components are held together with pins or rivets. Two different kinds of punches may be needed to remove a pin or rivet. A **starter punch** is used to break the pin loose

Shop Manual
pages 47–52

Tapered shank

Straight shank

Figure 3-56 A starter punch.

Figure 3-57 A pin punch.

(Figure 3-56). Then a **pin punch** smaller than the hole is used to drive a pin out of a hole (Figure 3-57). A starter punch also called a "drift punch," is made with a shank that is tapered all the way to its end. The tapered shank of a starter punch is stronger and will withstand greater shock from hammer blows than a pin punch. The starter punch is used to start the removal of pins and rivets.

Drifts are usually made of brass and are used to drive out hard or highly finished parts without damaging them. They are not always tapered. Long, tapered, steel drifts are often used to align two parts so that a bolt or screw can be installed.

Pin punches have straight shanks and are used to complete the removal of pins and rivets that have been started with a starter punch. The thin, straight shank of the pin punch is more likely to break under heavy blows from a hammer. The starter punch is used only to start a rivet or pin moving. The tapered design of its shank will prevent the starter punch from going all the way through. The pin punch is used to complete the job. Always use the largest starter and pin punches that will fit the hole (Figure 3-58).

The **center punch** is a short, steel punch with a hardened conical point ground to a 90-degree angle (Figure 3-59). A center punch is used to mark the centers of holes to be drilled. If the center punch is used correctly, the depression it makes in the metal will guide the point of the drill bit and make sure the hole is drilled in the correct spot.

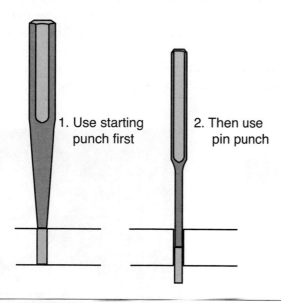

1. Use starting punch first

2. Then use pin punch

Figure 3-58 A starter punch starts the pin out of the bore, then a pin punch completes the operation. The punches are driven by a ball-peen hammer in most cases.

Figure 3-59 A center punch.

Figure 3-60 A cotter pin puller.

Cotter Pin Pullers

Diagonal cutting pliers are often used to pull out cotter pins after the ends have been straightened out or cut off. The **cotter pin puller** is a special tool that does the job more easily (Figure 3-60). This tool has a hook that engages the loop of the pin and a handle with which to pull it.

Pneumatic- and Electric-Powered Tools

Shop Manual
pages 36–42

Hand-operated wrenches have one basic disadvantage. They can be slow to use when there are many fasteners to remove. The technician's job can be done much faster using a wrench powered by electricity or compressed air.

Air-Operated Ratchet Drivers

An **air-operated ratchet driver** (Figure 3-61) is connected to an air line. Pulling the trigger causes the air to rotate a socket attached to the drive on the wrench. A reversing lever allows the technician to loosen as well as tighten. Many air wrenches are designed with an impact feature.

Impact Wrenches

A **center punch** is made of steel and used to mark the centers of holes to be drilled.

An **impact wrench** (Figure 3-62) is an air-operated wrench with a powerful driver. An impact wrench not only drives the socket but also vibrates or impacts it in and out. The force of the impact helps to loosen a bolt or nut that is difficult to remove. They are often used to remove and replace wheel lug nuts.

Most impact wrenches are not torque controlled. This means the technician must use a torque wrench to finish tightening after using the impact wrench.

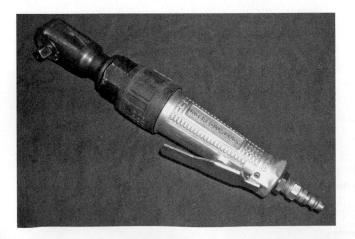

Figure 3-61 An air motor drives a reduction gear that in turn spins the socket driver. The rotation of the socket driver can be reversed for fastener removal/installation.

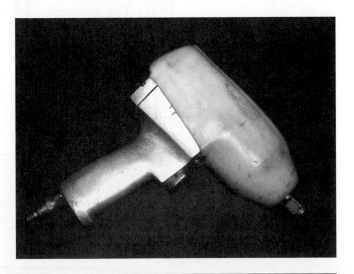

Figure 3-62 An impact wrench is handy and quick for removing fasteners. It should never be used to torque a fastener. Always use an impact socket with this wrench.

Figure 3-63 Designed for use with an impact wrench, it is a softer material than the chrome sockets. It can be hand driven with a ratchet.

Special heavy-duty impact sockets must be used with impact wrenches. Impact sockets are thicker and usually not chrome plated, as shown in Figure 3-63. Standard sockets cannot withstand the forces; they can break and cause serious injury.

Wheel Torque Sockets

Wheel torque sockets are commonly known as *torque sticks*. They are used with impact wrenches to deliver a specified torque to a fastener. They are similar to an extension in shape, but are used to prevent overtorquing a fastener while using an impact wrench for its speed. Tire stores are the primary users because of the high number of lug nuts that must be torqued each day.

Each stick is made for a certain torque and most toolmakers color them for that torque (Figure 3-64). The technician need not read the torque printed on the stick: just pick the right color for the desired torque and mount it to the wrench.

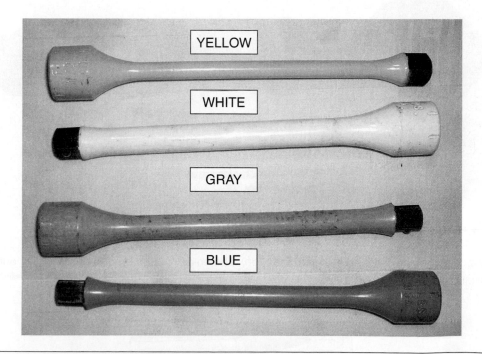

Figure 3-64 Color-coded torque sticks are designed to absorb torque above their designed limit. Some torque sticks have a fixed socket molded to one end sized to fit common wheel nut sizes.

The torque stick will deliver the wrench's output up to the stick's limit. Once the limit is achieved, the wrench's output is absorbed by the stick and no torque is delivered to the fastener. However, a torque wrench is still the best method to torque fasteners.

Some tool manufacturers offer adjustable torque impact wrenches. This wrench can be adjusted to provide accurate torque to many types of fasteners. Others offer an impact wrench that is preset to a maximum torque. Impact wrenches of this type work best on larger fasteners such as wheel lug nuts.

Metalworking Tools

Shop Manual
pages 47–52

Many automotive repair jobs require the technician to do some metalworking. Jobs like drilling holes to install accessories or cutting sheet metal to repair body rust damage are typical examples. The common metalworking tools are described here.

Chisels

Chisels are often used for cutting sheet metal, cutting off rivet heads, and splitting nuts that cannot be removed with a wrench. The most common chisel is called a *flat chisel* (Figure 3-65); these are often called "cold chisels." They are forged from round, square, rectangular, or hexagonal bars of tough carbon steel. A cold chisel is driven with a ball-peen hammer; the heavier the chisel, the heavier the hammer.

The *cape chisel* (Figure 3-66) is another common chisel. The cutting edge of a cape chisel is relatively narrower than the cutting edge of a flat chisel. The cape chisel is forged so that the cutting edge is slightly wider than the shank so that it will not bind when used to cut a narrow groove. Cape chisels are used for cutting narrow grooves such as keyways.

The cutting edge of a *round nose chisel* (Figure 3-67) has only one bevel. It is used mostly for cutting semicircular grooves and inside rounded-off corners.

Figure 3-65 A cold chisel with a holder (top). The holder can be used to grip chisel or punches while helping to protect the hands from missed hammer blows.

Figure 3-66 A cape chisel.

Figure 3-67 A round nose chisel.

Diamond-point chisels (Figure 3-68) are forged with a tapered shank of square section that is ground on an angle across diagonal corners. This results in a diamond-shaped cutting edge. Diamond-point chisels are used for cutting V-shaped grooves and for squaring up the corners of slots.

Always wear eye protection when using a chisel. Make sure the part to be cut is held securely in a vise. Hammering sometimes curls over or mushrooms the upper end of a chisel (Figure 3-69). The technician must grind the head smooth because the chips may fly off a mushroomed head and cause injury.

Twist Drills

The technician will often use **twist drills** mounted or chucked in an electric drill motor to drill a hole. The four parts to a twist drill are shown in Figure 3-70. The end of the drill is called the *point*. The spiral portion is made up of the *body* and the *flute*. The part that fits in the electric drill motor is the *shank*. A straight shank drill is commonly used in portable drill motors.

Twist drills are made in four different size groups: (1) fractional sizes from $1/64$ inch to $1/2$ inch and larger in steps of $1/64$ inch, (2) letter sizes from A to Z, (3) number sizes 1 to 80, and (4) millimeter sizes. Drill sets, or indexes, are sold in each size group. Drill size charts are available that list each drill and give decimal and metric equivalent sizes.

The size of a twist drill is stamped on the shank. But, after a great deal of use, the stamp may be difficult to read. A drill may be measured with a drill gauge. The drill gauge shown in Figure 3-71 is a metal plate with holes identified by size. The drill to be measured is placed in the holes until it is found which size hole best matches the drill.

Figure 3-68 A diamond-point chisel.

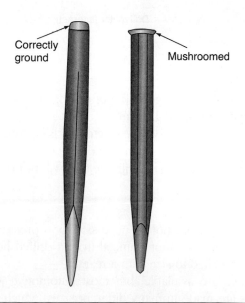

Correctly ground

Mushroomed

Figure 3-69 The mushroomed end of a chisel must be ground off.

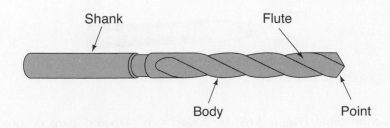

Figure 3-70 Parts of a twist drill.

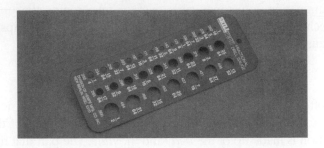

Figure 3-71 A drill gauge for measuring the size of a drill.

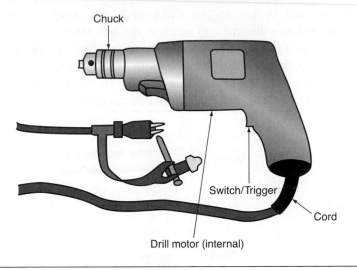

Figure 3-72 Parts of a portable electric drill motor.

A twist drill must be sharp to do a good job of cutting. The point of the drill is sharpened with a grinding wheel, usually in a special fixture.

Twist drills are most often used in portable electric drills (Figure 3-72). There are three common sizes of portable drills based on the size of the drill chucks. These are $1/4$, $3/8$, and $1/2$ inch. The size of the chuck determines the largest drill size that will fit inside.

Reamers

A **reamer** has cutting edges used to remove a small amount of metal from a drilled hole.

A **reamer** is used when a very precise hole is necessary for a precision fit. A reamer has cutting edges designed to remove a small amount of metal from a drilled hole. Many automotive parts have bushings that must be finished to size with a reamer.

Machine-driven reamers are available, but most automotive jobs require a hand-driven reamer. Like drills, reamers are made in many different sizes, which are stamped on the shank.

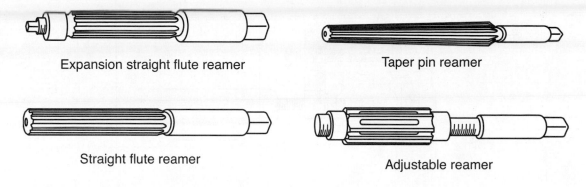

Expansion straight flute reamer

Taper pin reamer

Straight flute reamer

Adjustable reamer

Figure 3-73 Reamers are used to fit a bore to exactly the right dimension for the component or fastener being installed. Reamers are most commonly used for piston pin bushings and certain suspension/steering bushings.

Reamers are also made that adjust to many different sizes. Some common types of reamers are shown in Figure 3-73.

A reamer should be turned with a wrench or with a tap wrench in a clockwise direction. Turning a reamer backward will quickly dull its cutting edges.

Hacksaws

Hacksaws are made to cut metal. A technician may use a hacksaw to cut exhaust pipes and other metal parts that are made during a repair job. They consist of two parts: the *frame* and the *blade* (Figure 3-74). Almost all hacksaw frames are made so that they can use 8-, 10-, or 12-inch blades. The hacksaw blade installed in the frame is the part that does the cutting. Blades are made in different lengths and also with different numbers of teeth per inch (TPI). The common TPI are 14, 18, 24, and 32. The number of teeth is marked on the hacksaw blade. For almost all automotive work, the 18- and 32-tooth blades are best. The 18-tooth blade is used for all sawing except thin metal, such as sheets or tubing, which should be sawed with 32-tooth blades.

When installing a hacksaw blade in a frame, always have the teeth pointing toward the front of the frame and away from the handle (Figure 3-75). Hold the saw with both hands and push forward and down to cut. Release the pressure to back the saw up for the next stroke. Take about one stroke per second.

A **hacksaw** is made to cut metal.

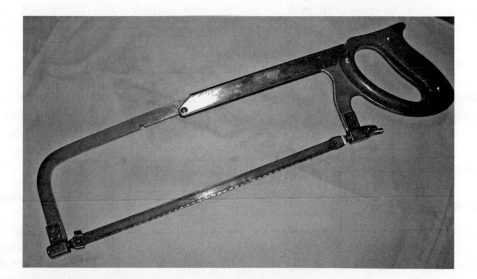

Figure 3-74 A hacksaw is used to cut metal only. It is commonly used on exhaust systems.

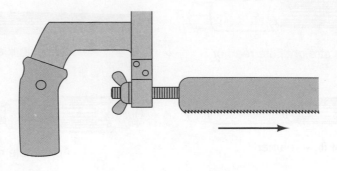

Figure 3-75 The teeth point forward or away from the handle.

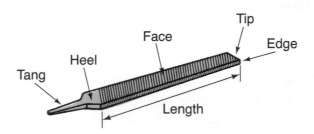

Figure 3-76 The parts of a file.

Files

A **file** is a hardened, steel tool with rows of cutting edges used to remove metal for polishing, smoothing, or shaping.

A **file** is used to remove metal for polishing, smoothing, or shaping. The parts of a file are shown in Figure 3-76. The *face* is the larger part of the file with the cutting teeth. The *tang* is the part that goes into the handle. The tapering part between the tang and the face is the *heel*. Files are made in different lengths, measured from the *tip* to the heel. The tang is shaped to fit into the handle. A handle must always be attached when filing to protect the technician from the sharp tang. The handle is set tightly on the file by striking the handle on a workbench, as shown in Figure 3-77.

Files with cutting edges that run in only one direction are called *single-cut files*. Files made with cutting edges that cross at an angle are referred to as *double-cut files* (Figure 3-78). The cutting edges may be spaced close together or wide apart. The wider they are spaced, the faster the file will remove metal. A file whose cutting edges are closer together will remove less metal and can be used to smooth or polish a metal surface. Files are also available in different shapes: flat, half-round, round, triangular, and square (Figure 3-79).

When using a file, grip the handle with one hand. Push down on the file face with the other hand. Because a file is designed to cut in only one direction, raise the file on the return stroke. Dragging it backward dulls the cutting edges. Mount small parts in a vise for filing. The correct way to hold a file is shown in Figure 3-80. When the teeth on the file become clogged with metal filings, remove them by tapping the file handle or brushing the teeth with a file card.

Threading Tools

Shop Manual
pages 50–52

A **tap** is used to cut internal threads.

Many repair jobs involve repairing or replacing the threads on automotive parts. A **tap** is a metalworking tool (Figure 3-81) designed to make or repair inside threads. A tap of the correct size is installed in a holding tool called a tap wrench (Figure 3-82). The tap is then turned in the hole to make a new thread or to repair damaged ones. Taps are available for all the common thread sizes.

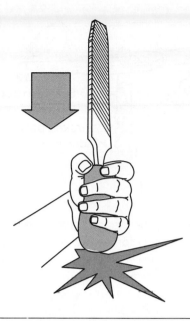

Figure 3-77 The file handle is held on only by the V-shaped portion of the file. The handle can come loose during file usage and the V portion can stab deep into the hand or wrist. Check the security of the handle often during use.

Figure 3-80 Correct way to hold a file.

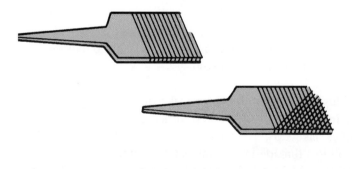

Figure 3-78 Single- and double-cut files.

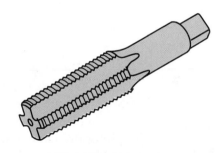

Figure 3-81 Taps are used to make internal threads in a bore or repair internal threads. An additional use is to clean internal threads. The tap's thread pitch and diameter must be matched exactly to the fastener being used.

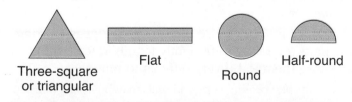

Three-square or triangular   Flat   Round   Half-round

Figure 3-79 Different types of file shapes.

Figure 3-82 A typical tap driver or tap wrench.

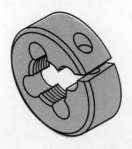

Figure 3-83 A die performs actions similar to those of a tap except it is used on external threads.

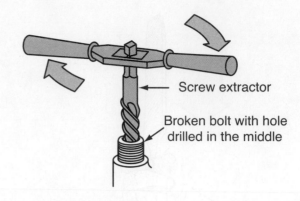

Screw extractor

Broken bolt with hole drilled in the middle

Figure 3-85 The extractor reverse (left-hand) flutes grip the fastener for removal. A hole must be drilled into the fastener for the extractor's insertion.

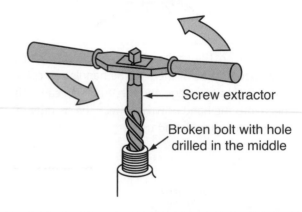

Screw extractor

Broken bolt with hole drilled in the middle

Figure 3-86 Using a screw extractor to remove a broken screw or stud.

Figure 3-84 A typical die stock or die wrench.

A **die** is used to cut external threads.

A **screw extractor** is used to remove broken screws, bolts, or studs from automotive parts.

A **gear puller** is a tool used to remove gears, bearings, shafts, and other parts off shafts or out of holes.

A **die** (Figure 3-83) is a metalworking tool used to repair or make outside threads. A die is installed in a tool called a *die stock* (Figure 3-84). The die is then turned down over the part to make new threads or repair damaged ones. Like taps, dies come in a variety of thread sizes.

When a bolt or screw has broken off in a threaded hole, the technician must remove it. One tool for this job is a **screw extractor** (Figure 3-85). A hole of the proper diameter is drilled into the broken screw. A screw extractor is inserted into that hole. A wrench is used to turn the screw extractor counterclockwise. The reverse threads dig into the broken screw and allow the technician to unscrew it as shown in Figure 3-86.

Gear and Bearing Pullers

All technicians should be able to set up and use a **gear puller** or bearing puller. Most shops have a large selection of these pullers (Figure 3-87). These tools are designed for hundreds of common service jobs on the automobile. Although the number and variety of puller tools is huge, they are all designed to do three simple jobs (Figure 3-88).

1. Pulling something off a shaft: A common service job is to pull a gear, bearing, or pulley off a shaft. These parts are made to fit tightly together. Rust and corrosion on the shaft can cause them to be very difficult to remove.

2. Pulling something out of a hole: Parts like bearing cups and seals may be installed inside holes with a press fit. These parts would be very difficult to remove without a puller.

3. Pulling a shaft out of something: Shafts are installed in many transmission components. These parts are very difficult to remove except with the correct puller.

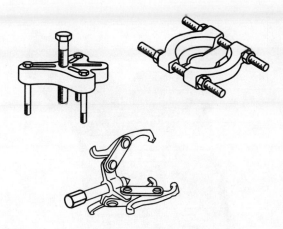

Figure 3-87 Pullers come in many shapes and sizes. Shown are the most common types used in automotive repair.

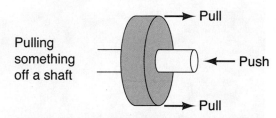

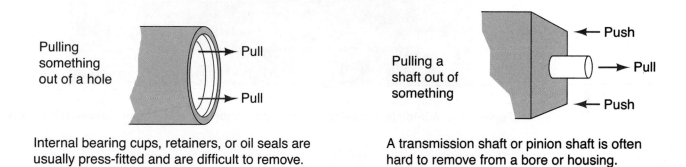

Figure 3-88 A puller can be used to remove a gear from a shaft (top), remove a bearing from inside a bore (bottom left), or remove a shaft from a bore (bottom right).

Equipment

Shop Manual
pages 33–36

There are many different types of equipment in an automotive shop. They range from large $^3/_4$-inch and 1-inch drive socket sets to automotive lifts capable of lifting up to 9,000 pounds safely.

Jack Stands, Wheel Blocks, and Jacks

Jack stands come in different heights with different load capabilities. The smallest can extend to about 24 inches in height and will support 2 to 4 tons of weight (Figure 3-89). Jack stands are used to support the vehicle or one end so the technician can move under it on a creeper. Most under-vehicle work can be performed using this type of stand.

Figure 3-89 A 3-ton jack or safety stand used to support a vehicle during undercarriage work.

Other stands are much taller and designed to support a portion of a vehicle raised on a lift (Figure 3-90). For instance, a stand may be used to hold a rear-axle housing as the suspension system is being repaired. This type of stand is usually rated at a 1-ton capacity.

Wheel Blocks

Wheel blocks are placed forward and rearward of the wheels to prevent the vehicle from rolling. Chocks should always be used if only one end of the vehicle is being raised. Place the blocks forward and rearward of at least one wheel left on the floor (Figure 3-91).

Jacks

The most common shop jack is a **floor jack** (Figure 3-92). This type of jack may be rated from 1 ton to 10 or more tons. However, the type found in most automotive shops is rated between 2 and 4 tons. Floor jacks are designed to roll as the vehicle is lifted. This keeps the lift plate in position under the vehicle, but it *can* present a problem. On an incline, the jack and vehicle could roll down the slope. This is the reason for using wheel blocks.

Wheel blocks are also known as "wheel chocks."

Floor jacks are designed to roll as a vehicle is lifted, thereby keeping the lift plate in position.

Figure 3-90 A tall jack stand is commonly used to help steady a vehicle on a lift during removal of heavy components. Also used to hold components in alignment for connections with other parts and for installation of fasteners.

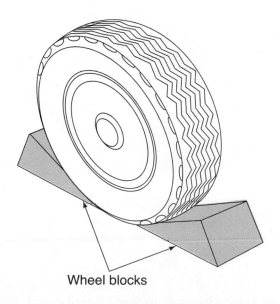

Wheel blocks

Figure 3-91 Blocks should be rearward and forward of the wheels before jacking.

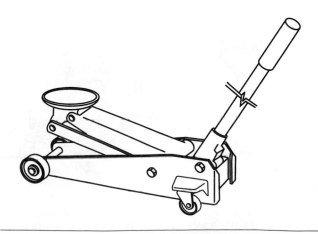

Figure 3-92 Floor jacks are designed to travel under the lift point as the vehicle is being raised.

There are also **transmission jacks.** They can be tall enough to fit under a lifted vehicle or low enough to be used with the vehicle on jack stands (Figure 3-93). This type of jack has attachments to secure the transmission to the jack.

Lifts

Automotive lifts make the technician's work much easier. The vehicle can be raised so almost any under-vehicle repair can be easily accessed and accomplished. The entry-level technician will use a lift to change oil and inspect or replace tires.

Single Posts

Single-post lifts are older types of lifts. The hydraulic cylinder is buried in a floor cavity. The lift arms and the top of the cylinder present a blockage to an otherwise smooth floor (Figure 3-94). In addition, many of the under-vehicle components like the transmission and exhaust system are blocked by the arms when the vehicle is raised. The floorboard and exhaust can also be damaged if the lift arms and pads are not positioned correctly. There is usually only one safety lock positioned at the lift's full extension. Any position between *completely down* or *completely up* relies on the hydraulic and air control valves to hold the vehicle in place.

Double Posts

The **double-post lift** is also known as the "above-ground lift" (Figure 3-95). It is designed to lift a vehicle and allow the technician to perform almost any repairs needed under the car. Many lifts of this type have a crossover between the two posts at the top. This provides a clear work area under the vehicle.

Most lifts of this type use two equalizing hydraulic cylinders to raise the vehicle. Cables or flat chains run over pulleys and attach to each cable, ensuring that the vehicle is lifted evenly. There are locks at different heights so the vehicle can be safely locked in place at several working heights.

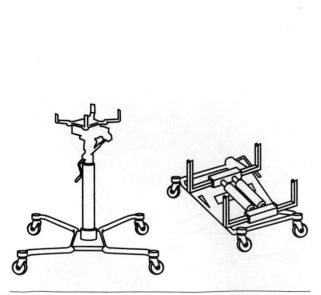

Figure 3-93 Transmission jacks are used to support the transmission during removal and installation.

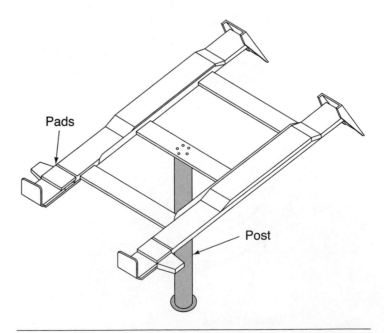

Figure 3-94 Single-post lifts have the post extending from and retracting into the floor.

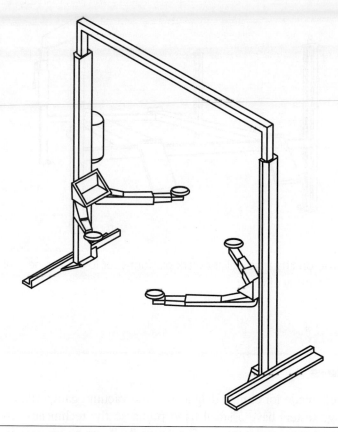

Figure 3-95 Double-post lifts with an overhead crossover leave a clean floor and work area. (Courtesy of Snap-on Tools Company)

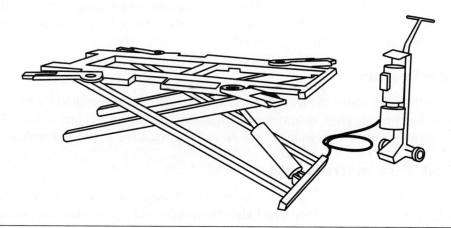

Figure 3-96 A scissor lift works very well for brake and tire repairs. (Courtesy of Snap-on Tools Company)

Scissors and Drive-Ons

Scissor lifts are best used for brake or tire service. They fit under the vehicle and have limited height (Figure 3-96). As such, there is almost no room to work under the vehicle.

Drive-on lifts allow the technician full access to under-vehicle components (Figure 3-97). Unless equipped with secondary jacks, the vehicle rests on its tires for the duration of the repairs. Drive-on lifts with secondary jacks are generally used for alignment. The vehicle can rest on its tires for alignment or the axles can be raised for access to the wheel and brake assembly.

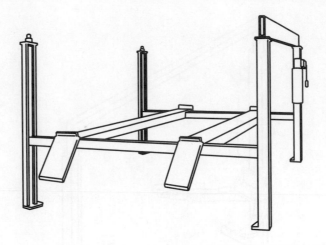

Figure 3-97 Drive-on lifts are excellent work platforms for lower engine, transmission, and exhaust repair. (Courtesy of Gray)

Testing Tools

Vacuum Testers

One of the best testers made for engine diagnosis is the vacuum gauge (Figure 3-98). Most automotive vacuum gauge testers have a small hand pump so the technician can supply a vacuum source to test vacuum motors and controls. The gauge itself can be used to diagnose the internal workings of an engine by reading the engine's intake vacuum under different operating conditions. Even on today's computer-controlled engine the vacuum gauge is still one of the best tools to diagnose internal engine mechanical problems. The vacuum gauge/pump can be used to measure or apply vacuum to any vacuum-operated components on the vehicle, such as the blend door motors on heating and air conditioning systems.

Pressure Testers

Like the vacuum gauge/pump, the pressure tester is a tool that has been around for awhile (Figure 3-99). It is used to measure pressure in systems such as the transmission and fuel systems. About the only change to an automotive pressure tester is the ability to read higher pressure without damage.

Electrical Test Instruments

Shop Manual
pages 90–92

Electrical theory is covered in Chapter 4, "Automobile Theories of Operation." Electrical measurements have become a way of life with today's technician and are essential test devices. Some of the electrical terms used here are also explained in Chapter 4, where they are discussed within the context of electrical theory.

Test Lights

AUTHOR'S NOTE: Test lights are now on the market that use light-emitting diodes (LEDs) and can be used to check electronic circuits. Before using one, ensure it can check electronic circuits without damaging them.

Even though the *test light* is old technology, it is still valid in many situations (Figure 3-100). One end, the *ground clip,* is attached to a metal portion of the vehicle. The other end, the *probe,* is used to probe the conductor. The test light will indicate voltage, but not the amount. If voltage is present, the lamp will light. It is used on heavy current circuits like the headlight and cooling fan motors.

Figure 3-98 One of the oldest types of engine testers. This vacuum gauge can also be used to measure pressure, up to about 10 psi, for mechanical fuel pump testing.

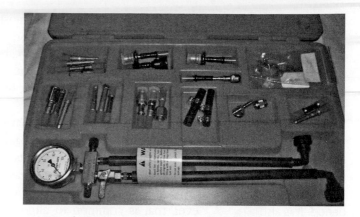

Figure 3-99 A typical pressure gauge set for testing fuel pressure on fuel-injected engines. Note the maximum reading of 100 psi on the gauge.

Multimeters

The multimeter is used to check almost any circuit on the car, but caution must be used when selecting one for purchase. It should be an automotive meter of **high-impedance** and preferably digital (Figure 3-101). A high-impedance meter is used to protect electronic circuits during testing. It will have two regular leads: a black and a red. Automotive multimeters have additional leads for temperature, engine speed, and other measurements exclusive to the

High-impedance is basically the same as high-resistance. It is required to obtain an accurate measurement of electricity.

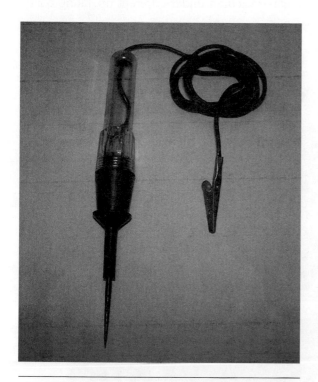

Figure 3-100 The test light on the left is simple to use, but should never be used on electronic devices. The LED light on the right can be used on electronic devices. Both will only show voltage is present, but not how much.

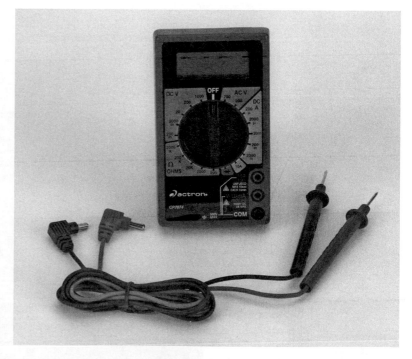

Figure 3-101 An automotive digital multimeter (DMM) is used to test electronic devices and some engine data. It can be used to perform simple voltage, resistance, and current measurements.

automobile. The controls on most multimeters include selections for the three basic electrical measurements: voltage, current, and resistance. Some meters require the operator to select the range of measurements. The better and more expensive ones have automatic ranging. *Ranging* is the scale of measurement. For instance, if voltage is being read, the operator may have to select a 0-to-20-volt scale or a 20-to-100-volt scale. Automatic ranging meters do this automatically.

Automotive multimeters have selections for revolutions per minute, temperature, **frequency,** and others depending on the make and model of the meter. Additional leads are required to make the most of these measurements.

There are meters that measure only resistance (ohmmeter), voltage (voltmeter), or current (ampere). Meters of this type are not common in the automotive field. There is one meter, however, that is common to almost every tire, parts, and repair facility. It is a *volt/ampere tester* (Figure 3-102). The VAT 40 is dated technology because it is an *analog* type of meter, but many still exist and work as good as new. An analog meter is similar to the speedometer on a car where a moving needle points to a number for the measurement. Its replacement, the VAT 60, performs the same tests, but the display data is in a digital format. Both are used to test batteries and charging systems. This is one automotive meter that is capable of reading high current. On a starting system, the current may be as high as 400 to 500 amperes. Most multimeters can read no more than 10 amperes.

AUTHOR'S NOTE: None of the VAT testers will test a dead battery. There is a test with the VAT 40 that allows the battery to be charged for 3 minutes and then tested. This particular test does not mean the battery is good, but it can tell the technician if the battery will accept a charge. A battery should have at least 12 volts before testing to achieve a correct test result.

Figure 3-102 A volt amps tester (VAT). The top unit is a digital VAT 60. This unit has a battery charger mounted in the bottom of the frame and is sold as the VAT 45. Note the battery tools on the mid-shelf.

Frequency is the number of times an action or repeated action happen within a given time period. It is also known as "duty cycle."

Computerized Test Equipment

Computerized Data Storage and Retrieval

Paper service manuals are rapidly being replaced by **computerized data** storage (Figure 3-103). A series of CD-ROM discs contain all of the information found in many different conventional service manuals. The data can be retrieved and printed quickly and easily. Most of the programs are user friendly and require no training other than some hands-on experience. Fax, e-mail, Internet, and direct satellite uplinks can also transmit computerized data. In addition, storage is not required except for the space required for the computer terminal. There are aftermarket data systems such as Mitchell's On-Demand and Snap-on's Alldata.

The computer data system is less expensive after initial setup than purchasing the required paper manuals. The system is designed for a shop. With the correct software and computer hookup, the shop and even the technician can ask, receive, and give technical assistance over the Internet. Groups like the Coordinating Committee for Automotive Repairs, International Automotive Technician's Network, and the National Institute for Automotive Service Excellence provide on-line tips and conduct forums for automotive representatives to discuss repairs and the business in general. Many automotive businesses have Internet sites along with libraries and schools that provide technical assistance and general information. An Internet search for automotive sites can be made using the keyword "automotive."

There are many different types of test equipment on the market today. They are designed to communicate with the vehicle's on-board computer or with a specific system. Most are known as *scan tools* or *hand-held lab scopes* (Figure 3-104). Computerized test equipment retrieves data and **diagnostic trouble codes (DTCs)** for the technician to use in diagnostic procedures. They will only collect data on electrical troubles. Mechanical problems will not show up as a problem, but may cause a sensor to transmit erroneous information to the computer. This data may show as a sensor problem on the scan tool. Correct interpretation of the information is essential for correct diagnosis.

A larger, more expensive test instrument is the **digital storage oscilloscope (DSO)** (Figure 3-105). Shops with the customer base and funds to purchase a DSO can diagnose a fault down the last loose screw. The small, hand-held type will perform many of the functions of the larger scopes, but the large unit usually performs a more detailed readout of several data streams from the vehicle's computer at once. They are also quicker, which provides a better readout of what is happening within the vehicle's electronic systems. Most are equipped with pickup tubes to sample

Shop Manual
page 53

Diagnostic Trouble Codes (DTCs) are numerical data that can be retrieved from the vehicle's computer with a scan tool. Each code designates a specific fault.

The **Digital Storage Oscilloscope (DSO)** is a large, expensive piece of equipment. It is capable of storing and displaying enormous amounts of data in graph, wave, and numerical forms.

Figure 3-103 Paperback manuals are almost nonexistent in many automotive shops, but are common for the do-it-yourself mechanic.

Figure 3-104 Shown is a common scan tool/graphing instrument used to collect electronic data from the vehicle's computers and assist in pinpointing the faults in the vehicle systems.

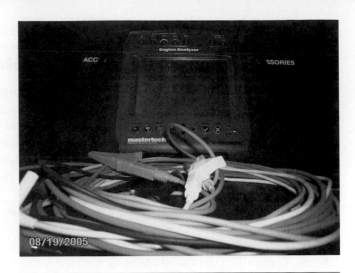

08/19/2005

Figure 3-105 This engine analyzer is dedicated to more electronic graphing, but not the full digital data retrieval capabilities of other analyzers. Most shops have versions of both types.

the exhaust and provide a total readout of four or five gases. By analyzing the amount of oxygen (O), carbon monoxide (CO), carbon dioxide (CO_2), hydrocarbons (HC), and nitrogen oxides (No_x), gases within the exhaust provide the technician with information on the operation of the engine and its systems.

Service Information

Before you assume that all of the tools studied to this point are all you will ever need, dream on. There are probably hundreds of different types of tools and equipment needed for today's cars and trucks. Tools range from an exhaust wrench (a curved box-end wrench) to specialized screwdrivers and sockets, and from equipment providing 25 tons of force to remove bearings to noid lights that test electrical impulses.

Service information identifies the vehicle, its major components, diagnostic checks and procedures, repair procedures, labor times, and much more. This data may be available in a paper book, on a computer hard drive, or through the Internet. Even procedures that seem simple, such as changing a tire, may become a problem if the vehicle is equipped with devices to measure tire inflation. Either now or in the near future, everything on the vehicle will be monitored or controlled by electrical/electronic devices except possibly the operator. To repair almost any passenger vehicle requires some access to service information. Some of the terms in this section have been covered earlier, so let's take a typical repair order and walk through the steps a technician follows to perform a "fixed right the first time" job.

An operator tells the service writer there is a shift problem with his six-speed automatic transmission. The writer gathers more details from the owner and starts a repair order. The assigned technician receives the repair order and realizes he has not worked on this type of vehicle, but is somewhat experienced in transmission diagnosis. After determining the vehicle is safe to operate, it is taken on a short test drive to verify the customer's complaint. A problem definitely exists with the transmission shift pattern. But is it caused by the engine, transmission, or some other defect? This is where access to service information is important.

Using the car's vehicle identification number (VIN), the technician locates the type of engine, transmission, and other items that may pertain to this vehicle's problem. The next step is to check the repair history of this vehicle to see if similar repairs have been accomplished. No similar repairs were found except for a replacement of the vehicle speed sensor (VSS) several weeks earlier. Since the VSS has an impact on transmission operation, the technician makes a note and continues the research.

A search of recall or technical service bulletins may point to the manufacturer having experienced this transmission problem with other similar vehicles. If any such reports are found, the technician's job may be half done at this point. Assuming no reports were found, the technician accesses the diagnostic procedures for this problem. He may retrieve DTCs from the vehicle computer and use them as a starting point for diagnostic steps. Usually on this type of problem, the "check engine" lamp will be illuminated, indicating a DTC. The technician can now pull up procedures under the heading "Testing with Codes" or a similar heading. This will lead to the diagnostic procedures that are most likely to find the fault. If no codes are present, go to the heading "Testing without Codes." This will lead the technician back to the "Codes" steps in many instances.

With this information, the actual repair can begin. The diagnostic steps will tell the technician what test tools to use and what the correct readings should be. The steps are arranged in a diagnostic tree, which is nothing more than a chart indicating "do this step, if the data is this; then go to next step or go forward to a specific step." This will eventually lead to the faulty part or system.

Another source of service information rests in the technician's head. If he understands how a typical modern transmission/transaxle works and how it interfaces with the other systems of the vehicle, then the problem is half solved. In this case, a problem with the engine may cause the computer to have the wrong data for proper shifting of the transmission. Or it could be a simple problem with thick, nasty transmission fluid that doesn't flow like it should. Either way, an understanding of what is supposed to happen can lead to a quick, accurate repair. Typical time for an average technician to access the service information and begin the diagnosis is perhaps 10 minutes.

Summary

❏ Wrench sizes are determined by the size of the bolt head or nut they fit. Sizes are given in either the U.S. (English) or metric system. Common wrench types are open-end, box-end, combination socket, and adjustable.

❏ Pliers are used as an extension of the technician's fingers. Common pliers are combination, channel lock, diagonal cutting, and vise grips.

❏ Screwdrivers are used to drive screws. There are many different types of screwdrivers to fit the many types of screw heads. Slotted screws are driven by the standard screwdriver. Recessed-head screws—such as Phillips, Reed and Prince, and clutch type—require a different screwdriver.

❏ Hammers are used to drive tools or position parts. The ball-peen hammer is used for driving punches and chisels. Softface hammers—such as brass, plastic tip, and rubber—are made to protect the parts being hammered on.

❏ Punches are used to remove parts held together by pins. Removing a pin requires both a starter punch and a pin punch. A center punch is used to make a mark for locating a hole to be drilled.

❏ Power wrenches are used for rapid removal or installation of bolts or nuts. Common power wrenches are the electric wrench, air-operated ratchet driver, and the impact wrench.

Terms to Know

Adjustable wrench
Air-operated ratchet driver
Allen wrench
Ball-peen hammer
Box-end wrench
Brass hammer
Center punch
Channel-lock pliers
Chisel
Combination wrench
Computerized data
Cotter pin puller
Diagonal cutting pliers
Diagnostic Trouble Codes (DTCs)
Die
Digital Storage Oscilloscope (DSO)

❏ Many repair jobs require the use of metalworking tools. Chisels are used for cutting metal. Twist drills are used to cut holes. Reamers are used to finish holes to an exact size. Metal is cut with a hacksaw. Files are used for removing, shaping, or polishing metal.

❏ Threads are made or repaired with several threading tools. Dies are used to make or repair outside threads. Taps are used to make or repair inside threads. If a screw is broken while still in a part, it can be removed with a screw extractor.

❏ Jacks that come with the vehicle for tire changing should not be used for any other repair work.

❏ Jacks are designed to lift only one end or one side of a vehicle.

❏ Lifts may be powered by air rather than hydraulics when air is used to operate under-floor controls for the hydraulic system.

❏ Service manuals may be on paper or computerized.

❏ Computerized data banks can contain data for many years, makes, and models of vehicles.

❏ Web sites on the Internet can provide assistance on vehicle repair or information on buying a vehicle.

❏ Computerized test equipment can interface with the PCM to display sensor and actuator data.

❏ Large DSOs display more data than hand-held scanners and scopes. They can also be equipped to measure certain exhaust gases.

❏ The most common electrical measuring instrument is the multimeter.

❏ Multimeters must be high-impedance to prevent damage to electronic circuits.

❏ VAT 40 and VAT 60 meters are used to test starting and charging systems and the battery.

Review Questions

Short Answer Essay

1. Describe how to determine the size of a wrench to use on a nut or bolt.
2. Explain how to safely pull and push on a wrench in a tight space.
3. What determines the size of a ratchet handle?
4. What is the purpose of a socket extension?
5. List the three main types of torque wrenches.
6. Describe how to safely use an adjustable wrench to loosen a nut.
7. Explain how to safely use an adjustable wrench to tighten a nut.
8. Explain why special heavy-duty sockets are used with an impact wrench.
9. What is the purpose of a die?
10. What three basic electrical measurements can be made with a high-impedance multimeter?

Fill-in-the-Blanks

1. The two different measuring systems for wrench sizes are _____ and _____.

2. A wrench with a box-end and an open-end is called a(n) _____ wrench.

3. Torque can be measured in U.S. (English) units called _____-_____. Torque can be measured in metric units called _____-_____.

4. When using an adjustable wrench, care must be taken to apply the load to the _____ jaw.

5. When using an impact wrench, be sure to use _____ sockets.

6. A(n) _____ hammer is used to drive a chisel.

7. The _____ will measure resistance only.

8. A file should never be used without a(n) _____.

9. A drilled hole can be finished to a specific size with a(n) _____.

10. A broken screw can be removed with a(n) _____ _____.

ASE-Style Review Questions

1. Two technicians are discussing a wrench with "size 10" stamped on the handle.
   Technician A says this is a metric-size wrench.
   Technician B says the 10 indicates the maximum amount of torque this wrench can handle without breaking.
   Who is correct?
   - **A.** A only
   - **B.** B only
   - **C.** Both A and B
   - **D.** Neither A nor B

2. The use of a wrench to remove a tight nut is being discussed.
   Technician A says an open-end wrench provides better contact between the fastener and the wrench.
   Technician B says the box-end wrench is a better choice.
   Who is correct?
   - **A.** A only
   - **B.** B only
   - **C.** Both A and B
   - **D.** Neither A nor B

3. Two technicians are discussing which of two socket wrenches to use on a tight nut.
   Technician A says using a twelve-point socket always provides twice as much contact surface between the fastener and socket.
   Technician B says a six-point socket is best on a hex-head bolt.
   Who is correct?
   - **A.** A only
   - **B.** B only
   - **C.** Both A and B
   - **D.** Neither A nor B

4. Two technicians are discussing a torque reading that is specified in U.S. (English) units.
   Technician A says a low torque, such as 12 foot-pounds, is best applied with a box-end wrench.
   Technician B says there are 12 inch-pounds in a foot-pound.
   Who is correct?
   - **A.** A only
   - **B.** B only
   - **C.** Both A and B
   - **D.** Neither A nor B

5. A torque specification calls for a reading of 144 inch-pounds.

   Technician A says this is approximately the same as 12 foot-pounds.

   Technician B says torque this large should always be converted to foot-pounds and applied with a heavier torque wrench.

   Who is correct?

   A. A only   **C.** Both A and B

   B. B only   **D.** Neither A nor B

6. A torque specification calls for a reading of 72 inch-pounds.

   Technician A says applying 72 inch-pounds is best done with a box-end wrench.

   Technician B says this torque would probably be used on a small fastener.

   Who is correct?

   A. A only   **C.** Both A and B

   B. B only   **D.** Neither A nor B

7. Two technicians are discussing an Allen wrench with an 8 stamped on the handle.

   Technician A says the wrench fits metric Allen-head screws.

   Technician B says this Allen wrench will also properly drive a U.S. (English) Allen-head screw.

   Who is correct?

   A. A only   **C.** Both A and B

   B. B only   **D.** Neither A nor B

8. Two technicians are discussing the use of an impact wrench.

   Technician A says heavy-duty impact sockets must always be used with an impact wrench.

   Technician B says standard sockets are designed to split open if used with an impact wrench to reduce possible damage or injury.

   Who is correct?

   A. A only   **C.** Both A and B

   B. B only   **D.** Neither A nor B

9. The removal of a pin from a part is being discussed.

   Technician A says a starter punch is used to begin the operation.

   Technician B says the final removal of the pin is done with a pin punch.

   Who is correct?

   A. A only   **C.** Both A and B

   B. B only   **D.** Neither A nor B

10. Lifts are being discussed.

    Technician A says dual-post lifts may extend from under the floor.

    Technician B says a drive-on lift is normally used for brake repair.

    Who is correct?

    A. A only   **C.** Both A and B

    B. B only   **D.** Neither A nor B

Measurements and Precision Measuring Devices

Upon completion and review of this chapter, you should be able to:

❏ Describe United States Customary (USC) system of measurement.

❏ Describe the metric system of measurement.

❏ Explain how to convert measurements between the two measuring systems.

❏ Explain when and how to read rulers.

❏ Discuss the need for feeler gauges and their general use.

❏ Describe how to read measurements made with micrometers.

❏ Describe dial indicators and calipers.

❏ Explain typical uses of vacuum and pressure gauges.

Introduction

In the latter section of Chapter 3, we discussed collecting repair data for the vehicle. Many times, the technician must collect information about the vehicle for comparison with manufacturer data. The method of collecting the data depends on what is being measured and what measuring devices are used. This chapter will discuss some of the most common automotive measuring devices. Your shop may send components out for machining based on your diagnosis. If you do not check for excessive wear or damage, you may be wasting the customer's money. Also starting in this chapter, you may see vehicle terms, such as *valve guide*, used to explain when a tool may be used. Detailed definitions of some of the terms will be held until a later chapter when the terms can be placed in context with other vehicle components.

Measuring Systems

The **measurement** commonly used in the United States comes from the English system and is known as United States Customary (USC). Its scale is based on the inch (2.54 centimeters). USC measurements have been given terms for a certain number of inches. A *foot* equals 12 inches, a *yard* equals 36 inches or 3 feet, and on up the scale.

Automotive USC measures **linear** distances in **decimals** (0.000) or **fractions** ($^1/_2$, $^7/_8$) of an inch (Figure 4-1). Fractional measurements are usually used for overall vehicle length, tire or brake sizes, and other components where the **tolerances** are not required to be so precise. Engine measurements such as *oil clearance* and *valve guide size* are expressed in thousandths (.000) or ten-thousandths (.0000) of an inch. The tolerances in tightly fitted and **meshed components** must be extremely precise to reduce wear and prolong the life of the vehicle.

The metric system of linear measurements is based on the **meter** (39.37 inches) or a portion of a meter. A kilometer (1,000 meters) is roughly five-eighths ($^5/_8$) of a mile (5,280 feet) (Figure 4-2). The meter can also be divided into smaller units. The most common units used on the automobile are *centimeter (cm)* or one-hundredth ($^1/_{100}$) of a meter and *millimeter (mm)* or one one-thousandth ($^1/_{1,000}$) of a meter. Technicians and automotive designers use the millimeter to measure close tolerances in the same manner as when using the decimal units of inches.

Since the early 1980s, there has been an effort to change the United States from the USC system to the metric system. Almost the entire world uses the metric system, and with global trade established it is becoming necessary to use a single system of measurement. Presently, most automotive measurements show the USC and its matching metric measurement together

---

When recording **measurements,** always include the whole number before the decimal or fraction, for example, 0.002 or 1.05.

Linear measures are measurements of distance such as a mile or a meter.

Decimal numbers are expressions of whole numbers in hundreds, thousands, or ten thousands.

Tolerances set the limit that a component can be different from a perfect size.

Shop Manual pages 69–71

Meshed components touch or engage each other all the time, such as two meshed gears.

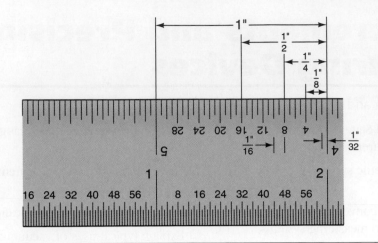

Figure 4-1 USC uses fractions of an inch to measure distances.

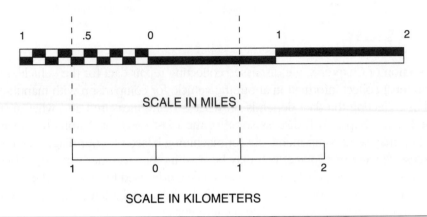

SCALE IN MILES

SCALE IN KILOMETERS

Figure 4-2 The metric system uses tenths as the base divider or multiplier.

(Figure 4-3). For example, the speedometer is scaled both in miles per hour (mph) and kilometers per hour (kph). Another point that makes metric measurements a viable alternative to USC is the math required to change from one unit of measure to the other. With the USC system, the technician must divide or multiply by some hard-to-use numbers like 12, 3, or 5,280. In addition, inches can be expressed as decimals or fractions. The metric system uses 10 for dividing or multiplying to change units of measure and the system does not use fractions. As an example, see which of the following can be calculated the quickest.

One mile (5,280 feet) = _____ inches
One kilometer = _____ centimeters

Component	Torque
Backing plate to axle flange	109 N m (80 ft.-lbs)*
Brake hose bracket	45 N m (35 ft.-lbs)*
Wheel cylinder to backing plated	23 N m (17 ft.-lbs)*

*Note: Not true torque specifications.

Figure 4-3 Most torque specifications give metric and USC data.

The answer to the first statement is:

$$5{,}280 \times 12 \text{ (inches in one foot)} = 63{,}360 \text{ inches}$$

The answer to the second statement is:

$$1{,}000 \times 100 \text{ (centimeters in a meter)} = 100{,}000 \text{ centimeters}$$

The first equation requires some math. The second one requires the addition or subtraction of zeros.

In the automotive field, almost every vehicle and part is made to metric specifications. However, there are still many vehicles on the road based on the USC system and those vehicles will be here for years to come. Eventually, the USC will be dropped and the metric system will be used worldwide. Until then, it is necessary to convert some measurements from USC to metric or the reverse to properly perform daily tasks.

Measurement Conversion

At times, the data found may not be in the measurement format needed. This may happen because a technician who has been in the field for awhile does not like the metric system or the tool being used only measures in one system. A prime example is brake disc thickness. Most new brake discs have the minimum thickness marked in millimeters (Figure 4-4). Most older shops use a brake disc caliper that measures in USC. To properly check the disc, the technician must first convert the metric data to a USC measurement. Sometimes a chart is readily available for quick conversion or the technician may have to perform some minor math. The following section will give a brief overview of converting measurements.

Figure 4-5 shows a chart to help make conversions. Notice that there is a **conversion factor** for going from the metric system to the USC and another for USC to metric. When making conversions, ensure that the correct conversion factor is being used. One of the most common conversions done in automotive repairs is converting inches to millimeters or centimeters or the reverse.

A measurement that may require conversion is shaft **end play** (Figure 4-6). This may be measured with a dial indicator, which may be marked in thousandths of an inch. However, the specification may be listed as metric. Using the data that follows, compute the measured end play into a metric unit.

Measured end play is 0.0008 in.
Specified end play is .2 mm to .3 mm.

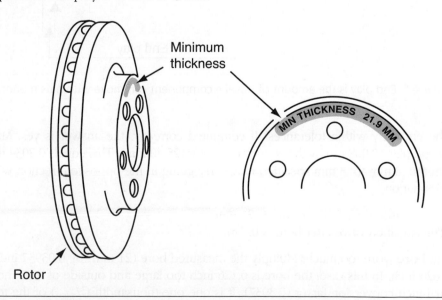

Minimum thickness

MIN THICKNESS 21.9 MM

Rotor

Figure 4-4 Most rotors give the minimum thickness on the disc component.

To Find		Multiply	x	Conversion Factors
millimeters	=	inches	x	25.40
centimeters	=	inches	x	2.540
centimeters	=	feet	x	32.81
meters	=	feet	x	0.3281
kilometers	=	feet	x	0.0003281
kilometers	=	miles	x	1.609
inches	=	millimeters	x	0.03937
inches	=	centimeters	x	0.3937
feet	=	centimeters	x	30.48
feet	=	meters	x	0.3048
feet	=	kilometers	x	3048.
yards	=	meters	x	1.094
miles	=	kilometers	x	0.6214

Figure 4-5 Conversion charts make switching between measuring systems easy.

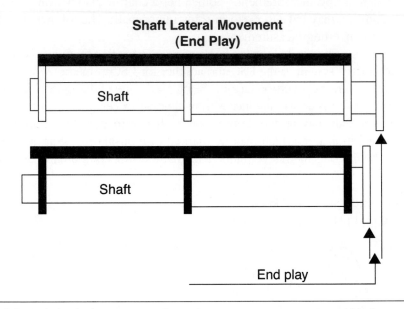

Figure 4-6 End play is the amount of travel a component can move within its mount.

Is the end play within tolerances? If computed correctly, the answer is yes. Multiply the measured end play (0.008 in.) by the conversion factor (25.40 mm). This equals 0.2034 mm, which falls within the .2 mm to .3 mm specified range. Try going in the opposite direction with the following information:

> The measured valve lifter bore is 21 mm.
> The specified valve lifter bore is 0.7 in.

Is the **bore** worn too much? Multiply the measured bore (21 mm) by 0.03937 inch. The answer is 0.826 inch. In this case, the bore is 0.126 inch too large and outside of tolerances. Notice the decimal-inch conversion factor (0.3937). It is one one-thousandth ($^1/_{1,000}$) of the total inches (39.37) in a meter.

A **bore** is usually considered to be a precise, machined hole. A hole is not machined.

Rulers and Feeler Gauges

Rulers

Rulers are simple devices used to measure straight-line distances where tolerances are not a major factor. They may be marked in USC or metric units. The ruler may have measurements on one side, both sides (Figure 4-7), or may be in the shape of a triangle with each side representing a different scale (Figure 4-8). This triangular ruler is used by drafters and engineers to change the scale of their drawings.

Rulers can be made of wood, plastic, or metal. The flat, wooden ruler is the most-used type. The typical USC ruler is 6 inches or 12 inches long. A single-sided ruler may have measurements

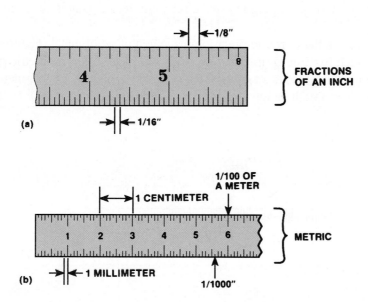

Figure 4-7 One scale has division of $^1/_{16}$ of an inch. The other is $^1/_{32}$ of an inch.

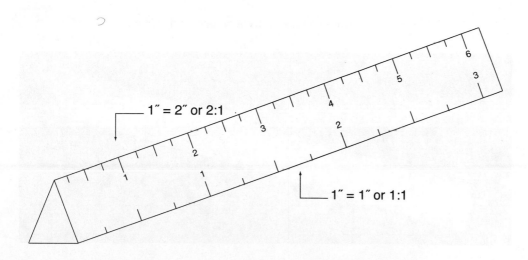

Figure 4-8 This type ruler is used by drafters and engineers for blueprinting.

along one or both edges. The scale will be subdivided or graduated in fractions of an inch. The graduations are normally in $^1/_8$ or $^1/_{16}$ of an inch. Rulers with scales on both edges may have their second scale in $^1/_{32}$ of an inch (Figure 4-9). USC rulers may have graduations along both sides; however, this is not a common practice. Most rulers have a thin metal strip embedded in the graduated edge to act as a straightedge for drawing or marking the work (Figure 4-10).

A metric ruler is made very similar to the USC version. The only real differences are the measuring system and overall length. Metric rulers are usually 20, 25, or 30 centimeters long. Metric rulers are graduated in millimeters, making it easier to interpret the measurement (Figure 4-11). There are no fractions on a metric ruler. Sometimes, a metric ruler will have the 0.5-millimeter marks labeled.

Feeler Gauges

At one time, **feeler gauges** were one of the most common measuring devices in an auto repair shop. Ignition points and spark plug gap were set using feeler gauges, as were other precision measurements.

A feeler gauge is a flat or round precision-machined piece of metal (Figure 4-12). USC sizes range from 0.001 inch (0.02 mm) upward to about 0.080 inch (1.6 mm). Metric gauges are graduated in tenths of a millimeter. Feeler gauges may be marked in either USC or metric, but the common practice is to stamp each gauge with both measurements (Figure 4-13).

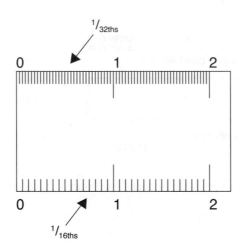

Figure 4-9 Rulers with $^1/_{32}$-inch readings can be very hard to decipher.

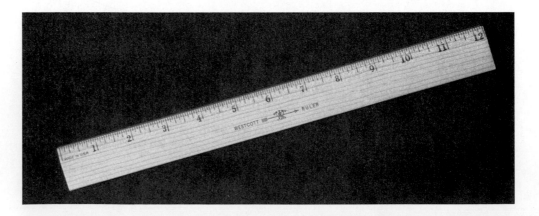

Figure 4-10 The metal edge is provided to help draw a straight line.

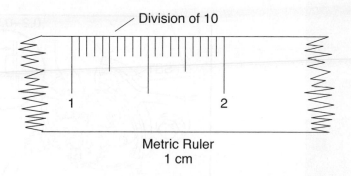

Division of 10

Metric Ruler
1 cm

Figure 4-11 Metric measurements are easier to transfer because they use no fractions.

Figure 4-12 Round-type feeler gauges are used to measure spark plug gap.

Figure 4-13 Most feeler gauge sets are marked in USC and metric measurements.

Figure 4-14 Shown is a typical feeler gauge set that can measure most gaps.

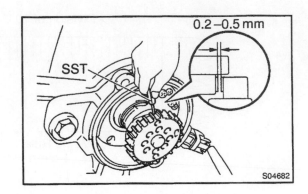

Figure 4-15 Air gaps should be checked with a nonmagnetic flat feeler gauge.

Gap usually refers to the gap between two nonelectronic components.

Air gap usually refers to the space between two magnetic components on an automobile.

Usually the term "feeler gauge" refers to a set of ten to twenty individual gauges (Figure 4-14). Feeler gauges are used to measure the distance between components. Flat gauges are used to check the **gap** between adjacent parts or measure the **air gap** between electronic components (Figure 4-15). **Round feeler gauges** are used almost exclusively to measure spark plug gap. Usually, the gauges are made from steel, but special gauges may be made from a nonmetallic material so they can be used to measure magnetic components air gap (Figure 4-16). Some gauges are built with an angle to make it easier to measure in tight places (Figure 4-17).

The proper use of a feeler gauge is fairly simple. However, the method of determining the point at which the measurement is correct relies on the technician's touch and feel. A feeler gauge is placed between two adjacent components. The object is to adjust the components until they are exactly x-thousandths of an inch apart.

Instructions for the use of a feeler gauge state that it must slide back and forth between the two components until a *slight drag* is felt (Figure 4-18). Technicians must develop the touch to find the slight drag required for a correct measurement.

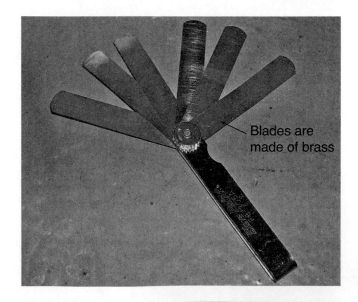

Figure 4-16 Typically, nonmagnetic feeler gauges are tan in color to indicate that the metal used is not steel or iron but usually brass.

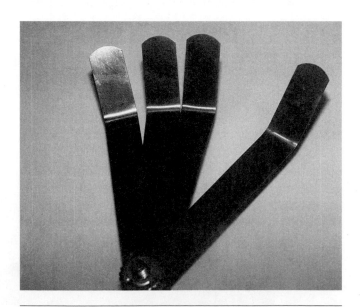

Figure 4-17 Angled feeler gauges assist in measuring gaps in tight places.

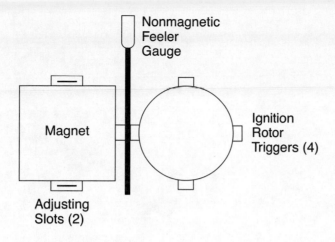

Figure 4-18 The gap must be adjusted until there is a slight drag on the feeler gauge.

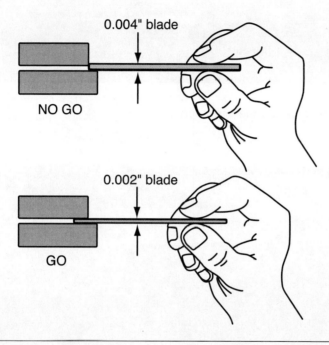

Figure 4-19 Two blades of different thickness with the correct dimensions can be used as go/no-go gauges.

In addition to regular feeler gauges, there are go/no-go gauges or **stepped feeler gauges.** This gauge has a tip that is 0.002 inch (0.05 mm) smaller than the rest of the gauge. If the tip slides into the space between the components and the remainder of the gauge will not slide in, the gap or space is correct (Figure 4-19). Two separate blades can be used to do the same check. For instance, a 0.005 inch (0.05mm) can be used as the "go" gauge and a 0.004 inch (0.010 mm) can be used as the "no-go" gauge.

Micrometers and Associated Gauges

Micrometers are used to precisely measure the different shapes of a component. There are outside, inside, and depth micrometers, each built with a specific purpose. Each type has a measuring scale, an adjustable measuring face, a fixed measuring face, and a frame (Figure 4-20). Because of the limited space on the scales, micrometers are usually set up to measure either USC or

A **micrometer** may be referred to as a "mike."

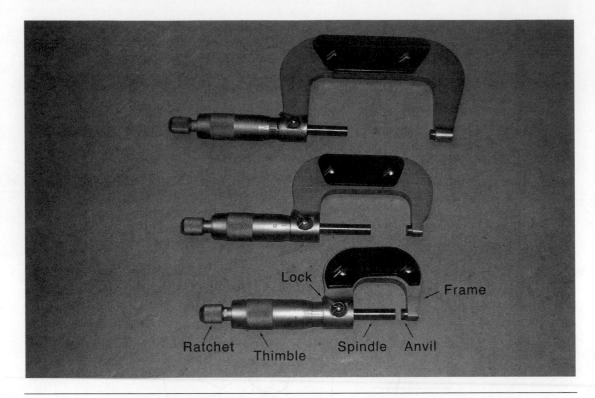

Figure 4-20 Parts of an outside micrometer.

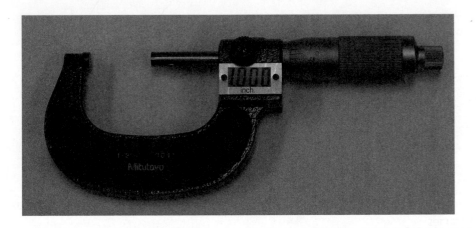

Figure 4-21 Electronic micrometers are quicker and easier to use and provide a push-button means for measurement conversion.

Shop Manual
pages 72–77

An **outside micrometer** is normally used in the automotive business to measure shaft external bearing journals.

metric. However, with the introduction of electronic and digital micrometers, the user can select either USC or metric measurements or the scale (Figure 4-21).

Outside Micrometers

An **outside micrometer** resembles a clamp that can measure linear distances. The *frame* supports the working components and is sized to measure within a limit, usually 1 inch or 25 mm. Metric scale will be discussed later. A 1- to 2-inch USC micrometer will measure a component that is between 1 and 2 inches in diameter or thickness (Figure 4-22). The *fixed face* or *anvil* provides a point from which the measurement is made (Figure 4-23). The other measuring face

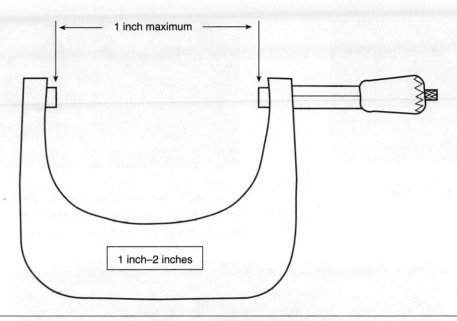

Figure 4-22 The micrometer frame shows the limits of its measurement.

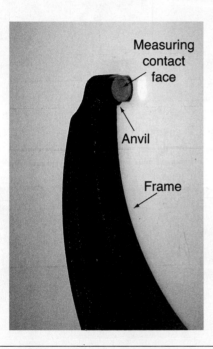

Figure 4-23 The anvil face is the nonmovable or fixed measuring face of a micrometer.

is on a *spindle* that can be extended to or retracted from the anvil by turning the *thimble* (Figure 4-24). The outer portion of the thimble is knurled for grip. At the outer end of most thimbles is a smaller, knurled protrusion that acts as a force release (Figure 4-25). The protrusion has a ratchet device that slips when the anvil and spindle are tight enough to make an accurate measurement. The ratchet also prevents excessive force from being applied to the spindle. The thimble rotates around a sleeve and extends from the frame in the opposite direction of the spindle. Most micrometers use some type of lock to hold the spindle and thimble in place (Figure 4-26). This allows a technician to remove the micrometer from the work to better see the measurement markings.

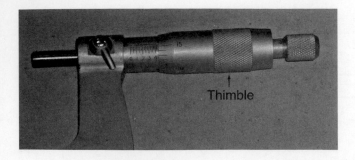

Figure 4-24 The thimble acts as a "wrench" to adjust the spindle.

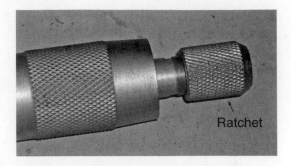

Figure 4-25 The thimble ratchet reduces the chance of micrometer damage and sets the correct contact force on the component being measured.

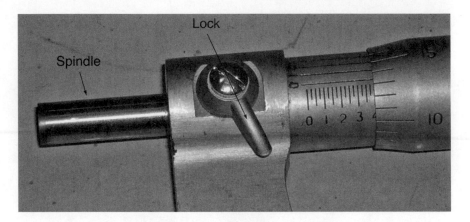

Figure 4-26 The lock fixes the spindle in place so the micrometer can be moved without changing the measurement reading.

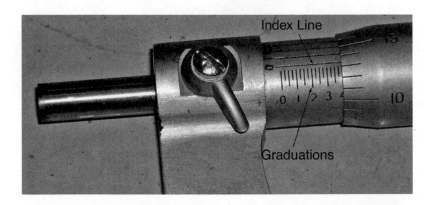

Figure 4-27 The index line is the point from which measurements are read.

The sleeve has an **index line** along its length. Above the line are numbers ranging from 0 to 10 at $^1/_{100}$-inch graduations (Figure 4-27). Below the lines are markings at 0.025 inch. The thimble markings are laid rotationally around its lower edge (Figure 4-28). Thimble graduations are 0.001 inch each and are numbered each 0.005 inch up 0.025 inch (Figure 4-29). One complete rotation of the thimble will move the spindle 0.025 inch. Some micrometers have an additional scale on the sleeve called a **vernier scale.** The vernier scale is laid rotationally around the base of the sleeve starting at the index line. Each line parallels the index line and is equal to 0.0001 inch (Figure 4-30). Vernier scales are used to read measurements in one ten-thousandths ($^1/_{10,000}$) of an inch.

Figure 4-28 USC thimbles are graduated each 0.001 inch and numbered each .005 inch up to a total of 0.025 inch per complete revolution.

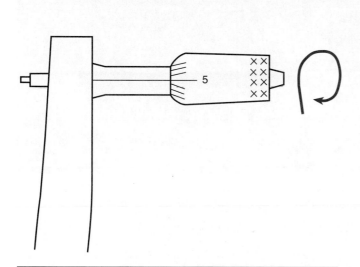

Figure 4-29 Each mark represents 0.001 inch and are numbered every 0.005 inch.

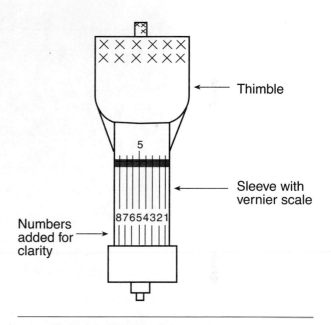

Figure 4-30 The vernier scale measures 0.0001 inch per line up to 0.0010 inch.

Micrometer Calibration

In order to have accurate measurements, the micrometer must be accurate itself. Any time the micrometer is removed from storage it should be calibrated. It should also be calibrated when used in extreme hot or cold climates, especially if stored or protected at normal room or body temperatures. The calibration is performed using a **standard gauge** and a tool, usually a small screwdriver (Figure 4-31). The standard gauge is a shaft that is precisely machined to an exact length. Micrometers usually have the gauge and a small tool, usually a screwdriver, included when the micrometer is purchased.

Gauges come in sizes that match the distance the micrometer can measure. A 1- to 2-inch micrometer will have a 1-inch gauge, and a 5- to 6-inch micrometer will be calibrated with a 5-inch gauge. Metric micrometers use a comparable system. Inside and depth micrometers can be calibrated by using an outside micrometer that has already been calibrated. The calibration

Figure 4-31 The five standards shown are used to calibrate outside micrometers from sizes 0–1 inch to 5–6 inch.

procedure is simple but essential. Chapter 4 in the Shop Manual will explain the procedure for calibrating an outside micrometer.

Reading an USC Outside Micrometer

Reading any micrometer is easy after a little practice. Place the anvil against the work. Rotate the thimble to move the spindle to contact the opposite (180 degrees) side of the work (Figure 4-32). Use the ratchet knob to ensure sufficient contact is made with the work for an accurate reading. When the ratchet slips, lock the spindle and *remove the micrometer* from the work.

For this example, we will say the component is somewhere between 2 and 3 inches thick. This means a 2- to 3-inch micrometer is selected (Figure 4-33). With the micrometer removed from the work, the index line is checked first. Notice that 0 and 1 are exposed (Figure 4-34). Since 1 is the highest number exposed, the component is between 2.100 and 2.200 inches thick. Below the index line there are three graduations past the 0.100 mark that are visible (Figure 4-35). Since each of these lines is equal to 0.025 inch, 0.075 inch is added to the 2.100 inches computed previously. The component is now at least 2.175 inches thick. Following the index line to where it disappears under the thimble, it is found that the third mark above zero on the thimble aligns with the sleeve's

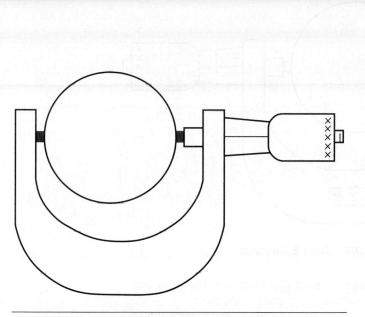

Figure 4-32 Ensure that the two measuring faces are opposite each other.

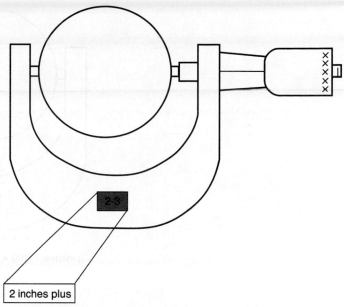

2 inches plus

Figure 4-33 The frame contains the measurement limits of the micrometer.

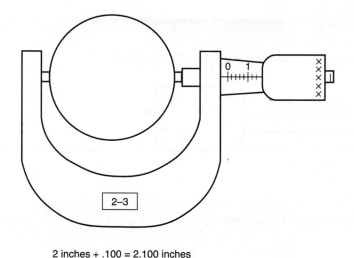

2 inches + .100 = 2.100 inches

Figure 4-34 Use the highest visible number to start your decimal measurement.

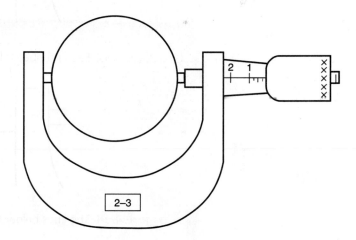

2 inches + .100 + .075 = 2.175 inches

Figure 4-35 The thimble is just past the 0.075-inch mark.

index line (Figure 4-36). This is the fourth number in the measurement. Since each line on the thimble equals 0.001 inch, 0.003 inch is added to the 2.175 inches from the last calculation. The final measurement for the thickness of this component is 2.178 inches.

Reading a Metric Outside Micrometer

A metric micrometer is laid out the same as a USC micrometer except for the size of the scales. Above the index line are 1-millimeter markings ranging from 0 to 25. They are numbered at each 5 mm (Figure 4-37). Below the index line are markings designated 0.5 mm each (Figure 4-38). The

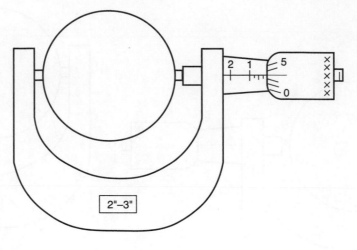

2 inches + .100 + .075 + .003 = 2.178 inches

Figure 4-36 The 0.003-inch line is in direct alignment with the index line.

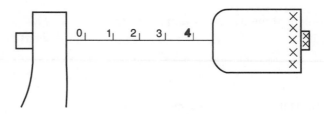

Figure 4-37 Each numbered line above the index line represents 5 millimeters.

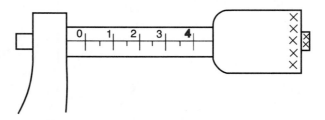

Figure 4-38 The lines below the index equal 0.5 millimeters.

thimble is marked from 0 to 50 or half of a millimeter (Figure 4-39). Each graduation is equal to 0.01 mm and is numbered each at 0.05 mm (Figure 4-40).

Reading a metric micrometer is very similar to reading a USC micrometer. Like the USC micrometer, the frame has a numerical value based on the total distance or length the micrometer is capable of accurately measuring. The frame on a metric micrometer is sized for a total of 25 mm (i.e., 0 mm to 25 mm, 50 mm to 75 mm).

The first digit to the left of the decimal is the lowest number on the frame, just as with a USC micrometer. For our discussion, a 25-mm-to-50-mm micrometer will be used as the example. On a 25-mm-to-50-mm micrometer this would initially be 25.00 mm. There are a few other differences between reading a USC and metric micrometer.

As stated, the first measurement (reading) is from the frame, in this case, 25.000. Look on the sleeve's index line and read the last visible figure. Figure 4-41 shows that the number five is visible. This is one change: Add the last visible number above the index line (5 mm) to the minimum size of the micrometer (25 mm). At this point the shaft is at least 30 mm in diameter.

25 mm (frame) + 5 mm (index) = 30.00 mm

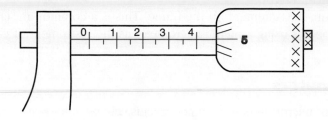

Figure 4-39 One rotation of the thimble equals half (0.5) of a millimeter.

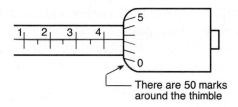

There are 50 marks
around the thimble

Figure 4-40 Each line on the thimble equals 0.01 millimeter for a total of 0.5 millimeter per thimble revolution.

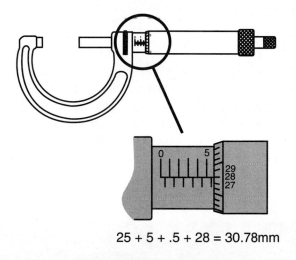

25 + 5 + .5 + 28 = 30.78mm

Figure 4-41 Similar to the USC micrometer, the measurement on a metric micrometer is a matter of adding those numbers visible on the frame and index lines.

The next step determines the first number to the right of the decimal. This number comes from the last visible marking on the index line under the sleeve. In Figure 4-41 the last visible mark under the index line is the 0.5-mm graduation. This is added to the previous reading and we now know the shaft is at least 30.50 mm in diameter.

30 mm + 0.5 mm = 30.50 mm

The last number in this measure is read from the thimble. Like the USC micrometer, find the graduation line on the thimble that most closely aligns with the index line on the sleeve. In Figure 4-41 this appears to be the 28 mark or the 0.28-mm graduation. Add this measurement to the previous step:

30.50 mm + 0.28 mm = 30.78 mm

The shaft in question has a diameter of 30.78 mm. When using a metric micrometer after a lifetime usage of USC micrometers, don't forget to add the last visible number above the index to the minimum measurement on the frame. This is a common mistake when an experienced technician begins using a metric micrometer as the instrument of choice for precision measurement.

Brake Micrometers

There are a couple of micrometers used almost exclusively by automotive technicians. They are used to measure brake rotor thickness and brake drum diameter. The brake rotor mike, as it is commonly called, works like any other outside micrometer except for its physical size, considering the size of measurement it can read and the speed of the spindle movement (Figure 4-42). A standard 0.00-inch-to-1.00-inch (0.0 mm to 25 mm) or a 1.00-inch-to-2.00-inch (25 mm to 50 mm) micrometer will fit comfortably in the palm of an average hand. The brake rotor micrometer is larger and is generally easier to handle. In addition, the internal gears that move the spindle allow for a quick adjustment, which saves some time during a brake repair job. The markings and numbers on this micrometer are also larger and easier to read. This allows the technician to make quick and very accurate measurements of a brake rotor's thickness and the parallelism of its machined braking surfaces.

The brake drum micrometer or drum mike is an inside micrometer sized up in the same manner as the rotor micrometer (Figure 4-43). Like the rotor mike, the technician can make quick and accurate measurements of the inside diameter of a brake drum. Use of an inside micrometer is discussed in the next section.

Telescopic Gauges

Telescopic gauges
have rounded tips at
the ends to better
match the curve of
the bore.

Telescopic gauges (Figure 4-44) and outside micrometers can be used to measure inside dimensions. Telescopic gauges come in various lengths and are T-shaped with the cross section consisting of two spring-loaded extensions. The leg of the gauge contains a rotatable handle that locks the extensions in place. The gauge is placed inside the bore, the handle is twisted to release the extensions, and the extensions spring out to the bore's walls (Figure 4-45). Once the technician is satisfied with the gauge's position, the handle is turned to lock the extensions. The gauge is removed from the bore and its extended length is measured with an outside micrometer (Figure 4-46). The

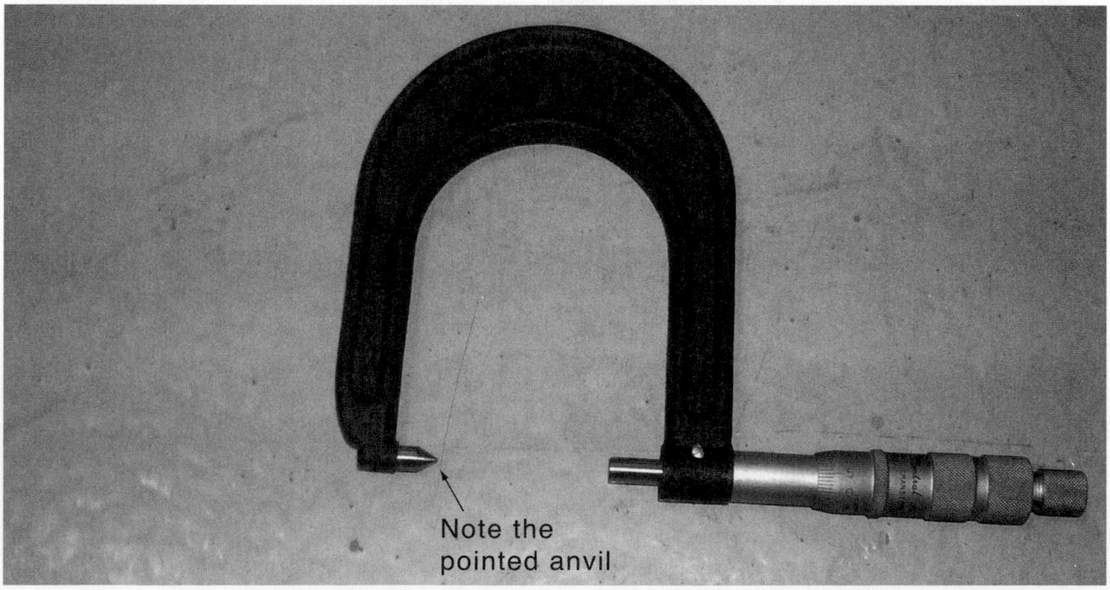

Note the
pointed anvil

Figure 4-42 A brake rotor micrometer is an outside micrometer with faster gears and a larger frame.

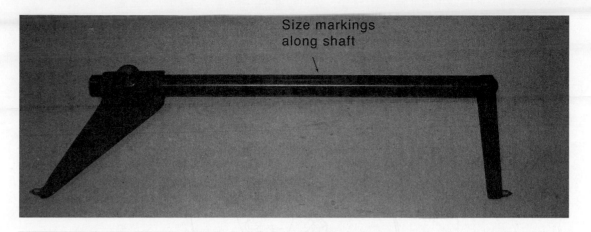

Size markings
along shaft

Figure 4-43 Drum micrometers are inside micrometers that, like the rotor micrometers, have faster gears and oversized frames.

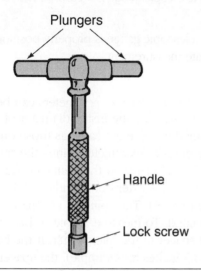

Plungers

Handle

Lock screw

Figure 4-44 The telescopic gauge has two spring-loaded extensions.

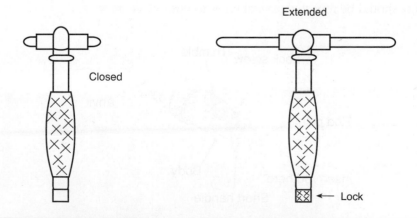

Closed

Extended

Lock

Figure 4-45 Springs push the extensions out to the cylinder wall.

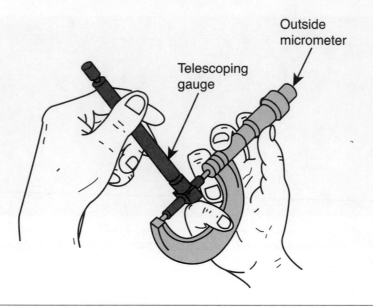

Figure 4-46 The removed telescopic gauge is measured with an outside micrometer.

technician must ensure that the telescopic gauge is properly positioned within the bore and the micrometer to guarantee an accurate measurement.

Inside Micrometers

Inside micrometers are used to measure the inside diameter of large holes (Figure 4-47). They can be USC or metric and the markings are the same as those on an outside micrometer. An inside micrometer looks like the outside micrometer minus the frame. An inside micrometer can only measure up to 1 inch or 25 millimeters without attachments (Figure 4-48). To make larger measurements, the technician must use extension spindles and a half-inch spacer supplied with the inside micrometer set (Figure 4-49). The spindles are made in different lengths and are attached to the micrometer as needed. To measure a hole or bore that is 4 inches in diameter, attach a 3- to 4-inch spindle and spacer to the micrometer. If the bore to be measured is between a whole inch and $1^1/_2$ inches (3.5 inches for example), the spacer would not be used. Once positioned in the bore, the measurement would be read the same as an outside micrometer.

The critical point in using an inside micrometer is its placement within the bore (Figure 4-50). The micrometer must be at a right angle to the vertical centerline of the bore. Because of the curved surfaces of the bore, the technician must shift the spindle back and forth until the largest measurement is shown on the micrometer. This skill comes with a good deal of practice, and all measurements should be repeated several times to ensure accuracy.

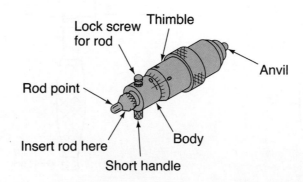

Figure 4-47 Inside micrometers have all the components except for the frame.

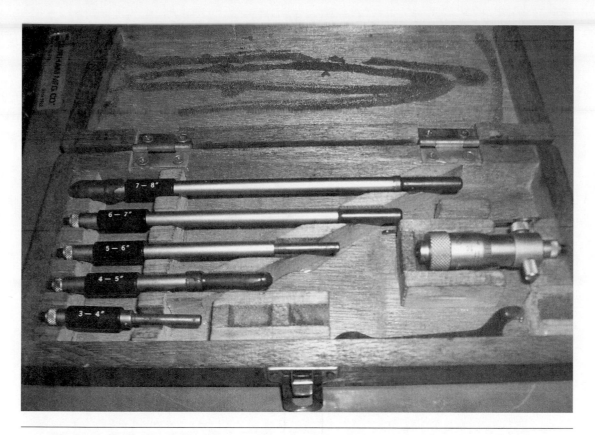

Figure 4-48 The extensions shown are used with an inside micrometer to measure diameters larger than the micrometer alone can measure.

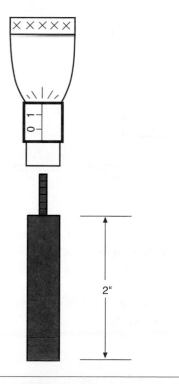

Figure 4-49 Extension spindles are attached to the micrometer to measure a large bore.

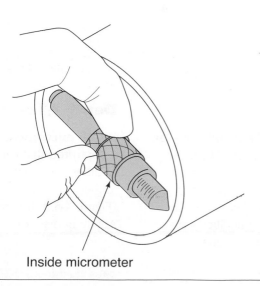

Inside micrometer

Figure 4-50 The micrometer must be carefully placed within the bore for accuracy.

Figure 4-51 A typical micrometer.

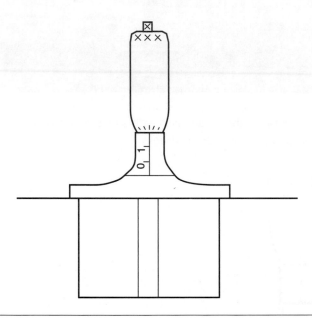

Figure 4-52 Place the base at the edge of the hole and extend the depth rod.

Depth Micrometers

Depth micrometers
are used by race
engine builders to
better position the
valves and pistons
for additional speed
and torque.

Depth micrometers look similar to inside micrometers. This type of micrometer has a base and a depth rod that can be extended like a spindle (Figure 4-51). The scales are the same as those of other micrometers. Place the base on a firm edge of the bore or hole and extend the depth rod to the work. This will provide an accurate measurement of how deep the work is within the bore (Figure 4-52). It should be noted that depth micrometers are used to make small measurements. Anything over an inch will require add-on spindle extensions.

Small Hole Gauges

Small hole gauges do what their name implies: they measure small holes. Technicians do not routinely use large bore gauges because micrometers are more practical. However, micrometers are not practical for small bores.

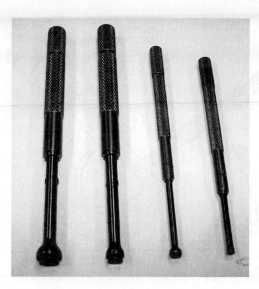

Figure 4-53 Small bore gauges are available in sets or individually. A common use is to measure the diameter of valve guide bores.

NASCAR is the National Association for Stock Car Automobile Racing.

AUTHOR'S NOTE: NASCAR and other automotive racing associations use fixed-size bore gauges. During race inspections, inspectors use the gauges to quickly check bores of carburetor air horns, cylinders, and other bored or circular components of the racecar.

A small hole gauge is straight with a screw handle on one end and a **split ball** on the other end (Figure 4-53). A wedge fits between the two halves of the split ball and extends into the handle. Turning the handle moves the wedge in and out of the split ball, causing it to expand or contract (Figure 4-54). The split ball end is slid into the bore where it is expanded by the wedge.

A **split ball** is a steel ball that has been precision cut in half and installed in machined holders.

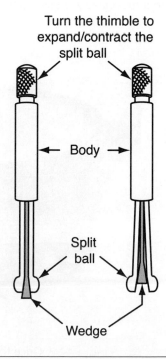

Turn the thimble to expand/contract the split ball

Body

Split ball

Wedge

Figure 4-54 The split ball of a small bore gauge can be expanded inside the bore and then removed for measuring.

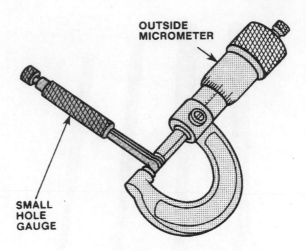

OUTSIDE
MICROMETER

SMALL
HOLE
GAUGE

Shop Manual
pages 81–83

A **dial indicator**
uses a plunger and
gear to move a
needle in proportion
to plunger
movement.

Runout is the
amount of side-to-
side movement in a
rotating device.
Runout may be
referred to as
"wobble."

Figure 4-55 The split ball is measured with a 0- to 1-inch outside micrometer.

Once the technician is satisfied with the fit, the gauge is withdrawn and the ball diameter is measured with an outside micrometer (Figure 4-55). Small hole gauges are used mostly in the automotive repair business to check valve guide bores.

Dial Indicators and Calipers

The **dial indicator** is used to check how far a component can move or is used to measure the distortion of a component, such as flywheel **runout.** A dial indicator must be mounted to a solid object for support during its use. The dial has a circular face with a scale around its edge (Figure 4-56). A moveable needle indicates the measurement much like a speedometer needle. Turning a knurled ring on the outside of the dial rotates the scale. There is a lock to hold the scale in place (Figure 4-57). The needle is geared to a moveable plunger extending from the bottom of the dial.

Figure 4-56 A dial indicator has a round dial with the graduations totaling 0.100 inch per one needle revolution.

Lock

Figure 4-57 Once the scale is adjusted to zero, it is held in place by the lock during measuring.

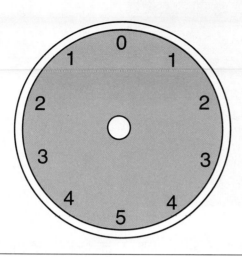

Figure 4-58 The balanced scale numbers meet at the midpoint of the face.

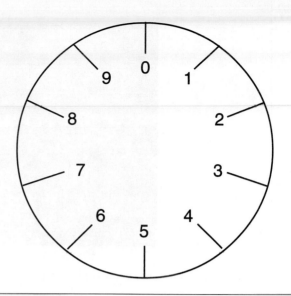

Figure 4-59 Numbers extend upward from zero on a continuous scale.

The scale can be in USC or metric measurements. The USC scale will use either $1/1,000$ or $1/10,000$ inch. The scale size will be noted on the face. Metric scales are marked in 0.01 millimeter. The face may be **balanced,** meaning the readings extend from zero counterclockwise and clockwise until they meet at the midpoint of the face (Figure 4-58). A **continuous** type has the markings start at zero and extend clockwise around the scale back to zero (Figure 4-59). Care must be taken with both types when taking large measurements. The plunger may be designed to move the needle more that one revolution around the scale. A single revolution is equal to 0.100 inch (2.54 mm).

Dial indicators can also be used to check shaft end play. Before setting up the dial, move the shaft as far as it can go in one direction. Mount the dial to indicate a zero reading at this point. The final step is to move the shaft as far as it will go in the opposite direction. The total movement will be shown on the dial scale.

Balanced scale indicators are usually used to measure runout. **Continuous scale indicators** are normally used to measure end play or lateral movement.

Dial Calipers

The **dial caliper** is a measuring device that can be used in place of almost all other automotive measuring devices (Figure 4-60). It can measure inside, outside, and depth dimensions. It has two

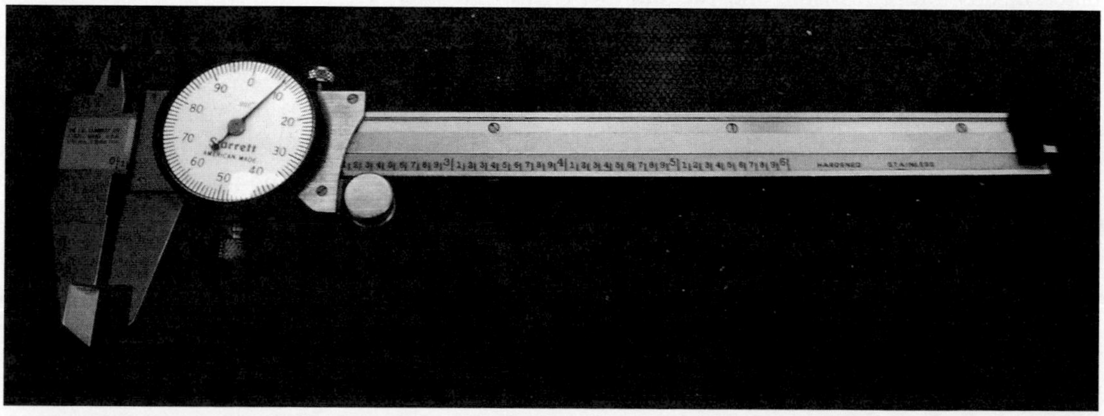

Figure 4-60 This nonelectric caliper is preferred by many technicians and machinists over the newer electronic version.

Figure 4-61 The dial face and scaling of a dial caliper are similar to the face of a dial indicator. The whole-inch measurements are read from the bar and added to the thousandths-of-inch readings from the dial.

pairs of jaws or calipers that fit inside or outside a component. The top half of each pair is fixed to the bar scale. The bottom half moves along the bar scale when the roll knob is turned. At the opposite end of the bar scale is the depth scale. The bar scale indicates the gross measurement of the caliper's movement. A dial indicator attached to the moveable caliper measures the small movement within its limits.

A typical USC dial caliper can measure between 0 and 6 inches. Its bar scale is divided in 0.100-inch graduations. The USC dial indicator measures in 0.001-inch segments (Figure 4-61). One complete rotation of the dial's needle equals 0.100 inch on the bar scale. A metric bar scale is in 2-millimeter divisions while its dial scale is in 0.02-millimeter graduations. One complete revolution of the needle is equal to 2 millimeters on the bar scale. There are electronic calipers on the market which will allow technicians to change between the two measuring systems.

Pressure and Vacuum Measurements

Shop Manual
pages 89–92

Various systems of the vehicle use pressure and vacuum theories to operate. Pressure and vacuum are used to control engine and transmission operations, brakes, and other systems. A detailed discussion of pressure and vacuum theory is in Chapter 7.

Pressure Gauges

Pressure is the amount of force being applied as measured in pounds per square inch (psi). If pressure is applied to a liquid in a closed, sealed system, the liquid can be used to transfer force. A hydraulic brake system is an example of pressure and liquid transferring forces.

Pressure gauges are fitted into the pressure line using adapters. The scales are marked (graduated) and numbered based on the size of the scale. A small scale gauge may be marked each 1 psi and numbered each 5 psi, while a larger scale may be marked each 10 psi and numbered each 50 psi (Figure 4-62). When the gauge is connected and the system energized, the scale will record the pressure being applied to the system.

Vacuum Gauges

Atmospheric pressure is considered to be 14.7 psi at sea level.

Vacuum properly defined as pressure below **atmospheric** and is generally measured in inches of mercury (in. Hg). This measure is based on how high (in inches) a vacuum source can raise a

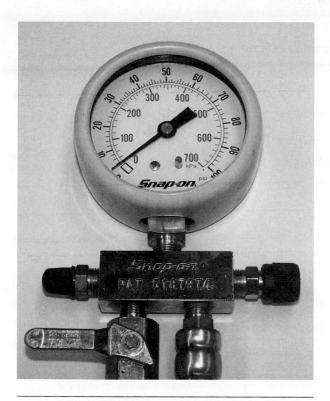

Figure 4-62 The amount of pressure a pressure gauge must measure determines the graduations and numbering of the scale. This gauge is capable of measuring 100 psi.

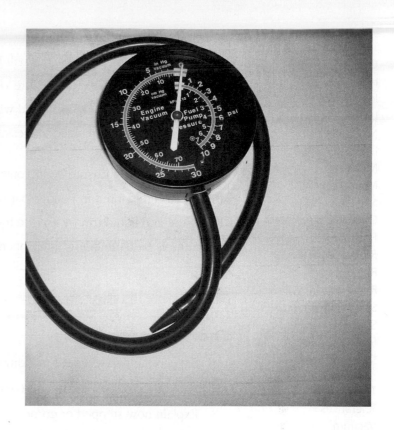

Figure 4-63 Almost all vacuum gauges used in automotive repair have a scale from zero to 30 inches of mercury.

column of liquid mercury in a tube. Obviously, technicians cannot carry tubes of mercury around their shops. The vacuum gauge's mechanical system is geared to move a needle a specified distance based on the amount of vacuum in its inlet tube. Vacuum gauges are usually marked in the same manner as pressure gauges except the markings indicate inches of mercury (Figure 4-63).

Summary

❏ The most common measurement system in the United States is the United States Customary or the English system.

❏ The base USC measurement is the inch.

❏ The base metric measurement is the meter.

❏ The metric system units are based on the number 10 and no fractions are used.

❏ USC and metric measurements can be converted to each other using a conversion chart and math.

❏ Rulers may be in USC or metric measurements.

❏ Feeler gauges are used to measure small gaps or distances between two parts.

❏ Micrometers are made for measuring the size of objects.

Terms to Know
Air gap
Atmospheric pressure
Balanced indicator
Bore
Continuous indicator
Conversion factor
Decimal
Depth micrometer
Dial caliper
Dial indicator
End play
Feeler gauge
Fractions

❏ Micrometers may be USC or metric.

❏ Small bores are measured with small hole gauges.

❏ Dial indicators are used to measure very small measurements.

❏ Shaft end play is usually measured with a dial indicator.

❏ Depth micrometers are used to measure the depth of a bore or the depth of an object within a bore.

❏ Dial calipers can measure depth, outside diameters, and inside diameters.

❏ Pressure is measured in pounds per square inch (psi).

❏ Pressure and liquid can be used to transfer force.

❏ Vacuum is pressure below atmospheric pressure.

Review Questions

Short Answer Essay

1. Explain the measurement markings on a USC micrometer.

2. List and describe the different components on a micrometer.

3. Explain how stepped or go/no-go feeler gauges are machined.

4. Explain how to convert 3 inches to millimeters.

5. Discuss the components of a dial caliper.

6. Discuss the difference between pressure and vacuum.

7. List and discuss the differences between a USC and a metric dial caliper.

8. Describe the means for measuring vacuum.

9. Explain how a micrometer is placed to measure a shaft's outside diameter.

10. Explain the metric system of measurement.

Fill-in-the-Blanks

1. The calipers on a dial caliper are moved by a(n) _____ _____.

2. Some micrometers have an additional scale called a(n) _____ _____.

3. One rotation of the spindle on a USC micrometer moves the spindle _____ _____.

4. A 3- to 4-inch micrometer can measure _____ inch.

5. A typical dial caliper will measure up to _____ inch(es) in _____-inch segments.

6. A stepped feeler gauge may be known as a(n) _____ gauge.

7. Large bores may be measured with a micrometer and_____ or a(n) _____.

8. A dial indicator may have a(n) _____ or a(n) _____ scale.

9. A micrometer has a(n) _____ to prevent an overtorque of the spindle.

10. The vernier scale is used to read _____ of an inch.

ASE-Style Review Questions

1. *Technician A* says pressure and liquid are used to make hydraulic brakes systems work.
   Technician B says vacuum helps the engine operate.
   Who is correct?
   - **A.** A only
   - **B.** B only
   - **C.** Both A and B
   - **D.** Neither A nor B

2. Dial calipers are being discussed.
   Technician A says the bar scale may be graduated in 0.100 inch.
   Technician B says a bar scale may be graduated in 2 millimeters.
   Who is correct?
   - **A.** A only
   - **B.** B only
   - **C.** Both A and B
   - **D.** Neither A nor B

3. Feeler gauges are being discussed.
   Technician A says a feeler gauge blade may be machined in two different sizes.
   Technician B says a standard feeler gauge may be used to check the air gap between a magnet and a pickup coil
   Who is correct?
   - **A.** A only
   - **B.** B only
   - **C.** Both A and B
   - **D.** Neither A nor B

4. *Technician A* says 2.54 centimeters is equal to 1 inch.
   Technician B says 1 inch is equal to 0.3937 millimeter.
   Who is correct?
   - **A.** A only
   - **B.** B only
   - **C.** Both A and B
   - **D.** Neither A nor B

5. *Technician A* says a micrometer may be used to measure the wobble in a gear.
   Technician B says a micrometer can be used to measure the diameter of an engine cylinder bore.
   Who is correct?
   - **A.** A only
   - **B.** B only
   - **C.** Both A and B
   - **D.** Neither A nor B

6. Small hole gauges are being discussed.
   Technician A says this type of gauge is sometimes used to measure bores that cannot be measured with an inside micrometer.
   Technician B says a wedge is used to spread two ball halves apart on this gauge.
   Who is correct?
   - **A.** A only
   - **B.** B only
   - **C.** Both A and B
   - **D.** Neither A nor B

7. *Technician A* says pressure is generally measured in inches of mercury.
   Technician B says vacuum is generally considered to be any pressure less than atmospheric pressure.
   Who is correct?
   - **A.** A only
   - **B.** B only
   - **C.** Both A and B
   - **D.** Neither A nor B

8. *Technician A* says a drum micrometer is basically an inside micrometer.
   Technician B says a rotor micrometer is used to measure diameter of a brake rotor.
   Who is correct?
   - **A.** A only
   - **B.** B only
   - **C.** Both A and B
   - **D.** Neither A nor B

9. Dial indicators are being discussed.
   Technician A says a dial indicator is part of a dial caliper.
   Technician B says a dial indicator is used to measure a small hole gauge.
   Who is correct?
   - **A.** A only
   - **B.** B only
   - **C.** Both A and B
   - **D.** Neither A nor B

10. *Technician A* says spindles are added to an inside micrometer to make shaft measurements.
    Technician B says telescopic gauges can be measured with a dial caliper.
    Who is correct?
    - **A.** A only
    - **B.** B only
    - **C.** Both A and B
    - **D.** Neither A nor B

Fasteners

Upon completion and review of this chapter, you should be able to:

❏ Describe the USC threaded fastener measuring system.

❏ Describe the metric threaded measuring system.

❏ Identify the grade markings of threaded fasteners.

❏ List and describe common thread repair tools.

❏ Identify and explain the purpose of common nonthreaded fasteners.

❏ Explain the importance of wire gauge sizes.

❏ Identify common types of electrical fasteners.

❏ Identify common multipin, electrical connectors.

Introduction

Automotive components are held together with **fasteners.** As a technician, you will spend much of your day removing and replacing fasteners. These small parts are very important. If fasteners are not used properly, the components they hold together can fail.

Many different types of fasteners are used on an automobile. Fasteners can be divided into two basic groups. Threaded fasteners use the clamping force from threads to hold parts together. **Nonthreaded fasteners** hold parts together without threads. Both types are discussed in the following sections.

Threaded Fasteners

Shop Manual
pages 103–107

A BIT OF HISTORY

Threads provide a mechanical advantage to hold parts together. The geometry behind threads was developed a long time ago, around 200 B.C. They were used by the ancient Romans to press grapes for wine.

Threaded fasteners are the most common fastener type. They use spirals called threads to wedge parts together. The common types of threaded fasteners are screws, bolts, studs, and nuts and they are shown in Figure 5-1.

Figure 5-1 Common types of threaded automotive fasteners.

Fastener Sizing and Torquing

For a fastener to work correctly it must have strength and size to handle the task. Fasteners are designed in different grade (strength) ratings for specific loads and applications. Replacement fasteners should meet the same requirement as the original item. A light-duty fastener may be replaced with a heavy-class fastener in most cases. However, some industry and vehicle parts have a certain class fastener that is designed to break or shear to prevent damage to components. A typical example would be the shear pin in a winch. To prevent overloading a cable or chain that may result in very serious injury and great damage, the shear pin is designed to shear apart when its limits are reached. Replacing a shear pin that "broke too quickly" with a stronger pin has resulted in at least one death to this author's knowledge.

Screws

The **screw** is one of the most common threaded fasteners. A screw fits into a threaded hole. Often there is a part without threads between the screw and the part with the threads, as shown in Figure 5-2. The threads on the screw engage the threads in the part. As the screw is turned, the two sets of threads work together to clamp the two parts together with a great deal of force.

Many different types of screws are used to hold automotive parts together. The most common types of screws used in automotive parts are shown in Figure 5-3. The hex head cap screw

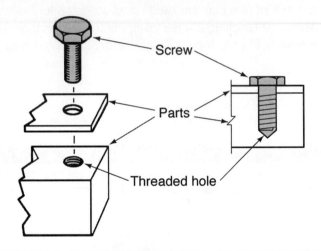

Figure 5-2 A screw fits through one part and into a threaded hole in a second part.

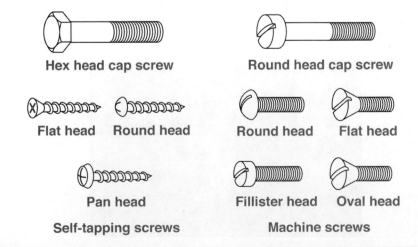

Figure 5-3 Common types of automotive screws.

is one of the most common types. It has a six-sided or hexagonal head and is driven with common automotive wrenches such as the box, open-end, combination, and socket. Hex head cap screws are used to hold large parts together.

Many other screws are driven with a screwdriver. They are used to hold small automotive parts together. These screws may have a slotted head or a recessed head like those discussed under screwdriver types in Chapter 3. They are often called *machine screws* because they are used to hold machined parts together. In addition, they are often classified by the shape of their head. The main head shapes are round head cap screw, round head (with threads that go all the way to the head), flat head, fillister head, and oval head.

Self-tapping screws do not fit into threaded holes. As they are driven into the part, they use their hard external thread to form, or tap, their own threads. Self-tapping screws are sometimes used in automotive body sheet metal parts.

Bolts

Many automotive parts are held together with **bolts.** Bolts use a nut instead of a threaded hole to hold automotive parts together. Usually the bolt fits through parts that do not have threads. The bolt and nut are driven together to clamp the two parts together as shown in Figure 5-4. Most bolts for automotive use have a hex head (Figure 5-5). The difference between this bolt and a hex head cap screw is that the bolt is used with a nut. The hex head cap screw is used in a threaded hole. There are bolts with heads that are not hex shaped, but they are not often used in automotive parts.

Since the early 1990s, a specific type of bolt has been used to hold high-pressure/force components together. The most significant are the bolts that hold cast-aluminum cylinder heads to cast-iron engine blocks. This type of fastener is known as a torque-to-yield bolt. They bear no outward visible differences from other same-size bolts. It is critical that they be used as engineered.

Standard bolts hold adjacent components together but may not provide exact, all-around clamping force to the components. This was the cause for many cylinder head failures in the 1980s

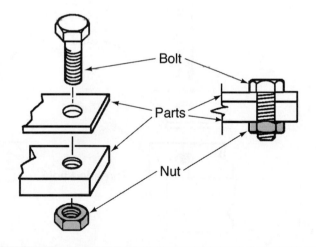

Figure 5-4 A bolt fits through both parts and into a nut.

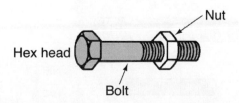

Figure 5-5 The hex head bolt is the most common fastener for automotive parts.

when this type of engine was being introduced. Torque-to-yield bolts helped to solve this engine problem. Torque-to-yield bolts must be replaced when removed; they are not reusable.

The primary reason for replacing torque-to-yield bolts is the method of tightening them to specifications. A typical torque specification for cylinder head bolts could be:

> 1st step tighten in sequence to 35 ft.-lbs
> 2nd step tighten in sequence to 45 ft.-lbs
> 3rd step tighten in sequence an additional 90 degrees

The third step is the key. This torque step actually stretches the bolt so it applies a very specific amount of clamping force to the cylinder head. The cylinder head is effectively sealed to the engine block with a special head gasket between them. It is also the reason torque-to-yield bolts must be replaced. Once stretched the bolt no longer has the tensile strength to properly clamp the components together.

Studs

A **stud** is a common type of automotive fastener that has no head. It has threads on both ends. One end of the stud fits into a threaded hole in the part as shown in Figure 5-6. A second part fits over the stud. Then the two parts are forced together with a nut on the other end of the stud.

Studs are used when getting the two parts in perfect alignment is important. A stud may have threads that run all along its length. The threads may be formed only on the ends, as shown in Figure 5-7.

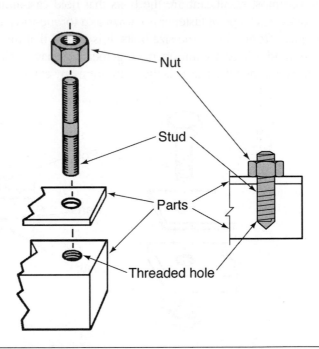

Figure 5-6 One end of a stud fits in a threaded hole and the other end has a nut.

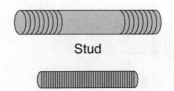

Figure 5-7 Stud threads may be on each end or along the entire length.

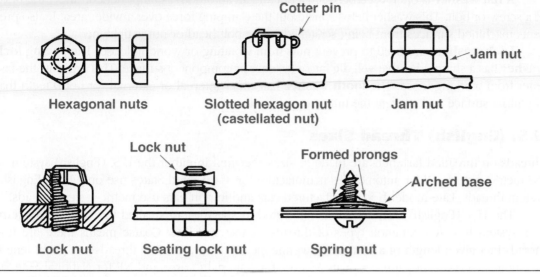

Figure 5-8 Common types of automotive nuts.

Nuts

Nuts are used with bolts and studs. Unlike other fasteners, they have their threads on the inside called internal threads. Many different types of nuts are used on automotive parts (Figure 5-8). The most common automotive nuts are hexagonal or *hex nuts*. Hex nuts have six sides and are made to fit box-end, open-end, or socket wrenches.

Several specialized nuts are available when parts may be subject to a great deal of vibration. Specialized nuts are also used where the part held by the nut is critical to safety. These parts might include the steering-wheel retaining nut or wheel spindle nut. The slotted or castellated nut is another important specialized nut. After the castellated nut is tightened onto the bolt or stud, a metal cotter pin is inserted through its slots and also through a hole that has been drilled in the stud or bolt. After the cotter pin is installed, it is bent around the nut.

Jam hexagon nuts are thinner than regular hex nuts. These are used in pairs and provide double protection against loosening. The lock nut has inserts of fiber or plastic. The inserts jam into the threads as the nut is tightened. This prevents the nut from getting loose during vibration. A **seating lock nut** is a thin nut with a concave surface that flattens when it contacts the top of the regular hex nut. The concave action binds the lock nut to the threads and prevents it from coming loose. **Spring nuts** are made from thin spring metal. They have prongs that fit into a sheet metal screw thread and an arched base that pushes down on the part to be retained. These nuts are used in trim and sheet metal parts.

Washers

Washers (Figure 5-9) are often used with other threaded fasteners for several reasons. First, they can help a nut distribute the clamping load on a part. They can also prevent a nut from getting loose. Finally, they can prevent a nut from digging into a machined surface.

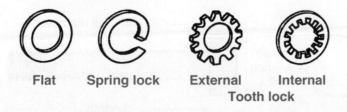

Flat   Spring lock   External   Internal
Tooth lock

Figure 5-9 Common types of automotive washers.

A **flat washer** is often used between a nut and an automotive component or under the head of a screw or bolt. This washer helps spread out the clamping force over a wider area. It also prevents machined surfaces from being scratched as the bolt head or nut is tightened.

Lock washers are used to prevent nuts from vibrating or working loose. The **spring lock washer** has a sharp edge that will dig into a fastener or component surface. This prevents the fastener from working loose. The **tooth washer** has either internal or external (or both) teeth that dig into a surface and prevent the nut from getting loose.

U.S. (English) Thread Sizes

Threads on threaded fasteners are made to sizes specified in either the U.S. (English) system or the metric system. Older automobiles manufactured in the United States use only U.S. (English) system threads. Late-model U.S.-manufactured cars and import cars use metric system threads.

The U.S. (English) threads are manufactured to specifications called the **Unified System.** This system has two common types of threads: coarse and fine. Coarse means there are few threads in a given length of a bolt or screw; fine means there are many threads in a given length of a bolt or screw. They are designated as NC for National Coarse and NF for National Fine, as shown in Figure 5-10. The difference between fine and course threads is easy to see. Coarse threads are used in aluminum parts because they provide greater holding strength in soft materials. Fine threads are used in many harder materials, such as cast iron and steel.

U.S. (English) threaded bolts and hex head cap screws are described by a number of measurements. Three important measurements are head size, shank diameter, and overall length, as shown in Figure 5-11. The **shank** is the part of the bolt between the head and the threads. Its diameter is part of the bolt's identification specification. Shank diameter sizes are available in most of the fractional divisions of an inch. Common sizes include $1/4$-, $5/16$-, $3/8$-, $7/16$-, and $1/2$-inch diameters.

The length of the bolt or cap screw is measured from the bottom of the head to the end of the bolt or screw. Bolts and hex head cap screws are commonly manufactured in $1/2$-inch increments. Typical length sizes would be $1/2$, 1, $1^1/2$, 2 inch, and so on.

Head size determines what wrench fits the bolt or hex head cap screw. The shank diameter is not the wrench size. The head size is determined by the distance across the hexagonal

Figure 5-10 Fine and coarse thread bolts with the same diameter.

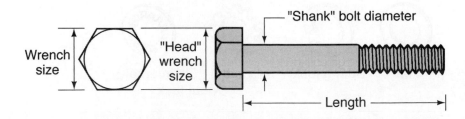

Figure 5-11 Typical measurements for a bolt or hex head cap screw.

flats on the head, as shown in Figure 5-11. A bolt with a $^1/_4$-inch shank diameter has a head size of $^7/_{16}$ inch. The common head sizes are shown in Figure 5-12.

The number of **threads per inch** is another important measurement in the Unified System. Each bolt and screw is identified by the number of threads in an inch, as shown in Figure 5-13. When a ruler is placed along the bolt shown in Figure 5-13, you can see that there are 20 threads in 1 inch. This bolt would be described as having 20 threads per inch (abbreviated to 20 TPI). When applied to a nut, these measurements refer to the size of the bolt that nut fits.

These measurements are used to give a callout or designation for a threaded fastener. If you were to go to an automotive parts store, you would find the fasteners in drawers with labels like $^3/_8 \times$ UNC $\times 1^1/_2$, as shown in Figure 5-14. The first measurement is the shank diameter of

Common U.S. (English) Head Sizes	
Wrench Size	Wrench Size
$^3/_8$"	1"
$^7/_{16}$"	$1^1/_{16}$"
$^1/_2$"	$1^1/_8$"
$^9/_{16}$"	$1^3/_{16}$"
$^5/_8$"	$1^1/_4$"
$^{11}/_{16}$"	$1^5/_{16}$"
$^3/_4$"	$1^3/_8$"
$^{13}/_{16}$"	$1^7/_{16}$"
$^7/_8$"	$1^1/_2$"
$^{15}/_{16}$"	

Figure 5-12 Common U.S. (English) bolt head sizes.

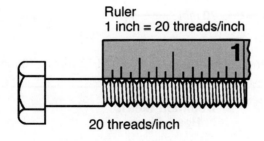

Ruler
1 inch = 20 threads/inch

20 threads/inch

Figure 5-13 A bolt with 20 threads per inch.

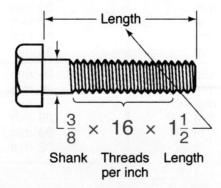

Length

$\dfrac{3}{8} \times 16 \times 1\dfrac{1}{2}$

Shank Threads Length
per inch

Figure 5-14 A complete bolt size designation.

$^3/_8$ inch. Each ✕ means "by." The second measurement is the 16 threads per inch. The UNC means that 16 threads per inch fits in the Unified National Coarse category. Finally, the $1^1/_2$ means that the bolt is $1^1/_2$ inches long.

Metric Thread Sizes

Fasteners manufactured in metric sizes have the same basic measurements as those described for the U.S. (English) system, but they are given in metric units. A common metric system fastener might be M12 ✕ 1.75, as shown in Figure 5-15. The M indicates that the fastener has metric threads. The first number is the outside diameter of the bolt, hex head cap screw, stud, or the inside diameter of a nut in millimeters. The second number, after the sign ✕ is the pitch. This is the distance between each of the threads measured in millimeters. After the second ✕ is the length of the fastener's shank in millimeters.

As you can see, there are no abbreviations such as NC or NF to identify fine and coarse threads. In the metric system, the pitch number distinguishes between fine and coarse. For example, a fine thread metric bolt may be M10 ✕ 1.0. A bolt of the same diameter with a coarser thread might be labeled M10 ✕ 1.25. The larger pitch number indicates wider spacing between threads. The length of metric fasteners is measured the same as in the U.S. (English) system, but the measurements are given in metric system units.

Like the U.S. (English) system, the metric fastener head size is not part of the designation or callout. The head sizes are manufactured in the common metric units. The common head sizes are made to fit common metric wrench sizes. The common head sizes for metric fasteners are shown in Figure 5-16.

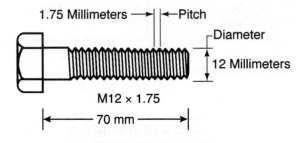

Figure 5-15 A metric fastener designation of a M12 ✕ 1.75 mm ✕ 70 mm bolt.

Common Metric Head Sizes	
Wrench Size	Wrench Size
9 mm	19 mm
10 mm	20 mm
11 mm	21 mm
12 mm	22 mm
13 mm	23 mm
14 mm	24 mm
15 mm	26 mm
16 mm	27 mm
17 mm	29 mm
18 mm	30 mm
	32 mm

Figure 5-16 Common metric head and wrench sizes.

SAE Grade Markings	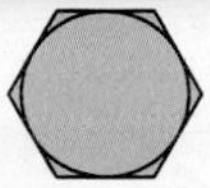	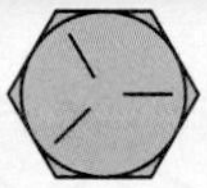	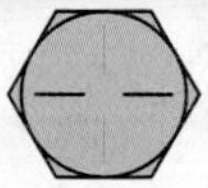	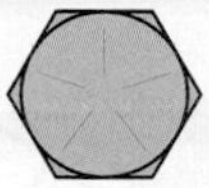	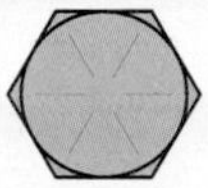
Definition	No lines: unmarked indeterminate quality SAE grades 0-1-2	3 lines: common commercial quality Automotive and AN bolts SAE grade 5	4 lines: medium commercial quality Automotive and AN bolts SAE grade 6	5 lines: rarely used SAE grade 7	6 lines: best commercial quality NAS and aircraft screws SAE grade 8
Material	Low carbon steel	Med. carbon steel tempered	Med. carbon steel quenched and tempered	Med. carbon alloy steel	Med. carbon alloy steel quenched and tempered
Tensile Strength	65,000 psi	120,000 psi	140,000 psi	140,000 psi	150,000 psi

Figure 5-17 Grade marking system for U.S. (English) bolts and hex head cap screws.

Hardness and Strength

Bolts and hex head screws used for different parts of the automobile have different strength requirements. A hex head cap screw used to hold on an engine flywheel must be much stronger than one used to hold on a headlight. On the other hand, the fasteners made to hold on the flywheel are much more expensive than the ones used on the headlight.

The strength or quality of fasteners is identified by **grade markings** on the fasteners. The markings are different for U.S. (English) and metric fasteners. The U.S. (English) grade markings for bolts and hex head cap screws are shown in Figure 5-17. The standards are set by the Society of Automotive Engineers (SAE). The system uses marks on the head of the bolt or the hex head cap screw. Unfortunately, the system is confusing because the number of marks and the grade number do not agree. A grade 0, 1, and 2 bolt has no markings on the head and is not very strong. A bolt with three marks is called a grade 5 bolt and is much stronger than a grade 1. The strongest bolt for automotive use has six marks and is called a grade 8. The chart in Figure 5-17 also describes the material each grade is made from and the **tensile strength.** Tensile strength is the amount of pressure the fastener can take before it breaks.

Metric fasteners use property class numbers to indicate their strength. These numbers are stamped on the head of the bolt or hex head cap screw. Typical metric bolt strength numbers are shown in Figure 5-18. The higher the property class number, the stronger the fastener. A 10.9 fastener is much stronger than one marked 4.6. A metric fastener without a number would be the same as a grade 0 U.S. (English) fastener. Metric studs have a marking system on the end of the stud that shows their property class number, as shown in Figure 5-18.

Nuts are also graded according to the same grading systems, as shown in Figure 5-19. Nuts must have the same grade as the bolts or studs with which they are used. English nuts have dots that represent grade markings. Three dots represent a grade 5. Six dots represent a grade 8. Metric nuts have numbers that represent strength. The number on the nut is its property class number. In both systems, the higher the number or the more dots, the stronger the nut.

Fastener Torque

Each threaded fastener on an automobile must be tightened just the correct amount. Untightened fasteners can loosen and cause parts to fail. Fasteners that are tightened too tightly can damage

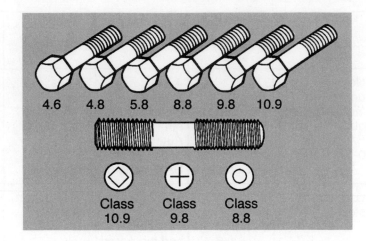

Figure 5-18 Grade marking system for metric bolts, hex head cap screws, and studs.

Inch System		Metric System	
Grade	Identification	Class	Identification
Hex nut grade 5	3 dots	Hex nut property class 9	Arabic 9
Hex nut grade 8	6 dots	Hex nut property class 10	Arabic 10
Increasing dots represent increasing strength.		Can also have blue finish or paint dab on hex flat. Increasing numbers represent increasing strength.	

Figure 5-19 Grade marking system for U.S. (English) and metric nuts.

the part or the fastener and cause the parts to fail. Overtightening causes fasteners to stretch. These no longer have the strength they had before the overtightening.

To avoid these problems, fasteners must be tightened with a torque wrench. There are torque specification charts for each important fastener in the service manual for the car you are working on. Always follow these specifications.

Pitch Gauges

When threads are damaged by using the wrong sizes or starting a fastener incorrectly with a wrench, we call this "cross threading" or "stripping."

CAUTION: Never use a regular impact wrench to tighten a fastener. The impact wrench can quickly exceed the required torque and damage the fastener or part.

There are two different thread systems in use, each with fine and coarse threads, which can cause a great deal of confusion. Metric threads cannot be used with English threads. Fine threads cannot be used with coarse threads. These different types of threads can make it difficult for the technician to tell one thread from another.

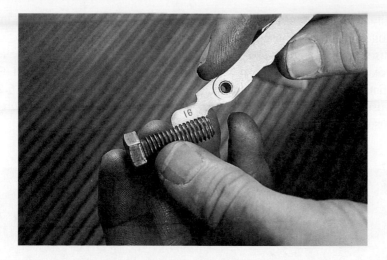

Figure 5-20 A pitch gauge is used to match threads for identification.

A thread **pitch gauge** (Figure 5-20) is used to identify fasteners. It has a number of blades with teeth. The thread size is written on the blade. By matching the teeth on the blade with threads on a fastener, the thread size can be determined. Pitch gauges are made for both metric and English threads.

Thread Repair Tools

Shop Manual
pages 108–113

There are several types of thread repair tools or tool sets on the market today. The tools come in metric and USC measurements and can help repair both internal and external threads.

Tap and Die Sets

Tap and die sets are designed to make threads instead of repairing them. However, careful use of taps and dies can be used to correct some minor thread damage without reducing the fastener's efficiency. A tap and die set is a combination of many different-size thread repair tools (Figure 5-21).

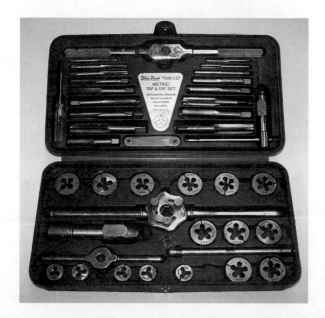

Figure 5-21 A typical tap and die set.

The set generally includes special wrenches, an external die, internal taper taps, internal flat taps, and a pitch gauge. Sometimes special dies and taps for pipes are also included.

A tap is used for making or repairing internal threads. Internal threads are located inside a hole. The taper tap is used for holes that extend completely through the material. Flat taps are designed with a flat tip to allow the tap to thread to the bottom of a hole (Figure 5-22).

A die is used to repair or make external threads on a shaft or rod (Figure 5-23). Special wrenches are used to turn the tap and die.

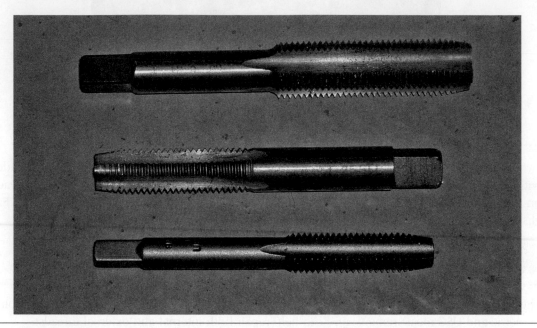

Figure 5-22 Bottom taps are used to thread holes that do not extend completely through the component.

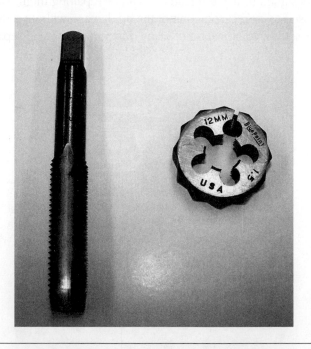

Figure 5-23 A tapered tap is used to thread holes that extend through the component. A die is used make threads on round stock iron or steel.

Helicoils

Basically, a **helicoil** is a threaded device that can be threaded into a hole. Once positioned, it will provide internal threads for a bolt (Figure 5-24). The helicoil is used to make a major thread repair but retains the same diameter size and thread type. It must be used in conjunction with a drill and tap that match the size of the helicoil. A small, single-size helicoil set can be purchased for one-time jobs (Figure 5-25). It consists of a drill bit, the tap, and five or ten helicoils. Larger sets can be purchased with several different-size helicoils and matching drill bits and taps.

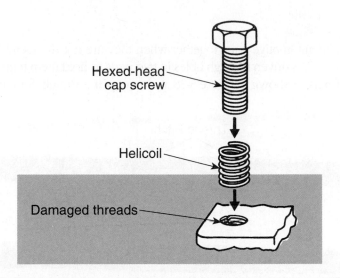

Hexed-head cap screw

Helicoil

Damaged threads

Figure 5-24 A helicoil is used to repair damaged threads.

Figure 5-25 Helicoil sets can be purchased in packets as shown or a helicoil can be purchased separately. However if purchased as individual coils, the correct drill bit, tap, and installation must be available.

Thread Restorers and Chasers

Thread-restoring files are similar to other files in their appearance and use (Figure 5-26). The cutting edges run over and between the damaged threads, which are smoothed and straightened by the file's actions. Thread files will only *repair* damaged threads.

Thread chasers fit over the damaged thread like a threaded fastener (Figure 5-27). The chaser is tightened onto the damaged threads and then turned back and forth over them until they can be used. Like the file, chasers can only *repair* threads, not replace them. This type of thread repair tool does a better job and will correct more damage than a thread file. Chasers are also easier to use.

Pins

A **pin** is often used to hold automotive parts together when they are not disassembled. A pin is a small, round length of metal. It is driven through holes in two parts to hold them together. There are several common types of pins, as shown in Figure 5-28. A **dowel pin** is straight for most of its length

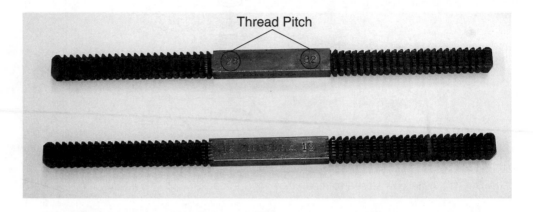

Figure 5-26 Thread files are used to straighten or clean threads. They cannot make new threads.

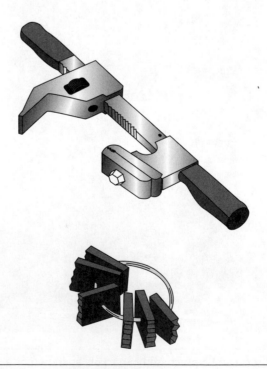

Figure 5-27 A thread restoring tool with insertable blades (lower). (Courtesy of Snap-on Tools Company)

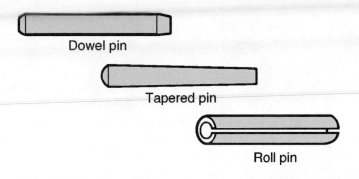

Dowel pin

Tapered pin

Roll pin

Figure 5-28 Three common types of pins.

but tapered on the very end. This tapered part helps to position two parts that fit together. **Tapered pins** are tapered for their entire length. The taper locks into the part for a tight fit. A **roll pin** is not solid, but rolled from a thin piece of metal. It is used where strong holding power is not required.

Snap Rings

A **snap ring** is often used to hold parts in place on a shaft. Because snap rings are used to retain parts in position, they are sometimes called "retaining rings." They are made from a high quality steel that allows them to be expanded or contracted by a tool for installation. Once in place the snap ring "snaps" back to its original size.

There are two basic types of snap rings as shown in Figure 5-29. The **internal snap ring** fits in a machined groove inside a hole. The **external snap ring** fits in a machined groove on the outside of a shaft. The external snap ring is expanded with retaining ring pliers for installation. The spring tension from the ring holds it in the groove. The internal snap ring is compressed with retaining ring pliers to fit it in the groove. When the tool is released, its spring tension will hold it in position.

Many snap rings have small holes on the ends, as shown in Figure 5-30. The ends of snap ring pliers are made to fit into these holes. Most snap ring pliers come with an assortment of ends to fit different sizes of snap rings (Figure 5-31).

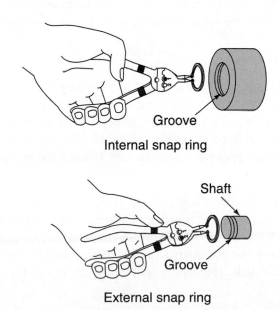

Groove

Internal snap ring

Shaft

Groove

External snap ring

Figure 5-29 Internal and external types of snap rings.

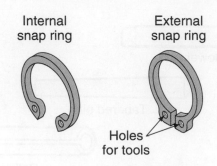

Internal snap ring   External snap ring

Holes for tools

Figure 5-30 Internal and external snap rings with holes for a snap ring tool.

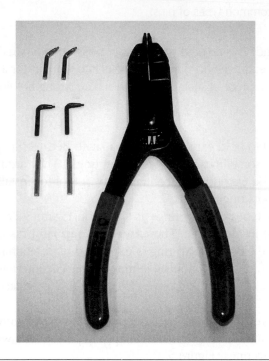

Figure 5-31 Snap ring pliers with replaceable tips are handy and relatively inexpensive, but they may not be able to survive the force sometimes needed to remove snap rings.

Keys

Another common way to retain a part on a shaft is to use a **key.** A key is a small, hardened piece of metal. They are often rectangular or half-moon shaped. As shown in Figure 5-32, the key fits into a slot called a **keyway.** When it is in position, half the key sticks up past the shaft. There is a matching keyway in the part to be retained on the shaft. The part to be retained in Figure 5-32 is a pulley. The keyway in the pulley fits over the key, holding the two parts in position.

Splines

Splines are another way to hold parts together. These are long teeth that can be cut on the inside or outside of a part. A set of splines is shown in Figure 5-33. The shaft in Figure 5-33 has external splines. The matching part has splines on the inside. The part with the inside splines will slip over the splines on the shaft. The mating splines hold them together.

Rivets

Rivets are used to hold parts together that are rarely disassembled. They are made from soft material such as aluminum. A rivet fits through a hole drilled in the two parts to be joined. The small

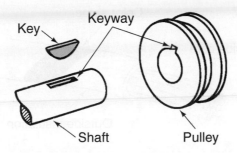

Figure 5-32 A key fits in a keyway on a shaft and pulley to hold the two parts together.

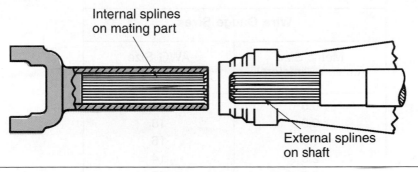

Figure 5-33 Splines can be used to hold a part on a shaft.

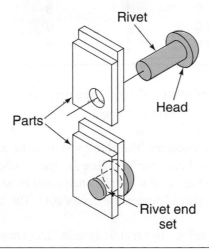

Figure 5-34 A rivet before and after installation.

end of the rivet is formed into a head with a rivet set or a ball-peen hammer. A rivet before and after installation is shown in Figure 5-34. A rivet is taken out by first removing the head with a drill or chisel. It may then be driven out with a punch.

Electrical Wire and Fasteners

A technician will be required to make repairs to the car's electrical system. This often means replacing damaged wires. Numerous types of wire and electrical fasteners are available to make these repairs. Wires, terminals, and connectors are considered to be the conductors of the electrical current within a vehicle.

When replacing a wire, always use a replacement that is at least as heavy as the one to be replaced. Automotive wire has two basic parts: an inside core of metal that conducts current flow and an outside of insulation material (Figure 5-35).

Shop Manual
pages 117–124

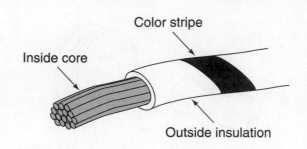

Figure 5-35 Automotive wire has an inside core and an outside insulation material.

Wire Gauge Sizes	
Metric Size (cubic millimeters)	AWG Size
0.5	20
0.8	18
1.0	16
2.0	14
3.0	12
5.0	10
8.0	8
13.0	6
19.0	5

Figure 5-36 Comparison of AWG and metric wire core sizes.

Wire sizes are based on two systems. The oldest system is the **American Wire Gauge (AWG).** When wire is made, it is assigned an American Wire Gauge number ranging from 0 to 20. These numbers specify the size of the wire center core diameter and cross-sectional area. The higher the number, the smaller the cross-sectional area of the center core. The largest cross-sectional core wire is 0 and the smallest is 20.

The newer system is based on the metric system. The cross-sectional area of the wire is measured in cubic millimeters. The wire size is then designated by a number such as 0.5 mm. This means the cross-sectional size of this wire core measures 0.5 cubic millimeters. A comparison of common AWG and metric wire sizes is shown in Figure 5-36.

The current-carrying ability of a wire is determined by its cross-sectional area. As shown in Figure 5-37, the greater the cross-sectional area, the more current can flow through the wire. The smaller the cross-sectional area, the more resistance exists and the less current can flow. The wire size of a circuit is carefully determined by automotive electrical engineers to meet specific requirements. For example, large battery cables are often no. 1 or no. 2 gauge because a larger flow of current is needed. On the other hand, a no. 16 gauge (1.0 mm^2) wire is often used with a headlight because less current is needed. As you can see, when a wire is replaced, it must be the same size as the one it is replacing.

New wire is sold in spools that are clearly identified by AWG or metric size. Wire size in various circuits on the automobile are often identified on the wiring diagrams. Two procedures can be used to discover the size of an unidentified wire. First, a wire gauge is available that has a series of graduated holes. Numbers are printed on the wire gauge to correspond with the hole

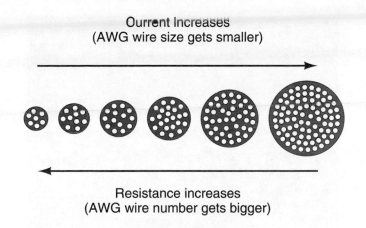

Current Increases
(AWG wire size gets smaller)

Resistance increases
(AWG wire number gets bigger)

Figure 5-37 Current flow increases and resistance decreases as the wire core gets bigger.

sizes. The center core of the wire to be identified is inserted into the various holes to determine the closest fit. If a wire gauge is not available, the diameter of the center core can be measured with an outside micrometer and then matched to a wire table to determine the size of the wire.

Wires are colored so that they can be traced on circuits on the automobile. Colors on the wire are also printed on the wiring diagrams for the car. You should always try to use the same color as the original when replacing wire.

Connectors

Soldering

Connecting two wires or a terminal to a wire is best accomplished using electrical-type **solder**. This type of connection adds little or no resistance to the conductor and is resin based. Regular solder is acid based so the soldered area is cleaned as the solder is being applied. Acid-based solder will damage the surface of the wire and may add resistance to the circuit. Only resin-based solder should be used for soldering wires together. Once soldered, the connection should be covered with heat-shrink material instead of tape whenever possible.

Butt Connectors

Butt connectors are used to connect two wires together (Figure 5-38). They are quick and easy to install but may add resistance to the total wire. Very-low-amperage circuits such as computer systems should be soldered.

Terminal Connectors

The **terminal connector** is a device fastened to the end of a wire that allows the wire to connect to or disconnect from a component (Figure 5-39). This allows components to be unplugged from the circuit without damaging the conductor or component. Most positive conductor terminals are covered in some type of insulator or installed in connectors to prevent shorting the circuits.

Molded and Shell Connectors

Molded and shell connectors come in many different sizes, shapes, and materials depending upon their designed use. In areas of the automobile where there are several connections in one place, the designers have, in many cases, installed connectors that can be matched according to shape,

Shop Manual
pages 120–124

Crimp-on terminal and butt connectors are called "solderless connectors."

Figure 5-38 Butt connectors are a quick easy way to connect two electrical wires. However, unless heat shrink is applied moisture and dirt can enter the connection.

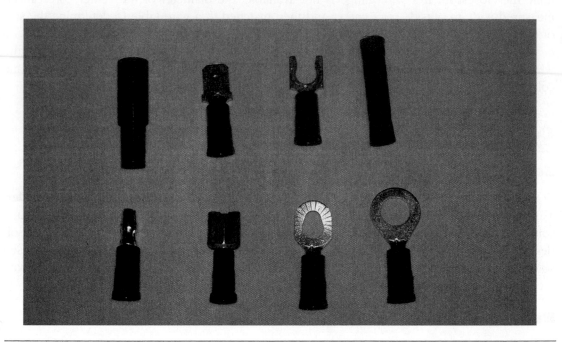

Figure 5-39 Terminals are installed at the end of the wire to form a way of connecting the wire to a permanent terminal like those on a switch or are mated with another terminal to connect two wires.

size, and color (Figure 5-40). This reduces the possibility of crossing circuits. Connectors can be made for a single wire or any number of wires installed into one shell or mold.

A molded connector imbeds wires inside a formed plastic or rubber coating (Figure 5-41). The wires cannot be removed from this connector. Replacement connectors with short wires can be purchased to replace the damaged connector. This part is called a **pigtail** and is made to exactly replace the original part, including the color of the wires.

Shell connectors are usually made of hard-formed plastic (Figure 5-42). With the right tool, damaged terminals can be removed, replaced, and reinserted into the shell. Shell connectors may have some type of seal to block moisture from entering the system. This is called a **weather-tight connector.** Most electronic circuits are connected this way.

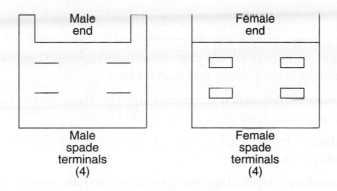

Figure 5-40 The male end will slide and lock into the female end.

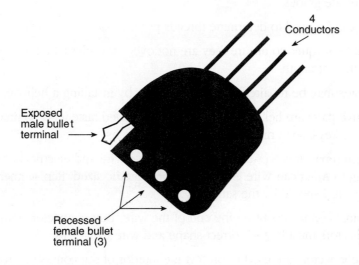

Figure 5-41 This is a typical molded connector. Note the bare terminal on the left. This is a four-wire trailer connector.

Figure 5-42 Multiple-circuit, hard shell connector.

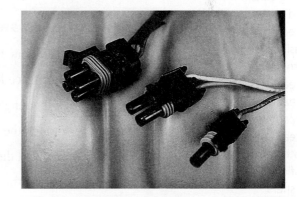

Figure 5-43 The seal is trapped between the mating connectors when they are plugged together.

Weather-Tight Connectors

Most of the electrical connections on late-model vehicles use the weather-tight design. This type of connector may be molded or a shell with rubber seals. As the two the connectors are plugged together, the seal on one side is trapped and clamped between the two (Figure 5-43). The clamped seal prevents moisture and dirt from entering the connector.

Summary

❑ Threaded fasteners use threads to hold automotive parts together. Common threaded fasteners include bolts, screws, studs, nuts, and washers.

❑ Threads are measured and classified according to the U.S. (English) and metric systems. U.S. (English) threads are made to standards of the Unified System.

❑ Unified System threads are classified according to the bolt shank diameter, length, and number of threads per inch.

❑ Metric system threads are classified according to shank diameter, length, and pitch.

❑ Both metric and U.S. (English) fasteners have grade markings to show fastener strength. U.S. (English) fasteners use marks on the bolt head to indicate grades. Metric bolts use property numbers to indicate grades.

❑ A pitch gauge can be used to determine threads per inch or thread pitch.

❑ Fasteners must be torqued to ensure they are not over- or undertightened. Overtightened fasteners lose their strength.

❑ Damaged threads may be repaired with a tap, die, or by installing a helicoil.

❑ Many automotive parts are held together with nonthreaded fasteners. Common no threaded fasteners include keys, snap rings, rivets, splines, and pins.

❑ Electrical system repair involves the use of automotive wire and electrical connectors. Wire is sized according to American Wire Gauge sizes or is metric sized. Replacement wires must be the same gauge and should be the same color.

❑ Electrical terminal connectors fit on the end of the wire. Butt connectors join two wires together. Connectors must be the correct shape and wire gauge size.

❑ Weather-tight connectors are used to protect the interior of the connector from moisture.

Review Questions

Short Answer Essay

1. Describe the difference between a bolt and a screw.

2. Explain the difference between a bolt and a stud.

3. Describe how the length of a bolt is measured.

4. Explain how to tell the strength of a U.S. (English) bolt.

5. Describe how to tell the strength of a metric bolt.

6. Explain how to tell the strength of a metric nut.

7. Explain the purpose of a pitch gauge.

8. Explain why a torque wrench should be used when tightening fasteners.

9. Why must the correct wire gauge size wire be used when replacing wire?

10. Explain the purpose of soldering wires and terminals.

Fill-in-the-Blanks

1. A measurement across the head of a bolt gives a(n) _____ size.

2. A measurement from under the head of a bolt to the end of the threads gives a _____ measurement.

3. Grade markings are found on the _____ of a bolt.

4. The number of threads on a metric fastener is called a(n) _____ measurement.

5. A replacement thread installed in a tapped hole is called a(n) _____.

6. A key fits in a(n) _____.

7. The two types of snap rings are _____ and _____.

8. The two types of electrical connectors are _____ and _____.

9. The _____ the inner core of a wire, the greater the resistance.

10. The _____ connector may add resistance to the circuit.

ASE-Style Review Questions

1. A $^3/_8$-inch bolt is being discussed.
   Technician A says a $^3/_8$ wrench fits the bolt.
   Technician B says the diameter of the bolt is $^3/_8$ inch.
   Who is correct?
   A. A only   **C.** Both A and B
   B. B only   **D.** Neither A nor B

2. A $^1/_4$ × 28 UNF bolt is being discussed.
   Technician A says this fastener is $^1/_4$ inch in diameter.
   Technician B says this fastener has 28 threads per inch.
   Who is correct?
   A. A only   **C.** Both A and B
   B. B only   **D.** Neither A nor B

3. The grade markings on a U.S. (English) bolt head are being discussed.
   Technician A says the more marks, the larger the bolt.
   Technician B says the more marks, the weaker the bolt.
   Who is correct?
   A. A only   **C.** Both A and B
   B. B only   **D.** Neither A nor B

4. The grade markings on a metric bolt are being discussed.
   Technician A says the larger the number, the weaker the bolt.
   Technician B says the larger the number, the stronger the bolt.
   Who is correct?
   A. A only   **C.** Both A and B
   B. B only   **D.** Neither A nor B

5. Coarse and fine thread bolts are being discussed.
   Technician A says a fine thread bolt has more threads per inch than a coarse one.
   Technician B says fine thread bolts will require a twelve-point wrench to install and tighten the bolt.
   Who is correct?
   A. A only   **C.** Both A and B
   B. B only   **D.** Neither A nor B

6. The installation of a hex head cap screw is being discussed.
   Technician A says always start the screw by hand.
   Technician B says once the screw is started it is best to use a speed wrench to tighten the screw to the torque point.
   Who is correct?

   A. A only **C.** Both A and B
   B. B only **D.** Neither A nor B

7. The installation of a hex head cap screw is being discussed.
   Technician A says it can be installed with an impact wrench.
   Technician B says it should be tightened with a torque wrench.
   Who is correct?

   A. A only **C.** Both A and B
   B. B only **D.** Neither A nor B

8. The use of washers is being discussed.
   Technician A says washers distribute the load on the part being tightened.
   Technician B says washers protect machined surfaces.
   Who is correct?

   A. A only **C.** Both A and B
   B. B only **D.** Neither A nor B

9. Automotive wire size is being discussed.
   Technician A says the AWG number includes the insulation around the wire.
   Technician B says the larger the center core, the smaller the AWG number.
   Who is correct?

   A. A only **C.** Both A and B
   B. B only **D.** Neither A nor B

10. Connectors are being discussed.
    Technician A says terminals are used to connect two wires.
    Technician B says a butt connector can do the same job as soldering two wires.
    Who is correct?

    A. A only **C.** Both A and B
    B. B only **D.** Neither A nor B

Automotive Bearings and Sealants

Upon completion and review of this chapter, you should be able to:

- ❏ Explain the purpose of bearings.
- ❏ Identify the different types, construction, and uses of automotive bearings.
- ❏ List the types of grease used in a typical light vehicle.
- ❏ Explain the different types, construction, and uses of gaskets.
- ❏ Explain the types and uses of chemical automotive sealants.
- ❏ Explain the different types, construction, and uses of seals.

Introduction

We now have some idea of the theories of operations and how the various components are held together in a vehicle. There are many moving pieces in a typical vehicle, many of which work against or with others. Each moving part requires some type of lubricant and a means to retain that lubricant on or in the assembly. In this chapter, the use of bearings and assembly sealing will be discussed. In addition, common usage of each will be covered. Use and lubrication will also be covered in various system chapters in this book.

Bearings and Bushings

When two components move against or with each other, **bearings** are used. Bearings are used between a moving part and a stationary part or between two adjacent moving parts. A bicycle wheel spins on two bearings that are mounted on and around a shaft or an axle. Bearings are used in the engine, driveline, suspension, steering, and in almost every system in a vehicle. Some components of the electrical system even use bearings.

Bearing Loads

Bearing loads are usually computed by the weight and movement of the load. The weight is used to determine the size, placement, and type of bearing to be used. The movement is either **radial load** or thrust load. Thrust load is also known as **axial load.** Radial load is the up-and-down movement of the load while axial is the front-to-rear (in and out) movement of the load. It should be noted that the term "load" not only applies to the weight being supported but also the direction and amount of force against that weight. The force can be thermodynamics in the engine or centrifugal force acting against a turning vehicle.

Bearing Journals and Races

Bearing journals are machined areas on a shaft (Figure 6-1). They have fine, smooth finishes to prevent damage to the bearing. Usually the area is hardened more than other parts of the shaft to withstand the loads. In most bearing failure cases, the journal will have to be machined and a smaller bearing used. There is a limit to the amount of machining before the shaft has to be replaced.

Shop Manual
pages 135–136

Bearings reduce the friction, heat, and wear of moving components.

Radial load is produced by the component pushing directly against the bearing from inside out or vice versa.

Axial load is produced by the component pressing on the sides of the bearing.

Journals may be hardened more than the rest of the shaft.

Figure 6-1 The highlighted areas are the bearing journals on this shaft.

Races

Bearing races serve the same purpose as the journal. However, races are usually purchased as part of the replacement bearing (Figure 6-2). A race can be assembled as part of a bearing assembly or it can be a loose part packed and shipped as part of the assembly. Many wheel bearings have a race that must be press-fitted into the wheel hub before the bearing is installed. If either a bearing or race fails, replace both.

Bearing Inserts

A BIT OF HISTORY

Crankshaft bearing inserts on the Ford Model T could be made from strips of leather.

Bearing inserts are known as "plain bearings" and are most commonly used in the engine as **rod bearings** or **main bearings** (Figure 6-3). The piston rod's lower end wraps around a portion of the crankshaft (Figure 6-4). Both the rod and crankshaft move. The bearing allows them to work with reduced friction and increased life. The main bearing is fitted between the rotating crankshaft and the engine block.

A bearing insert is so named because it is fitted into a mounting component. It is made from high-strength steel that is machined to a fine-surface finish, coated, and cut or stamped to the correct length and width. Bearing coating may be babbit, copper-lead, or aluminum alloy.

Two insert pieces are needed to make one complete bearing that is then curved to fit inside the rod and rod cap (Figure 6-5). One end is notched so the insert will lock into its mount. If the

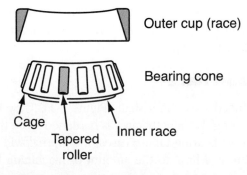

Figure 6-2 A tapered roller bearing is used almost exclusively on a vehicle's nondriving wheels. Shown is a complete tapered roller bearing. The outer race is separate and must be driven into the hub with a special tool.

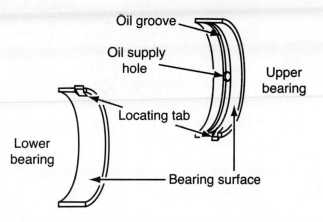

Figure 6-3 This type of bearing may be known as a bearing insert and is commonly used as a bearing for the crankshaft.

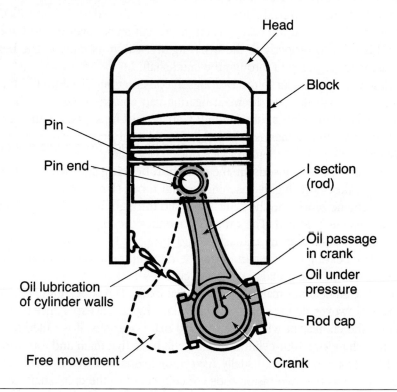

Figure 6-4 The lower end of the rod and rod cap holds a rod bearing in place.

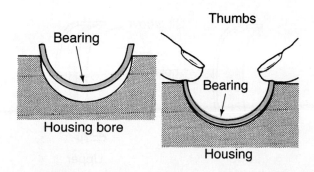

Figure 6-5 The bearing insert is pressed into the mount, usually a rod or main bearing cap, a rod, or the engine block.

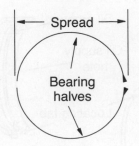

Figure 6-6 The two bearing halves are not in a perfect circle until installed. This is called "bearing spread."

two halves of an insert are placed together like a circle and measured, it will be determined that the circle is not perfectly round (Figure 6-6). This is called spread. But when the two halves are fitted into a rod and rod cap, they are forced into a perfect circle. If they do not, the insert will damage the crankshaft and fail quickly. Rod bearings are machined to fit a specified diameter on the crankshaft. A small length at the end of each insert half sticks over the rod and cap. This is called **crush** and is used to properly fit the insert tightly within its mount. The bearings are lubricated by pressurized oil pumped through the crankshaft.

Main bearings are very similar to rod bearings in construction. They are usually a little wider and a little longer. One half fits into the main bearing cap and the other into the engine block. The top half of a main bearing has a lubrication hole that must be aligned with the oil galley hole in the block (Figure 6-7). In this manner, the bearing can be lubricated with pressurized oil. Main bearings also have a lock to keep them in place during operation.

Two other types of bearing inserts are the **camshaft bearing** and **thrust bearing.** Camshaft bearings are shaped like short, highly machined metal tubes (Figure 6-8). They are press-fitted into cavities of the engine block and the camshaft is slid through from one end. There are usually four or more bearings per camshaft. The camshaft operates the engine's valves.

Thrust bearings are fitted at the crankshaft and are used to limit the amount of end-to-end crankshaft travel. The bearing may be part of one of the main bearings or separate pieces (4) that are fitted into the block next to a main bearing (Figure 6-9).

Main, rod, and cam bearings mainly support radial loads. Thrust bearings are made to support axial or thrust loads. However, during normal operation, an engine will experience both types of loads.

The areas on the crankshaft where the rod and main bearings fit is called the bearing journals. A typical crankshaft for a four-cylinder engine will have five main and four rod bearing journals (Figure 6-10). Most V-6 engines usually have four mains and three double-rod journals. A double-rod journal means that two piston rods are connected to the crankshaft side by side with a small space between them.

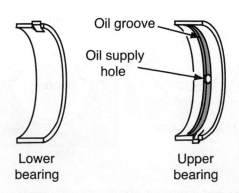

Oil groove

Oil supply hole

Lower bearing

Upper bearing

Figure 6-7 The upper bearing half or insert has an oil hole that aligns with the oil passage in the engine block.

Figure 6-8 Note the oil hole in each of the three camshaft bearings shown.

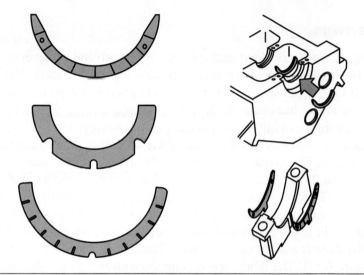

Figure 6-9 Thrust bearings are used to control or limit a shaft's forward and backward (in and out) movement within its mount. The figure illustrates the thrust bearings for a crankshaft.

Main journals

Rod journals

Figure 6-10 Crankshafts have a main bearing journal between each or between each pair of rod bearing journal(s) plus one at each end. There is a rod bearing journal for each piston.

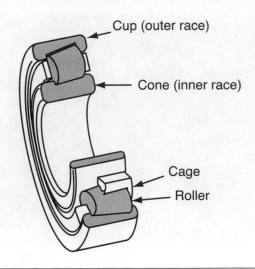

Figure 6-11 A roller bearing is made of high-strength steel and can support heavy loads.

Roller Bearings

All **roller** and **ball bearings** provide a rolling surface for the moving components so friction and wear are reduced.

A **roller bearing** is a set of machined shafts of steel. The length and diameter of the shafts or rollers are determined by the size of the load that the bearing is expected to carry. The typical roller bearing in a vehicle is fairly small and has several rollers trapped in a cage or separator (Figure 6-11). The cage holds the rollers in a circle. An inner race is fitted through the circle's center and over a shaft. The roller bearing assembly can now rotate around the shaft with each individual roller turning within the cage. An outer race is fitted around the outside of the bearing and into the rotating component like a gear. The contact area between the roller and race extends completely across the length of the roller. Roller bearings can be found in transaxles, electrical alternators, and other vehicle components. They are used to support radial loads and a small amount of thrust.

Many roller bearing assemblies are made with the same diameter at each end. Tapered roller bearings are made a little differently and usually support a wheel assembly. One end of the circle created by the cage is smaller than the other. This places the rollers at an angle and forms a cone shape (Figure 6-12). The inner race is commonly known as the *cup* and fits into a wheel hub. This race can be separated from the bearing assembly and fitted into the hub. The inner race or cone is part of the assembly and cannot be separated without destroying the bearing. It is made to slide onto the spindle or shaft around which the wheel turns. Tapered roller bearings can support axial and radial loads. When two sets of tapered roller bearings are used in the same component (like a wheel assembly), the bearings complement each other in handling the loads applied against the wheel assembly.

Needle bearings are not used for high-load conditions.

A third type of roller bearing is known as the **needle bearing.** Needle bearings are very small in diameter and may not be held within a cage. They fit into a bearing cup such as a driveline com-

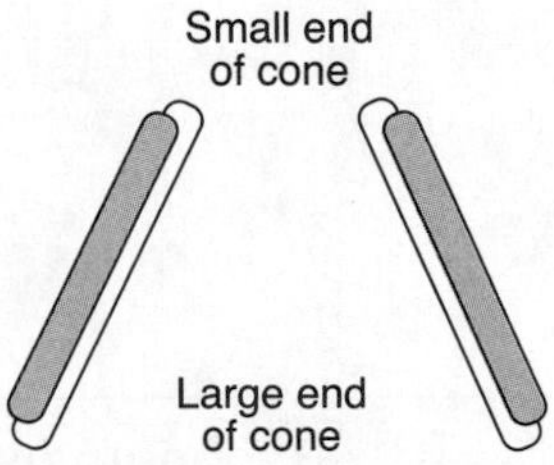

Small end
of cone

Large end
of cone

Figure 6-12 A tapered bearing forms a cone with the small end installed toward the inside of the component.

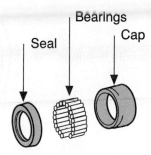

Seal   Bearings   Cap

Figure 6-13 Normally, needle bearings fit inside a cup but may be placed in a cage similar to the roller bearing.

ponent or into some older alternators (Figure 6-13). Improper disassembly of a component with needle bearings usually results in many needle bearings on the bench or floor, or they just disappear. If this happens, all of the bearings must be replaced. There can be no empty spaces between the needles. A needle bearing usually support only radial loads.

Ball Bearings

The construction of a **ball bearing** is similar to that of a roller bearing. The primary difference is the use of highly machined steel balls in place of rollers (Figure 6-14). The balls fit into a machined groove that keeps them aligned and distributes the load evenly to each ball. The two grooved sections form the cage that holds the bearing together and makes the inner and outer races. Contact between the ball and race is very small but the bearing can support axial and radial loads well. Some bearings use two rows of balls to support loads better.

On the front wheels of many front-wheel-drive vehicles, tapered roller bearings have been replaced with **ball bearings.**

Bushings

Bushings are made of thin steel tubing cut to specific lengths. Usually, they are copper coated, but other materials may be used for special situations. The bushing will not axially support loads very well since their purpose is to limit the radial movement of a shaft. Rotating shafts must be kept in a straight line to prevent damage to the shaft and component. Bushings form

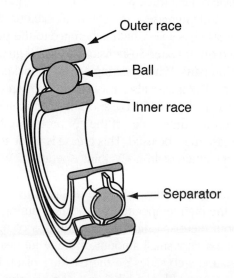

Outer race

Ball

Inner race

Separator

Figure 6-14 A ball bearing usually has the balls locked in a cage and comes in one piece. Ball bearings can support heavy radial (up/down) loads, but may not be sufficient for heavy axial (sideway) loads.

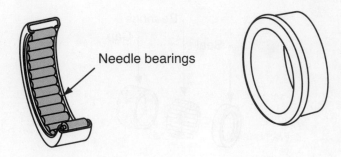

Figure 6-15 Bushings are normally used to support one end of a shaft that has little load placed upon it.

a journal on which the shaft can ride (Figure 6-15). However, bushings are not used in high thrust or radial load areas like the camshaft or crankshaft. They can support a transaxle shaft or the shaft in an alternator or pump. The bushing is press-fitted into a housing and the shaft slides through it.

Lubricants

Shop Manual
pages 138–141

Grease is a lubricant that reduces friction and smoothes movement between parts of components.

Grease is a lubricant used on roller and ball bearings and suspension system components. Many bearings, like those used in the engine, transmission, and differential, are lubricated by the oil stored and used within the component. Other components, like the wheels and suspension, require the parts or bearings to be packed with grease.

The National Lubricating Grease Institute (NLGI) (Figure 6-16) classifies automotive grease into two main categories: L for suspension components and U-joints on the drive shaft, and G for wheel bearings. The L classification is further broken into an A class for mild duty and B for mild to severe duty. Wheel bearing grease, G, has three subclasses:

A for mild duty
B for mild-to-moderate duty
C for mild-to-severe duty

A typical passenger vehicle or light truck would use LA for suspension and GB for wheel bearing. The difference between L and G grease is the chemical makeup that is designed to meet certain operating conditions. L grease has a very high adherence to the part that is lubricated. It has a high resistance to being washed off by water. Suspension or chassis grease will not break down from the movement of the suspension parts. Wheel bearing grease is thicker than L grease and is formulated for high resistance to heat. The grease also resists being thrown off by the spinning bearing.

With the advent of front-wheel drive (FWD), specially formulated grease was needed for the constant velocity (CV) joints on the drive axles. If the CV protective boot or the joint itself is being replaced, new CV joint grease must be used. This grease is made to withstand the high stresses of temperature and movement inherent in a CV joint operation. Wheel bearing grease should *never* be used in CV joints.

There are special lubricants for brakes. On disc brakes, this lubricant is formulated not so much for movement, but for the high temperatures generated during braking. It is applied to the slide pins or adapters that hold the disc caliper in place and is commonly referred to as caliper lubricant. Drum brakes require a very small amount of brake lubricant to be applied at the specific places where the brake shoe web slides on the backing plate. In most cases, caliper lubricant can be used. A detailed review of brake lubricants can be found in the *Today's Technician Automotive Brake Systems* Classroom and Shop Manuals.

While there are specific greases for the suspension, bearings, and U-joints, the most commonly used grease in an automotive shop is multipurpose grease. As its name implies, it can be

Figure 6-16 G denotes wheel bearing grease. L is for steering and suspension components. (Reprinted with permission from the National Lubricating Grease Institute)

used for different purposes. On a vehicle, multipurpose grease is used for suspension, U-joints, and wheel bearings. However multipurpose grease should not be used in all situations, and the service manual must be consulted for specific applications.

Gaskets

Gaskets are used to retain lubricants or to seal a chamber. In most cases, the lubricants are under low pressure but the sealed components may be stationary or moving. A gasket usually seals high-pressure areas and the space between two stationary components or parts.

Head Gaskets

The gasket that is under the most force is the cylinder head gasket (Figure 6-17). This gasket seals the extremely high pressure of the combustion chamber and seals water and oil passages between the engine block and the head. It must accomplish this task under high and low

Shop Manual
pages 155–157

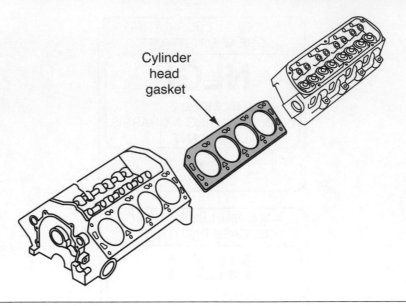

Cylinder head gasket

Figure 6-17 Cylinder head gaskets are one of the most critical and most stressed gaskets used on a vehicle. They must contain high temperatures and pressure while allowing adjacent components to expand and contract during temperature changes.

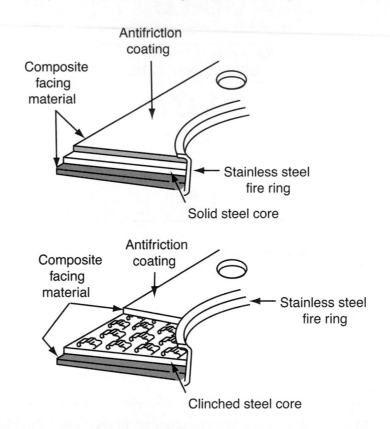

Antifrictlon coating

Composite facing material

Stainless steel fire ring

Solid steel core

Antifriction coating

Composite facing material

Stainless steel fire ring

Clinched steel core

Figure 6-18 The antifriction coating allows the sealed components to slide over the gasket without damaging it.

temperatures, expanding and contracting components, different materials, and exposure to chemicals including acids.

A typical head gasket is composed of layers of different materials bonded together (Figure 6-18). Older head gaskets were stamped or embossed steel sheets. They worked well in many applications but they are not recommended for vehicles produced after the early 1980s. The use of aluminum cylinder heads mounted on cast iron blocks produced some engineering problems on sealing this area. Aluminum expands twice as quickly as cast iron. Also, current engines and heads are lighter and more flexible. Head gaskets must expand and flex with the materials above and below them. To correct this, the manufacturers redesigned the components, selected new materials, and created new manufacturing processes to produce workable gaskets.

The modern head gasket has a steel core made of solid or clinched steel plate. A dense **composite** facing is placed on each side of the core and covered with a low- or antifriction material. Holes are punched through the gasket to line up with the engine cylinders and coolant and oil passages. The rim around each cylinder hole has a stainless steel fire ring to help protect the gasket material from burning air and fuel in the combustion chamber. The antifriction coating is usually Teflon or a silicone-based material. Some types of head gaskets have a silicone bead placed around the oil and coolant holes to prevent liquid leakage (Figure 6-19).

Similar to head gaskets are intake and exhaust gaskets (Figure 6-20). They differ in construction and material and must perform well after being exposed to high temperatures and chemicals. Sometimes they are placed between two different metals and must be able to expand properly with both.

> **Composite** refers to a mixture of materials to form a more durable gasket.

Other Gaskets

Cork gaskets are used to seal low-pressure areas like the valve covers and oil pans. When heated, cork will become brittle and crack if overtightened. Rubberized cork gaskets are often used as

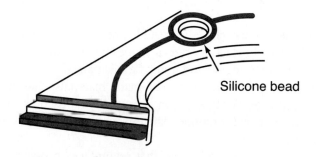

Silicone bead

Figure 6-19 The smaller holes in a cylinder head gasket allow for movement of liquids, coolant, and oil between the block and cylinder head.

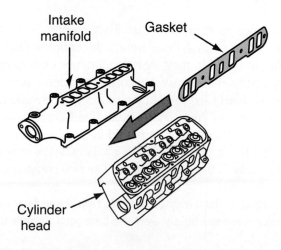

Intake manifold   Gasket

Cylinder head

Figure 6-20 This type of gasket is used to seal two components that create high temperatures. (Courtesy of Fel-Pro Incorporated)

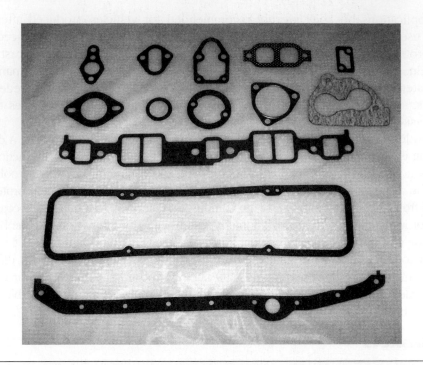

Figure 6-21 Neoprene gaskets can be reused if handled carefully during removal and installation.

valve cover seals. This type of cork gasket is treated with a reduce-shrinking process and maintains resiliency (the ability to return to its original shape) when clamped and released. Cork gaskets may also be found on older engine and transaxle oils pans.

Synthetic rubber gaskets are made from man-made materials. The material is usually neoprene and the gasket is reusable. For this reason, they are often used as valve cover gaskets because valve covers require removal for routine maintenance (Figure 6-21). They may also be found as oil pan gaskets and in other areas.

Paper gaskets are made of treated paper that retains low-pressure, low-temperature areas between two stationary components. A typical application is between the different housings and covers on a manual driveline or on some water pumps. Paper gaskets may be relatively thick for some applications, like the water pump, or they may be as thin as a regular piece of paper.

Chemical gaskets are used on many new cars. They are installed by robotic equipment capable of laying an almost perfect $1/8$-inch bead around the entire sealing area, thereby making a formed-in-place gasket. Because of this, they are difficult to use during repairs since a poorly laid bead will leak. Aftermarket part vendors may use cork or neoprene gaskets for repairs.

Gasket sealers are sometimes used to hold the gasket in place and help it seal. Care should be taken when using gasket sealers. Instructions should be followed, as many gasket manufacturers do not want any sealer on their gasket or placed in certain amounts at specific places. The sealer may be applied with a brush or pressed from a tube. Gasket sealers are anaerobic chemicals, meaning they will cure only in the absence of air. The two components have to be fitted and torqued before the sealer will cure.

A gasket sealer or thread sealant may be required on some fasteners that extend into the cooling passages. The sealer is spread lightly over the fastener's threads to seal the area between the threads.

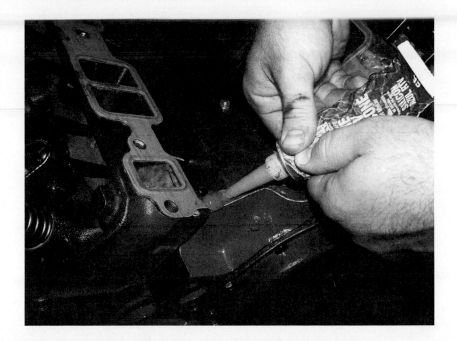

Figure 6-22 Use silicone only where directed by the gasket manufacturer. Silicone applied improperly can prevent the gasket from sealing. (Courtesy of Fel-Pro Incorporated)

Chemical Sealants

Chemical sealants include the sealing materials mentioned earlier and a popular sealer known as **room temperature vulcanizing (RTV)** (Figure 6-22). RTV is an aerobic sealant and may be known as "silicone rubber." Aerobic means RTV cures when exposed to air. It can be used to seal two stationary components such as an oil pan and an engine block. It cannot be used as a head gasket or an exhaust manifold (high heat and pressure). It cannot be used on fuel systems either because fuel deteriorates RTV. RTV will set in about 15 minutes in warm temperatures and high humidity. It cures in about 24 hours. It will not set or bond to a gasket. RTV comes in different colors based on application and temperature ranges.

Seals

Seals are normally used between moving components and their housings to retain liquids. Seals can be used in high-pressure and high-temperature areas and are usually made of butyl rubber or neoprene within a metal ring. The ring provides the connection to the stationary component.

A **lip seal** is the type most commonly used on vehicles (Figure 6-23). The seal has a lip that is installed facing the lubricant. In this way, the pressure of the oil against the inner side of the lip forces the lip edge tightly around the moving part. The lip may have a **garter spring** behind it so it is held to the shaft better (Figure 6-24). Some seals have a smaller lip called a dust lip that faces away from the lubricant. It does as its name implies, keeping dirt from entering the component. Seals are found at the front and rear of the engine crankshaft, front and rear of a transaxle, and around drive axles.

Shop Manual
pages 146–148

Chemical sealants and sealers should be checked for their use with electronic engine controls.

Shop Manual
pages 145–146

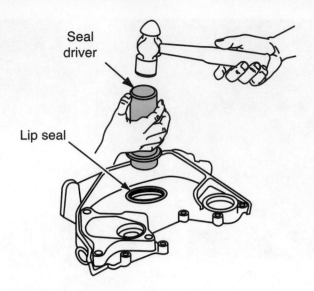

Figure 6-23 This lip seal fits into a cover plate.

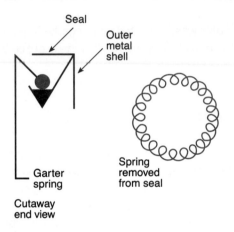

Figure 6-24 The garter spring is a small bracelet that fits behind the lid of a seal.

O-Rings

An **O-ring** is so named because of its shape and the manner in which it is used. The O-ring is shaped like a rubber bracelet without a latch (Figure 6-25). It fits into a groove, either in the moving component or in the stationary component. The ring extends above the groove and is pressed against the second component (Figure 6-26). O-rings are commonly used on high-pressure, hydraulic fittings such as power steering, brakes, and air-conditioning systems. A specially cut type called a *square-cut O-ring* is used on disc brake systems (Figure 6-27).

A third type of seal is used primarily to protect rather than seal a component. These seals are usually referred to as boots. A common one now in use is the boot for the CV joint. It does seal to some extent, but its main purpose is to keep debris and moisture from entering the joint. A common assertion is that if the CV boot is split, the CV joint will need to be replaced after about 8 hours of operation. Though this assertion is general in nature, it does point out how fast a component can fail when its seal or gasket fails.

Boot or dust boots are also used on the drum brake's wheel cylinders. In this case, they only protect the internal pistons and bore from dirt and brake dust. Other seals within the wheel cylinder perform the task of keeping the brake fluid inside the bore.

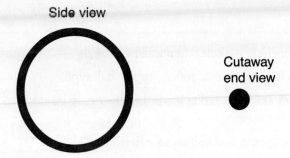

Side view

Cutaway
end view

Figure 6-25 An O-ring.

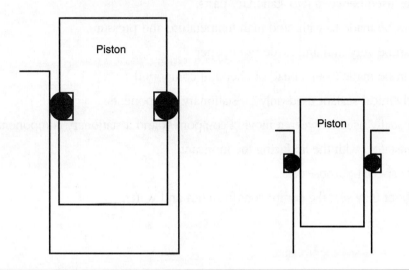

Figure 6-26 The O-ring may move with the piston or be installed in the wall so the piston moves over the ring.

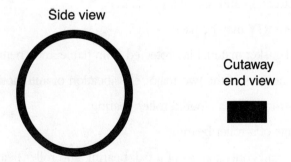

Side view

Cutaway
end view

Figure 6-27 This type of O-ring is square cut. When bent and released, it tends to return to its original shape.

Summary

❏ Automotive sealants include gaskets, chemicals, and seals.

❏ Automotive bearings may be inserts, rollers, or the ball type.

❏ Bearings must support axial and radial loads, but they may be designed for one specific type of load.

❏ Generally, automotive grease is classified as either L or G.

❏ L grease is for suspension and A is for wheel bearings.

❏ A special grease is used on CV joints and brake systems.

❏ Gaskets are used between two stationary parts.

❏ Gaskets can be made to withstand high temperature and pressure.

❏ Gaskets can be very thin and made from paper.

❏ Gaskets can be made from a bead of chemical compound.

❏ RTV is a chemical sealant used only on stationary components.

❏ Seals are usually used between a moving component and a stationary component.

❏ Seals are installed with the lip facing the lubricant.

❏ O-rings are special-purpose seals.

❏ A rubber boot may seal the component from dirt and water.

Review Questions

Short Answer Essay

1. Explain the construction of a typical head gasket for newer engines.

2. Describe the construction and use of a lip seal.

3. List the areas where RTV may be used.

4. Explain how head gasket material is protected from flame and chemicals.

5. List and describe the uses of the two major classification of automotive grease.

6. Describe the construction of a tapered roller bearing.

7. List the components of a roller bearing.

8. Describe how the load contact areas of a ball bearing and roller bearing differ.

9. Describe bearing inserts.

10. Explain how a rod bearing is installed.

Fill-in-the-Blanks

1. A(n) _____ _____ is used to support the thrust load of a crankshaft.

2. A type of _____ _____ is used to support the wheel assembly.

3. L type grease may fail if used on _____ _____.

4. The oil pan may be sealed by a(n) _____ or _____ _____.

5. High pressure within the oil pan may cause the crankshaft _____ _____ to leak.

6. The head gasket has holes punched through it to allow passage of _____ and _____.

7. Gaskets may use a(n) _____ to hold it in place.

8. A seal is installed with the dust lip _____ from the lubricant.

9. An O-ring is installed in a(n) _____ in the moving component.

10. Air cures a(n) _____ sealant.

ASE-Style Review Questions

1. Seals are being discussed.
   Technician A says the lip faces the lubricant.
   Technician B says a garter spring may be used on some seals.
   Who is correct?
   A. A only
   B. B only
   C. Both A and B
   D. Neither A nor B

2. The head gasket is being discussed.
   Technician A says the steel ring is placed around the cooling passage.
   Technician B says silicone may be used to seal the oil passages.
   Who is correct?
   A. A only
   B. B only
   C. Both A and B
   D. Neither A nor B

3. *Technician A* says a gasket can be used to replace a seal.
   Technician B says a chemical can be used to replace a gasket.
   Who is correct?
   A. A only
   B. B only
   C. Both A and B
   D. Neither A nor B

4. *Technician A* says lubricant pressure can be used to better seal the lip to the shaft.
   Technician B says a garter spring is used to better seal the dust lip.
   Who is correct?
   A. A only
   B. B only
   C. Both A and B
   D. Neither A nor B

5. RTV is being discussed.
   Technician A says RTV is an aerobic sealant.
   Technician B says the absence of air cures RTV.
   Who is correct?
   A. A only
   B. B only
   C. Both A and B
   D. Neither A nor B

6. Gasket sealers are being discussed.
   Technician A says a sealer can be used to replace the gaskets.
   Technician B says sealers are cured by the absence of air.
   Who is correct?
   A. A only
   B. B only
   C. Both A and B
   D. Neither A nor B

7. *Technician A* says anaerobic chemicals cure when the parts are assembled and torqued.
Technician B says aerobic chemicals cure completely in about 24 hours.
Who is correct?
   A. A only
   B. B only
   C. Both A and B
   D. Neither A nor B

8. The use of O-rings is being discussed.
Technician A says O-rings should be used in low-pressure systems.
Technician B says a square cut O-ring is used on drum brakes.
Who is correct?
   A. A only
   B. B only
   C. Both A and B
   D. Neither A nor B

9. Bearings are being discussed.
Technician A says needle bearings are a type of roller bearing.
Technician B says flat roller bearings are used to support a wheel.
Who is correct?
   A. A only
   B. B only
   C. Both A and B
   D. Neither A nor B

10. The use of bearings and bushings is being discussed.
Technician A says a roller bearing may be used in an alternator.
Technician B says bushings are used to support the camshaft.
Who is correct?
   A. A only
   B. B only
   C. Both A and B
   D. Neither A nor B

Automobile Theories of Operation

Upon completion and review of this chapter, you should be able to:

❏ Discuss the requirement to apply theories of physics in designing and operating vehicles.

❏ Discuss the first two laws of thermodynamics and their application in an engine.

❏ Define *energy* and describe the two types of energy.

❏ Discuss how pressure and vacuum help operate the engine.

❏ Discuss Newton's three Laws of Motions.

❏ Define electricity, voltage, current, and resistance.

❏ Discuss magnetism.

❏ Name the types of electrical circuits.

❏ Discuss the use of hydraulics in vehicles.

❏ Discuss other factors in designing and operating a vehicle.

❏ Discuss horsepower and torque.

Introduction

The automobile is made up of eight major interactive systems based on the application of certain physics theories. The seven basic systems are the engine, body and frame, electrical, brakes, steering and suspension, climate control, and driveline. The eighth is the computer system, which is not required to make a simple vehicle work but does provide the controlling element for the basic systems. The systems power the vehicle, deliver power to the wheels, and comfort and protect the passengers. Functional systems working together make it possible for the automobile to be what it is today: reliable, durable, and comfortable. In this chapter, we will discuss some of the theories of operations and examples of how they apply to the major systems. Understanding theories will assist the technician in diagnosing problems that do not respond to the conventional practice of electronic diagnostics and direct testing of suspected failures.

Thermodynamics

Engines use the theory of **thermodynamics** to generate power. The first law of thermodynamics is a law of energy conservation: Energy cannot be created or destroyed. The amount of internal energy in a system is a direct result of the heat transferred into that system and the work done. Heat and work are means by which natural systems exchange energy with each other. Basically, it is the relationship between heat and mechanical energy (Figure 7-1).

Shop Manual
page 153

Thermodynamics is the study of using radiant heat to perform mechanical work.

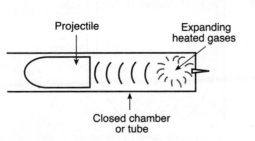

Figure 7-1 The force of hot, expanding gases is transformed into mechanical energy by the movement of the projectile.

The second law concerns the thermodynamic cycle. In a perfect heat engine, all of the heat produced would be completely converted to mechanical work. Nicolas Leonard Sadi Carnot, a nineteenth-century French scientist, proved that this ideal cycle could not exist. The best engine would lose heat and work because the engine must be exhausted. The exhaust would remove and waste a great deal of heat.

Liquids that are heated enough will change to a vapor and expand greatly in volume. A prime example of this is boiling water. As the water is heated, it expands in volume, which is the amount of space it occupies. Somewhere around 212°F, the water changes from a liquid to a gas, or vaporizes. The gas or steam greatly increases its volume. For example, place a lid on top of a container of boiling water. As the water boils more rapidly, the lid will begin to rattle and move about. We can capture the energy of the water by bolting the lid down and not allowing the steam to escape. However, if we do not in some way allow the steam to escape or use its energy, the volume of steam will continue to build until it ruptures the container and an **explosion** occurs.

Most lightweight vehicles use gasoline that is rated at about 110,000 **British Thermal Units (BTU)** per gallon. BTUs are used to measure the heat-producing capabilities of a chemical or mechanism. Burning vaporized gasoline in a closed chamber called a **combustion chamber** will result in high heat and a rapid expansion of gases (Figure 7-2). This rapid heating and expansion of gases creates a very high force that acts against the engine's pistons and produces useable power (Figure 7-3). The chamber is closed but it does have one side that can move: the piston. The energy of the expanding gas is transmitted against the piston, forcing it downward. This, in turn, rotates the crankshaft, producing a twisting force or *torque*. As the torque is transmitted through the vehicle's drive system, it moves the wheels and work is performed. Work in this instance is measured as **horsepower.**

However, the engine loses heat and therefore work through the exhaust and cooling system. The exhaust is required to remove spent gases while the cooling system must remove heat from the mechanical components. Roughly two-thirds of the heat in a gallon of gas is lost through exhaust and cooling. Only about 35,000 BTUs are left to operate the engine and vehicle. After friction and overcoming the weight, the mass of components, and the vehicle, only about 19 percent to 20 percent of the energy in a gallon of fuel is actually applied to the drive wheels. Vehicles made in the late-1990s through 2004 do somewhat better, but the nature of the beast prevents any large increase in the efficiency of current internal combustion engines.

An **explosion** is a sudden, violent release of energy. An engine uses a controlled burn to achieve power.

The **combustion chamber** is the open chamber at the top of an engine's cylinder. It is here that the fuel is ignited.

One **horsepower** is equal to the force needed to raise 33,000 pounds at one foot per minute.

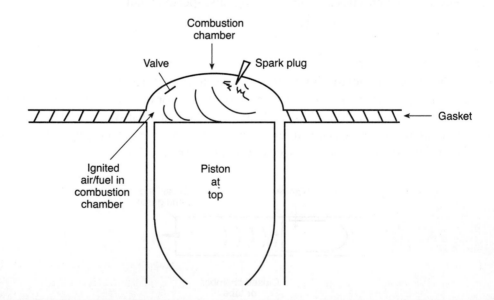

Figure 7-2 The air/fuel mixture burns in a controlled manner from the ignition point across the chamber.

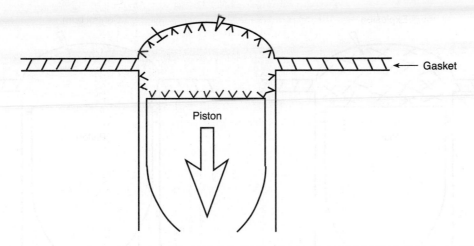

Figure 7-3 The force of the gases acts against the piston the same as it did against the projectile in Figure 7-1.

The application of thermodynamics in an engine requires fuel, oxygen, spark, and a combustion chamber. The engine will not operate or operate poorly if a fault occurs in any of these four elements. Almost nothing will burn without oxygen present. Considering that only about 20 percent of the earth's atmosphere is oxygen and the number of fires that occur worldwide, it is apparent that oxygen does not have to be present in large amounts. By the same token, almost everything will burn if enough oxygen is present. Steel will burn like a match if enough oxygen is available.

The control of the fuel and air (oxygen) to an engine is essential for a safe, controlled burn (Figure 7-4). While an automobile does not and cannot use pure oxygen, too much or too little upsets the **ratio** of air and fuel, resulting in a poorly operating engine. An example of upsetting the air/fuel ratio would be a dirty air filter. The filter will not allow sufficient air to enter the combustion chamber. With insufficient oxygen, the gasoline may or may not fire. If it does fire, it will not produce the necessary energy to perform the work needed.

Energy and Work

Energy is defined as the capacity of matter to perform work through its motion or position. The motion of an object produces **kinetic energy. Potential energy** is available energy based on the position of an object. Consider the action of a baseball. While the pitcher is holding the ball in his or her hand, the ball represents potential energy. When the ball is released toward the hitter, it has kinetic energy based on its weight and speed. The hitter's bat has potential energy until he or she swings it. When the bat and ball connect, the kinetic energy of both result in an

Ratio is the proportion or relation one element has to another. In this case, it is the relation of air to fuel.

Shop Manual
pages 166–168

Potential energy available is based on the mass and position of a stationary object. The speed of a moving object plus its mass creates the **kinetic energy** of the object.

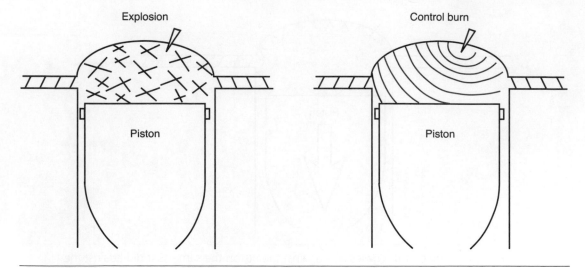

Figure 7-4 An explosion (left) creates uncontrollable energy radiating in unpredictable directions. A controlled burn radiates energy equally in all directions at once.

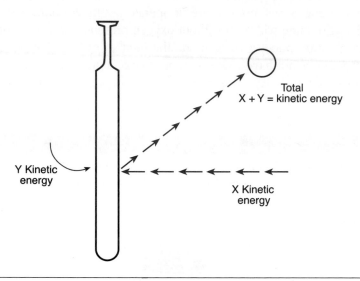

Figure 7-5 The kinetic energy created by merging or connecting a moving ball and bat creates a larger amount of kinetic energy in the ball.

expenditure of energy (Figure 7-5). The ball is driven in a different direction with great force. An equal amount of opposing force is absorbed by the bat and the hitter (see Newton's Laws of Motion later in this chapter).

When an object moves and energy is expended, work has been accomplished. This includes raising a hand or moving a house. **Work** is a unit of energy and may be measured in **foot-pounds (ft.-lbs).** Foot-pounds measure torque or twisting force. Work is measured by multiplying the **force** times the distance, or it is stated as one foot-pound will lift one pound the distance of one foot (Figure 7-6).

Gasoline is potential energy in the fuel tank. Its rapid expansion in the combustion chamber results in great amounts of kinetic energy. This energy is used to transform a resting vehicle with potential energy into a moving vehicle with large amounts of kinetic energy. This is proven by the amounts of kinetic energy released through actions occurring in a traffic accident.

There is one major drawback to all energy-producing devices now used. It is a fact that a machine or device cannot produce more energy than it consumes, in other words, there is no

Force is the pressure of one mass against another.

156

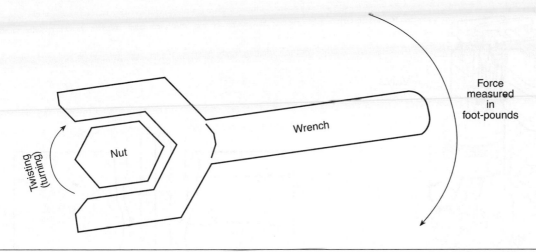

Figure 7-6 Torque is measured by determining the amount of twisting force being applied.

perpetual energy device today. A device that seems to contradict this statement is the alternating current generator. Without getting into depth on generator operation, the generator draws its power from the battery or its own electrical output and seems to produce more than it uses. However, when factoring the horsepower required for driving the generator's pulley there is definitely a loss of energy. *Today's Technician Automotive Electricity and Electronics* has the details for the generator's operation.

Pressure and Vacuum

Pressure is the amount of force used to hold a gas or liquid in place. The atmosphere applies about 15 pounds of pressure per square inch to every external inch of all objects on the surface of the earth including the human body. Atmospheric pressure can be used to achieve work in a vehicle or components and can be used to create pressure within a system to perform work.

Shop Manual
pages 157–162

Vacuum is defined as any pressure below atmospheric. The earth does not have sufficient natural vacuum to perform automotive work, but it can be created to some extent by certain components. One of those components is the engine.

As discussed in the section on thermodynamics, fuel and air must be present in proper amounts in the combustion chamber to create power. Since the chamber can be sealed, the downward movement of the piston increases the volume of the cylinder, creating a vacuum above it. This combination of vacuum and outside atmospheric pressure forces fresh air past a valve that is timed to piston movement (Figure 7-7). Fuel is pressure-injected into the airflow near the same valve or directly into the chamber during the intake of air. With the proper equipment, the amount of fuel can be regulated in correct ratio to the amount of air entering the chamber. Older systems used the movement of air passing through a ventura to create a low-pressure area and draw fuel from the carburetor fuel bowl (Figure 7-8). Now, almost all engines use electronic devices to control the fuel.

The force of the driver's foot creates pressure within a brake system (Figure 7-9). The force of the driver's foot is transferred by pressurized fluid to the brakes at each wheel. Engine vacuum and atmospheric pressure can be used to assist the driver in braking. A steering system uses a pump to supply high-pressure fluid to a power-assist unit controlled by the actions of the steering wheel. In both cases, pressure is used to assist the driver in the operation of the vehicle. Both systems are covered in later chapters.

Pressure is measured in **pounds per square inch (psi).** The use of pressurized fluid is covered later in this chapter in the paragraph on hydraulic theory. Vacuum is usually measured in **inches of mercury** or **inches of water.**

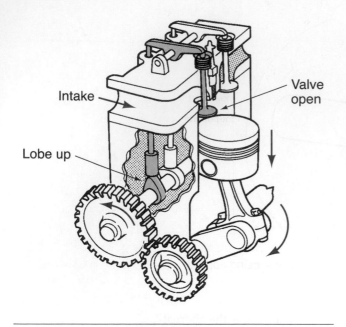

Figure 7-7 This valve opens to let the air/fuel into the cylinder.

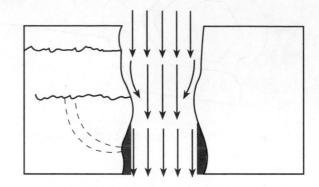

Figure 7-8 An object obstructing the airflow will create a low-pressure area downwind and behind the object.

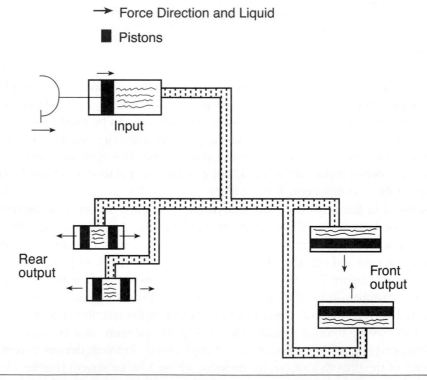

Figure 7-9 Applying force to a liquid in a sealed system will create pressure.

The most common vacuum measurement used in automotive repair is inches of mercury. Vacuum supplied against mercury in a closed tube will cause the mercury to rise based on the amount of vacuum (Figure 7-10). The amount of rise is measured in inches. However, since mercury is a deadly poison and a glass tube is not practical in the shop, calibrated mechanical gauges are used to register the measurement.

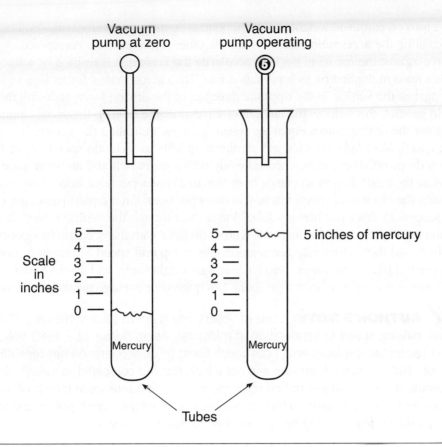

Figure 7-10 A given amount of vacuum will raise mercury or water to a specific height.

Newton's Laws of Motion

Shop Manual
pages 158–162

Newton wrote three basic Laws of Motion while proving that the earth revolved around the sun instead of the sun moving around the earth. The laws pertain to almost every system on the vehicle.

Newton's First Law deals with movement: An object at rest tends to stay at rest and an object in motion tends to stay in motion. This unchanging action is called **inertia** and it assumes that there is no outside force acting on or against the object. A stationary vehicle will not move unless something forces it to move. Parking a vehicle in neutral with the brakes off on level ground will result in a parked vehicle. However, parking the same vehicle on a hill, in neutral, and without brakes will allow gravity to act on the vehicle and cause it to roll down the hill, thereby overcoming rest inertia (potential energy) and generating motion inertia (kinetic energy).

Newton's Second Law: A moving object will travel in a straight line and in one direction. Try driving a car on a flat, level road. Assuming that the car is set up correctly, it will travel in a straight line. A force (steering) must act upon the car to change its direction. A second example is an inflated balloon. When an inflated balloon that has not been knotted is released, it immediately moves through the air in erratic directions. The balloon is not shaped to travel through air and it changes direction as the air acts against it. When all of the trapped air in the balloon is exhausted, the balloon falls to the earth. The balloon's shape and the loss of pressure are only two actions that work not only on the vehicle as a whole but on each moving component of the vehicle as well.

Within this law is a rule regarding acceleration of an object. The acceleration of a vehicle is directly proportional to the driving forces applied to it. This applied force must overcome the friction of the various moving components, tire against road, and the total **mass** of the vehicle and its load. Mass is not the weight of the object but the amount of matter in that object. A vehicle weighing 2 tons on earth will weight about 670 pounds on the moon, but its mass will be the same. When calculating the acceleration rate of a vehicle other forces must be considered. As the drive wheels force against the road and begin to accelerate the vehicle, it is resisted by a force equal to the vehicle's mass multiplied by its acceleration rate. Then aerodynamic forces begin to act simultaneously against the vehicle in the opposite direction of the driving force, reducing the acceleration rate. In general, this reduces the amount of force available to keep the acceleration rate high. At some point the driving forces equal the resisting forces, indicating the vehicle has reached its maximum speed. Most light vehicles will accelerate quickly to a certain speed before the rate of acceleration drops off. This can be checked easily with a stopwatch and an open, clear stretch of road. Measure the time taken to accelerate from zero to 25 miles per hour. Record the time but continue running the clock until speed reaches 50 miles per hour. On a typical passenger car or light truck the second 25 miles per hours will be slower even though the vehicle is already moving at the first time reading. Some individuals will tend to disagree with this scenario by quoting the 100-foot, 200-foot, and the quarter-mile acceleration rate or top end speed of dragsters. Consider this fact: Six thousand plus horsepower, aerodynamic lightweight body, and a vehicle designed to deliver maximum acceleration in midrange gears. Not quite your average pizza delivery vehicle.

AUTHOR'S NOTE: "One of Volvo's selling points back in the early 1980s was its passing speed or acceleration at midrange. As an owner of a 1980 Volvo 260 V-6 (don't remember displacement), I can vouch for its quick acceleration between 45 mph and 65 mph. Not real quick from stop and not a high top end compared to today's vehicles, but we would drive around just to find someone to pass. To quote some friends of mine, "it really screams in passing gear." This was not because of its great horsepower and torque, but the way the engine was engineered on when to deliver its power.

For every action there is an equal and opposite reaction—Newton's Third Law. Again, it is assumed that there are no outside forces in play. If a ball were thrown against a wall in outer space, it would return to the thrower with the same force and same speed. Doing the same on the earth would result in the ball slowing on its return journey. This is due to air resistance and gravity working against the ball so it is slowed, and its direction of travel is changed.

Earlier, we discussed kinetic energy in a vehicle accident. Consider, for example, that a 3-ton vehicle moving at 60 miles per hour with 100 tons of kinetic energy was suddenly stopped by an immovable object. The action of stopping 100 tons would immediately result in an opposite and equal reaction of 100 tons against the immovable object. That release of energy would not only affect the object but the vehicle as well. Since energy cannot be destroyed, the 100 tons of suddenly released kinetic energy must be dissipated or changed in some manner, in this case the destruction of the vehicle and object.

The movement, weight, and operation of all vehicle components are subject to the laws of motion. To move a piston, energy must be used. In turn, the actual movement of the piston in turn creates work. Each movement of every part requires energy and produces work. The engineering of a vehicle requires a balancing of the energy used versus the amount of work performed.

Electrical Theory, Measuring, and Circuits

Georg Simon Ohm (1787–1854) wrote the law regarding electrical theory and measurement. It states that it takes one **volt** to push one **ampere** of current through one **ohm** of resistance. A change in one of the three affects the other two values proportionally. In order to understand Ohm's Law, we must first look at **electricity.**

The basic element of any material is the **atom.** It is made up of three parts: neutrons, protons, and electrons (Figure 7-11). The center or nucleus of the atom is made of the neutrally charged neutrons and positively charged protons in equal numbers. An equal number of negatively charged electrons travel in orbits around the nucleus much as the planets travel around the

Shop Manual
pages 181–184

The **volt** is the standard measurement of electricity.

The **ampere** is the standard measurement of moving electrons.

Ohms are the standard measurement of electrical resistance.

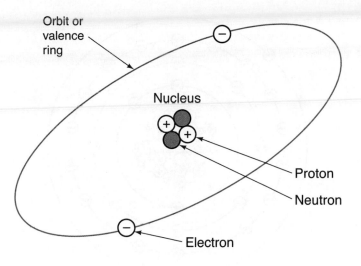

Figure 7-11 A balanced atom has equals numbers of protons, neutrons, and electrons.

sun. An atom that has equal numbers of electrons, neutrons, and protons is in balance. A balanced atom really does not produce any outside actions. If the atom can be unbalanced in some way, some type of energy can be created and used. Splitting the atom creates enormous amounts of energy, either uncontrollably, as in the case of nuclear weapons, or in controllable situations like nuclear power plants. However, if only a portion of the atom is moved and it basically remains intact, useable and controllable energy can be acquired.

Conductors

Causing an electron to leave one atom and move to another creates a chain reaction in which an extremely large number of electrons are moving (Figure 7-12). At this point, electricity or electrical current is present and can be used if controlled. However, there are only a few natural elements that can be used. In order for an element to be used for electricity, it must not have more than four electrons in the atom's outer orbit or the **valence ring.** The best element or **conductor** is copper (Figure 7-13). It has one electron in the valence ring. That electron can be forced out of orbit in a relatively easy manner. Elements with two, three, or four electrons in the valence ring can be used but require more voltage. For ease of understanding in this book, we will assume the conductor is copper.

> A **conductor** is any material that will carry electrical current.

Voltage

An electron is moved by **voltage**. Voltage acts like the pressure in a water hose and is a measurement of **electromotive force (EMF).** EMF is produced or is available when there is a difference in the numbers of free electrons at one end of a circuit compared to the other end (Figure 7-14). This

Conductor

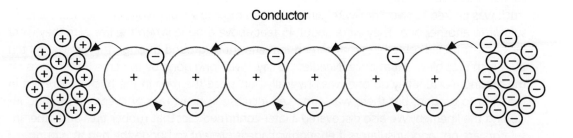

Figure 7-12 Electricity is the mass movement of electrons from negative to positive.

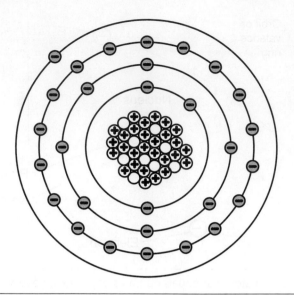

Figure 7-13 The valance ring is the outermost electron orbit. Electrons can only be stripped from this orbit for electrical current.

Conductor

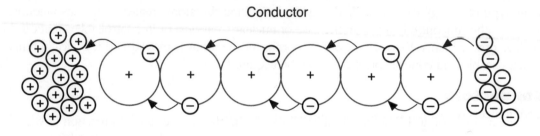

Figure 7-14 The difference between the number of electrons at the positive and negative sides creates voltage.

is known as **potential differential** or voltage. If there is no potential difference between negative and positive, then there can be no voltage. An example is a driver trapped in a car under a live, 13,500-volt electrical line. Many people might assume that the driver is protected by the rubber tires that act as insulators. The actual situation, however, is that the driver and the vehicle are the same voltage as the electrical line, hence there is no potential differential. When the driver attempts to leave the vehicle and places a foot on the ground, he or she has created a conductor between the line and the ground. Since the ground will have very little voltage, the potential differential is enormous and 13,500 volts of electricity will travel through the new conductor and usually result in death.

AUTHOR'S NOTE: An incident that brought the concept in the last paragraph home in a big way happened to me while employed as a technician at a large electrical utility company. I was changing a tire on a truck very near the spot where another bucket truck was parked as two "hot-wire" linemen were moving 25,000-volt live lines from one insulator to another one. They were about 45 feet above ground when the live wire they were moving contacted something that carried that current to ground through the vehicle. All ten 20.5-inch tires blew out at once. Needless to say, I was long gone by the time the tire rims hit the ground. Fortunately no one was physically hurt since the men in the bucket were not in the path to ground, but there was a marked increase in heart rate of everyone present, including the linemen. We also discovered a later-confirmed fact that rubber tires and other insulators are not good insulators if enough voltage is present to bridge the gap to ground.

As voltage pushes an electron out of orbit, two things happen at once. The atom becomes unbalanced and tries to regain balance by attracting another electron. The moving electron attempts to attach itself to another atom. A chain reaction has started. All of the electrons will move from the negative (ground) toward the positive because unlike polarities, negative and positive, attract each other. The movement of the electron is known as the **Electron Theory of Electricity.** When about 628 billion electrons pass a single point in one second, one ampere of current is available for use.

The Electron Theory of Electricity is actually what happens at the atomic level; electrons move from negative to positive. However, because of the way a vehicle is wired the **Conventional Theory of Electricity** is applied. In this theory it is *assumed* that the electrons flow from positive to negative. One reason for this is that automobiles are wired as negative grounded; the battery's negative post, through its cable, is connected directly to the vehicle's frame and other metal components. In this way, the entire vehicle becomes a source of negative-polarity electrons. When performing a test for voltage, the technician is not actually looking for the moving electrons, but for the positively charged atom or **ion** left behind. This type of wiring system is known as a **one-wire return.**

Current

Current is the amount of moving electrons expressed in an understandable term as amperes. A typical automobile has systems that use current ranging from milliamperes up to 30 amperes. Most electronic systems, like the engine computer and radio, use very low amperage while engine cooling-fan motors may use up to 30 amperes. The electrical current will continue to flow toward the positive until it meets resistance. It is wise to remember that voltage doesn't kill, current does.

Resistance

Resistance is anything that uses voltage and is measured in ohms (Figure 7-15). Elements that are nonconductive are resistant and classified as **insulators.** The only way to use electricity is to create a resistance within a system.

Resistance may be caused by **corrosion** on wires and connectors, small wire size, or it may be engineered into the system. Unintentional resistances like loose connections or corrosion disrupt the current flow in the system and devices may or may not work. Each electrical system is engineered to have specific resistance using specific amounts of current and voltage. Lamps, motors, and other electrical devices are engineered to use resistance and are known as loads. Loads

Corrosion can be found most prominently on battery cables, usually resulting from a lack of maintenance.

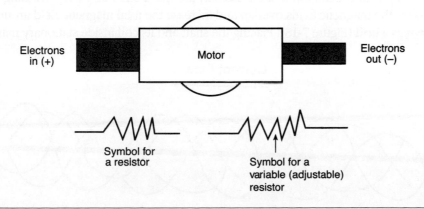

Electrons in (+)   Motor   Electrons out (−)

Symbol for a resistor

Symbol for a variable (adjustable) resistor

Figure 7-15 Resistance uses or stops current. Note the symbols for a resistor.

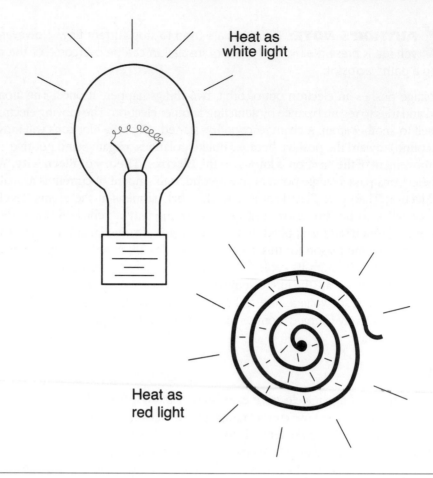

Heat as white light

Heat as red light

Figure 7-16 Depending on the type of resistance and circuit design, current can be used for light or heat.

Magnetism may be natural, such as magnetic rock or something electrical. Materials that are magnetized by electricity are known as "electromagnets."

are designed to use the electrical current for tasks. The lamp has a filament with a specific resistance (Figure 7-16). As current flows through the filament or resistance wire, it produces heat. The heat is seen as white light. An electric stove uses specific resistance coils for heating. The coils are heated to high temperature by current flow, which is usually seen as a red light. Electric motors must overcome the resistance of a shaft and attached devices.

Magnetism

Any time current is flowing, a **magnetic field** is created around the conductor (Figure 7-17). If the conductor is wound tightly in a spool fashion around a shaft and each winding is insulated from the others, the magnetic fields overlap and increase the total magnetic field around the shaft. This is known as a **coil** (Figure 7-18). Placing the shaft and its coil inside stationary magnetic fields

Current Flow

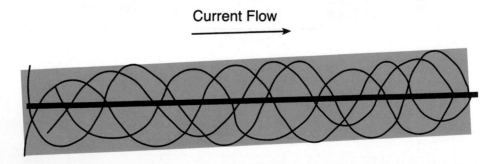

Figure 7-17 Current creates a magnetic field as it moves through a conductor.

Straight conductor, magnetic flux of 1

Conductor folded unto itself, magnetic flux of 2 plus

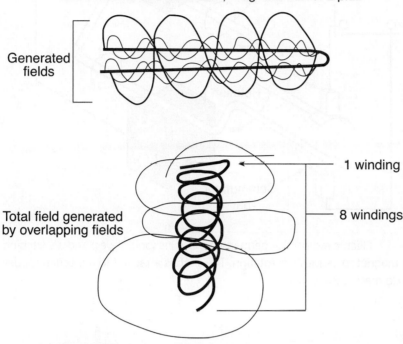

Generated
fields

1 winding

Total field generated
by overlapping fields

8 windings

Figure 7-18 Overlapping magnetic fields will create a larger field.

makes the shaft turn. Stationary magnets produce the second magnetic fields. The positive and negative polarities of the shaft's coil fields attempt to move toward their opposing polarity of the stationary fields (Figure 7-19). The result is a spinning shaft, which can be attached to a fan or other device. The polarity of the shaft's magnetic field can be continuously changed to continue the rotation. Magnetic fields can also be engineered to produce electrical current. Devices that use or produce electricity will be covered in other chapters in this book.

A BIT OF HISTORY

Michael Faradday (1791–1867) was an English chemist and physicist best known for his experiments with electromagnetism. Using the principle that electricity could be made by moving a magnet inside a coil of wire, he developed the first electric motor, generator, and transformer. He was responsible for several electrical terms in use today including *electrode, cathode, anode,* and *ion.* The unit of capacitance, the **farad,** is named in his honor.

Electrical Measurements

Since electricity can be used for definite purposes, there must be a way to measure it for control. Using electrical measuring instruments and service data as outlined in Chapter 3 and using Ohm's Law, we can calculate if a system is working properly. Since Ohm stated that one volt will push one ampere through one ohm of resistance, the mathematical formula for measuring and calculating

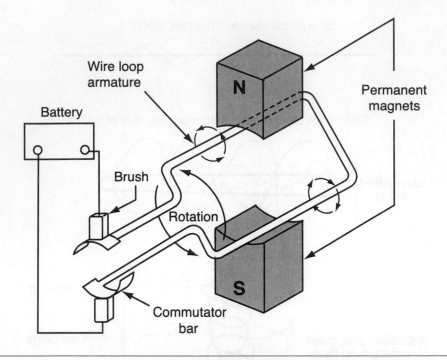

Figure 7-19 Electric motors use permanent magnets (pole shoes) and a switching polarity rotating magnet to cause shaft rotation. The shaft is attached to a mechanical device, such as a gear, to do mechanical work.

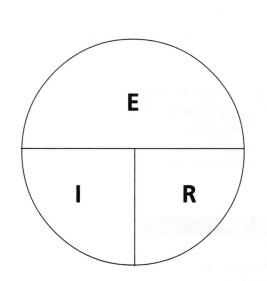

Figure 7-20 A method of remembering the mathematics formula for Ohm's Law.

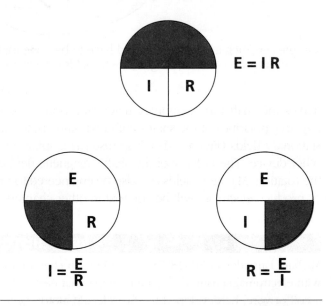

Figure 7-21 The equation can be used to find one unknown factor.

electrical values is E = IR, where E is EMF (voltage), I is current (ampere) and R is resistance (ohms) (Figure 7-20). If two of the values are known, the third can be calculated (Figure 7-21). Consider the following problem.

A system is using 12 volts to feed a current through 4 ohms of resistance (Figure 7-22). The missing measurement, current, can be determined using Ohm's mathematical formula. Since E = IR, then I = E/R or current is equal to voltage divided by resistance. In this case, I = 12/4 or I equals 3 amperes. If the system or device being checked uses 3 amperes to operate, then the system is

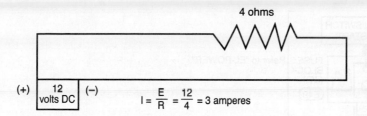

$$I = \frac{E}{R} = \frac{12}{4} = 3 \text{ amperes}$$

Figure 7-22 If a circuit can be measured and two known factors are found, Ohm's Law can be used to determine the third factor of the circuit.

functional. A device that uses more or less current would not work properly in this system. By the same token, if we measured this system with an instrument and found that only one ampere was being used, then the system is inoperative.

Consider a second situation.

A technician used a service manual to find that a system is supposed to have 12 volts supplied and 6 amperes of current (Figure 7-23). According to the formula, the working device should have 2 ohms of resistance (R = E/I). After taking measurements, it was found that only 1 ampere of current was present. Applying the formula again (R = E/I), the system appeared to have 12 ohms of resistance. Somewhere in the system there was too much resistance and the device did not work.

Circuits

Up to this point, we have discussed electrical systems. The systems are comprised of one or more electrical circuits. A **circuit** is the path that electricity follows from the negative side to the positive side. A circuit can work with a power source, conductor, and load. However, if those were the only things used, the operator would have to open the hood and connect the wires to the battery just to start the car. Such a procedure would have to be done for every device or system to be turned on or

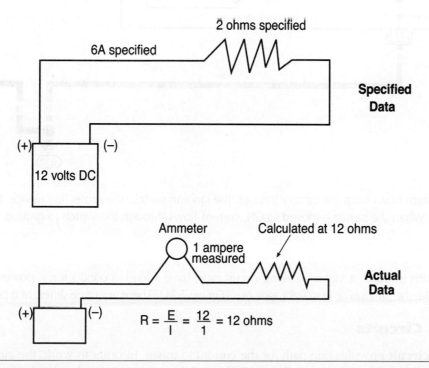

$$R = \frac{E}{I} = \frac{12}{1} = 12 \text{ ohms}$$

Figure 7-23 Measuring the actual voltage, current, and resistance of a circuit will assist the technician in diagnosis.

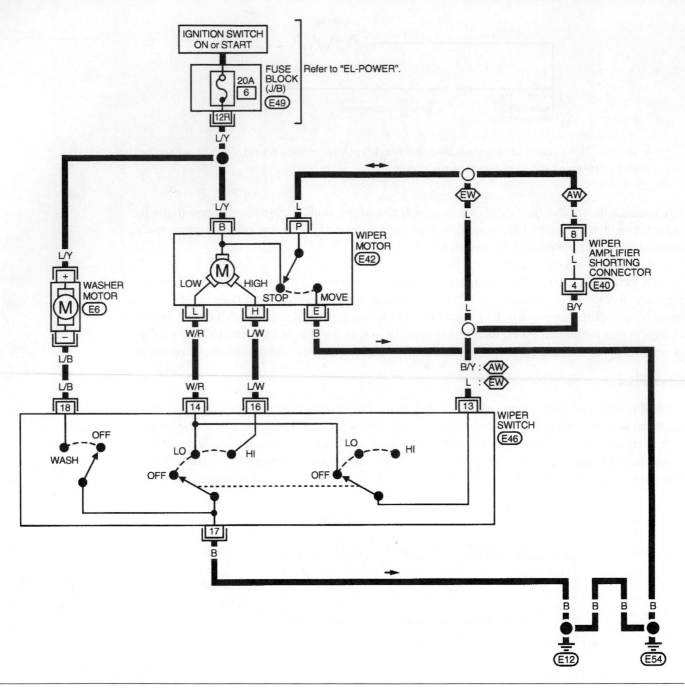

Figure 7-24 Current in this system flows from the battery through the ignition switch, the protection device, the motor, and finally to the wiper switch. When the switch is moved to ON, current flows through the switch to ground.

off. With this in mind, a viable working circuit must have a load, a conductor, a power source, and controls like the ignition or headlight switches (Figure 7-24). There are three different types of circuits.

Series Circuits

A **series circuit** provides one path for the current to travel. In order to work, the circuit must be intact from one end to the other (Figure 7-25). A defect anywhere in the circuit would affect the entire circuit. Older Christmas tree lights used series circuits. If one lamp went bad, then all of the other lamps would not work.

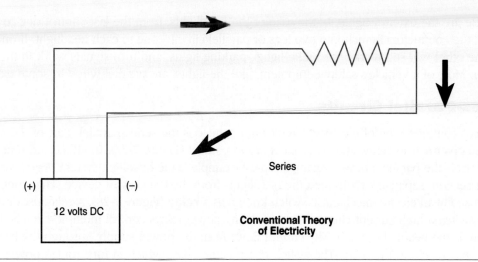

Series

Conventional Theory
of Electricity

(+) (−)

12 volts DC

Figure 7-25 A series circuit has one path for current to follow.

There is only one true series circuit on the typical vehicle and that is the starter control circuit. The current must flow from the battery, through the ignition switch, and finally through the park/neutral (clutch switch on manual transmission) to the starter relay. If the transmission is in a drive gear (clutch pedal up), then the park/neutral (clutch) switch is opened and the starter relay will not switch.

Parallel Circuits

A **parallel circuit** provides more than one path to ground and is common on vehicles (Figure 7-26). One portion of the circuit can work while the others may not. Typical examples of this circuit on

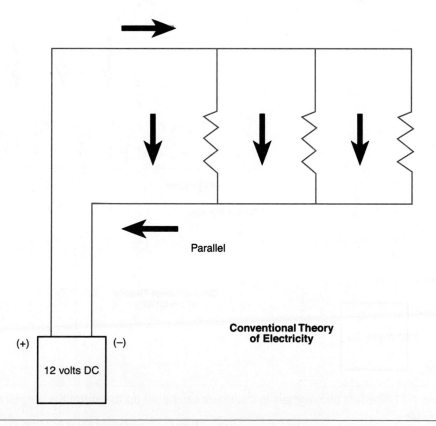

Parallel

Conventional Theory
of Electricity

(+) (−)

12 volts DC

Figure 7-26 A parallel circuit has multiple paths to ground.

a vehicle are the exterior lights. In the case of the headlights, from the last control device in the system, the conductors branch into two legs or parallel circuits, one to each headlight. If one light fails, the other will still work. The brake lights, parking lights, and turn signals work in the same manner. Most of a vehicle's safety equipment, like the lights, are parallel wired to some extent.

Series-Parallel Circuits

The most common type of electrical circuit on a vehicle is the series-parallel. Part of the **series-parallel circuit** is in series and other parts are in parallel (Figure 7-27). In effect, the series portion controls the parallel portion. Again, a typical example is the headlight system. As discussed in the last section, separate circuits feed the headlights from the last control device. That control device is usually an electric-mechanical switch known as a **relay** (Figure 7-28). A relay uses low current to switch a high current circuit on or off. Many newer relays control multiple circuits. Before the relay is the headlight switch. The current flows from the power supply, through the headlight switch, and finally to the relay. The switch controls the relay, which in turn routes power to the lights. When the switch is moved to the ON position, the relay closes and current moves from the power source through the relay to the headlights.

Circuit Protection

Circuits should be protected by some device that will fail if the circuit is overloaded. This protects other circuits and possibly the whole vehicle. There are three basic protection devices: **fuses, circuit breakers,** and **fusible links.** If a circuit protector fails, it is for a reason and the circuit must be checked to find the problem. Simply replacing the protector may damage the circuit or at the very least burn through it again.

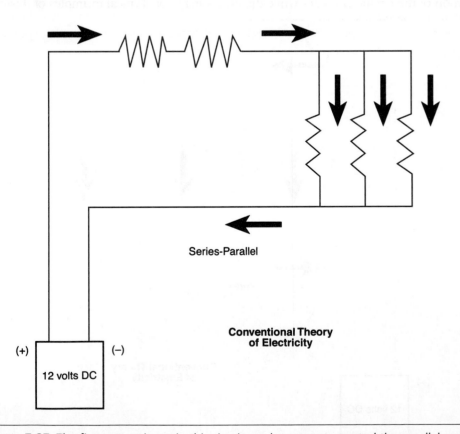

Series-Parallel

Conventional Theory
of Electricity

(+)   (−)

12 volts DC

Figure 7-27 The first two resistors in this circuit can be set up to control the parallel portion of the circuit.

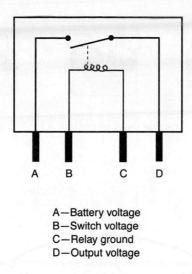

A—Battery voltage
B—Switch voltage
C—Relay ground
D—Output voltage

Figure 7-28 Current through the coil (B to C) will switch on the high-current circuit (A to D).

A fuse has a small internal wire through which all of the circuit's current flows (Figure 7-29). When a circuit current exceeds the capability of the fuse, the wire burns through, thereby stopping the current. The fuse can be removed from its holder and visually checked for failure. Fuses are rated in different amperes ranging from .5 to 60. The larger ones, 30 to 60 amperes, are used in circuits such as charging units, fuel pumps, and cooling fans. The fuse may be a cartridge-type or blade-type. The cartridge has been replaced with the blade for convenience and reliability. A fuse is not serviceable and must be replaced when it fails.

Circuit breakers open because of high current and are rated like fuses (Figure 7-30). Some circuit breakers will reset after they cool off. Others have to be replaced or manually reset. Breakers are better in some circuits because their design allows them to be slow acting compared to a fuse. This prevents an electrical surge from tripping the breaker. The problem with a circuit breaker is that most have to be checked with an ohmmeter or multimeter. There are usually no visible signs of failure.

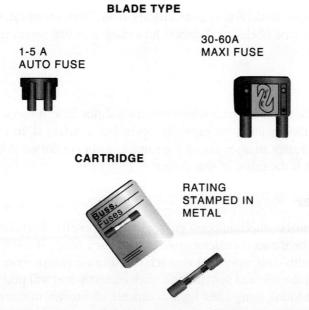

BLADE TYPE

1-5 A
AUTO FUSE

30-60A
MAXI FUSE

CARTRIDGE

RATING
STAMPED IN
METAL

Figure 7-29 Fuses are made to be effective, but they are also inexpensive and easy to replace.

Figure 7-30 A circuit breaker can be used to protect a high-current circuit or used as the flasher for the turn signal.

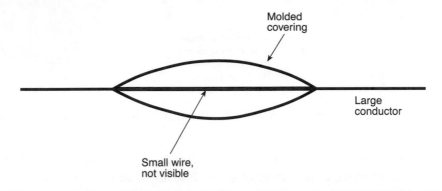

Figure 7-31 The fusible link is being phased out but can be found on trucks and older vehicles.

A fusible link is used on some heavy-current circuits. The link is a wire within a molded form (Figure 7-31). The wire is sized below the wire of the circuit and usually rated for low amperes. Again, excessive current will burn through the link. A major problem with this type of protection device is the diagnosis and replacement of it. Many times the link fails without any visible signs. They are usually checked with a test light, voltmeter, or multimeter and may be placed in locations difficult to access. Fusible links must be replaced when they fail. Fusible links have been replaced on newer vehicles with high-rated blade fuses.

Circuit Defects

There are only three possible defects in an electrical circuit. They are **open, short-to-power,** and **short-to-ground.** Any type of electrical device, including ones that use magnetism, can have one or more of the defects.

Open

An open circuit is an incomplete circuit where current will not flow. When a light switch is moved to the OFF position, the circuit is intentionally open, but a defect is an unintentional open. A burned-through fuse creates an open circuit the same as a cut conductor. A blown lamp is not the electrical defect, but it is the cause of that defect.

Short-to-Power

A short is an unintentional connection between two or more positive (hot) conductors (Figure 7-32). This could happen where two conductors rub against each other, thereby weakening the conductors' insulation. In this case, operating a switch could cause two or more circuits to switch on. In most cases, none of the affected systems will work correctly and will probably trip a protection device. A short can be found using a test light, voltmeter, ohmmeter, multimeter, or a short finder. A **short finder** is the common name for a tool that can be moved over a conductor for testing.

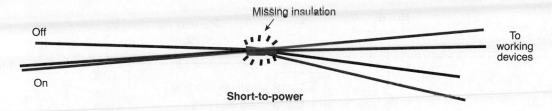

Short-to-power

Figure 7-32 In this short, current is being directed into a circuit that is actually turned off. The result will be a blown fuse or poorly performing electrical devices.

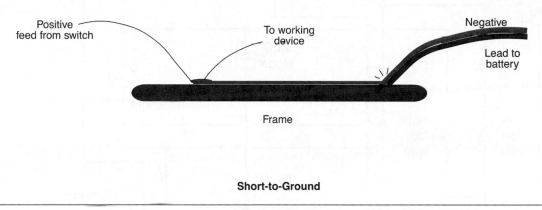

Short-to-Ground

Figure 7-33 If the protection device does not blow in this grounded circuit, the conductor's insulator will probably melt and possibly start a fire.

A moving needle indicates current and that a magnetic force is present, and a stationary needle indicates no current or no magnetism. The same tool can be used to locate a grounded circuit.

Short-to-Ground

A short-to-ground is an unintentional connection between a feed conductor (positive) and the ground of the vehicle (Figure 7-33). Since the ground is the negative side of the power source, and there will probably be no resistance at this time in the circuit, extremely high current will flow. The protection should trip to protect the circuit. If it does not, heat can damage the circuit very quickly or a fire could erupt, possibly causing damage and injury.

Electrical Service Information

Shop Manual
page 53

In order to diagnose an electrical circuit or system, information must be obtained on how the circuit is placed in the vehicle. It may also be necessary to trace a circuit without tearing the vehicle apart to chase wires. There are two items to help the technician in this area. They are the vehicle's wiring diagram and electrical component locator.

Wiring Diagrams

The **wiring diagram** is a paper or computerized document that shows every circuit or a single system circuit on a vehicle (Figure 7-34). It also shows controls, switches, loads, and other devices in each circuit. A single circuit may extend over several pages or screens, so tracing the circuit can be a time-consuming task. Wiring symbols are usually shown on the first page or screen of the wiring diagram section of the manual. Wire colors and sizes, along with circuit designators, are also shown on the diagram.

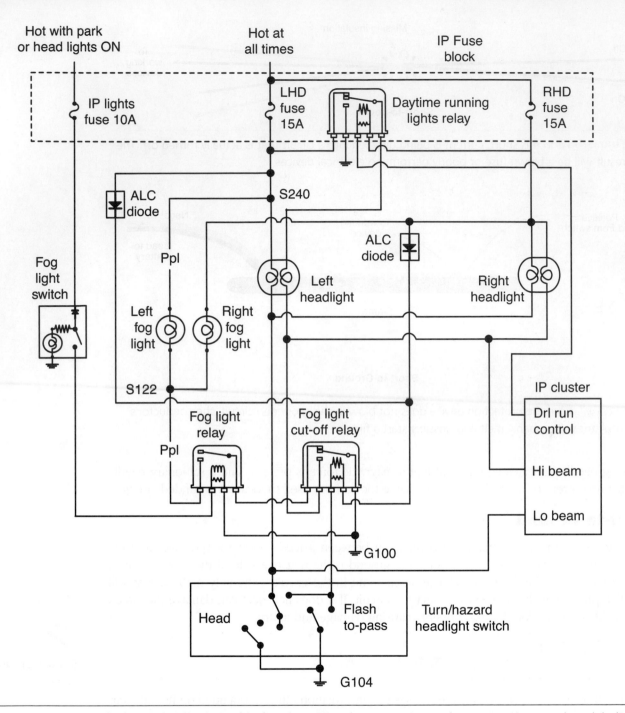

Figure 7-34 Circuit for a DaimlerChrysler antitheft system. Note the connection to the powertrain control module (PCM). A system fault may be related to a fault in one of the circuits in or connected to the PCM.

Newer vehicle wiring diagrams usually have separate drawings known as system diagrams (Figure 7-35). They show the wiring for a single system on one page and usually include all the information needed to trace the power flow. However, if a system diagram indicates the system is controlled by a relay or computer it may be necessary to find the diagram for that relay or computer to determine what electrical signal(s) are needed to cause the system in question to turn on. For instance, two possible sensor signals needed to switch on the cooling fan relay are the engine temperature and/or the air conditioner compressor control circuit. If one of those signals is in-

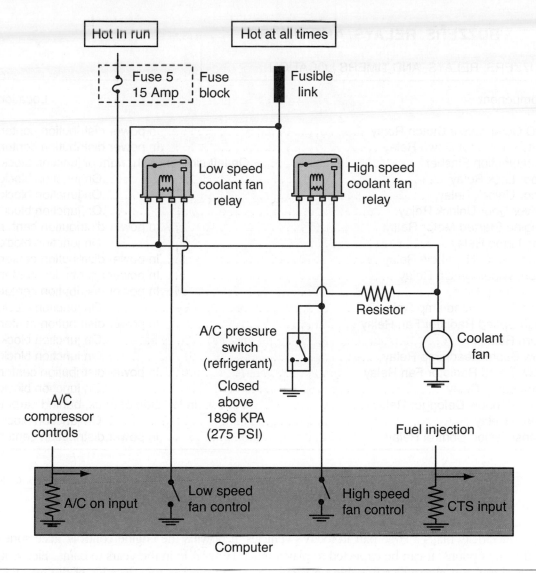

Figure 7-35 This system diagram shows the color of the conductor's insulation material. The first part is the primary color and the second is the tracer color (i.e., red/br is a red wire with a black tracer).

correct or missing, the cooling fan may not be energized, leading to engine overheating. Those signal wires may not be shown on the cooling fan system circuit. Even using a wiring diagram to locate a specific component or connection can be difficult.

Component Locators

A **component locator** shows the location of each electrical component on the vehicle. It can be a list of the items with their location spelled out, as shown in Figure 7-36. Some manufacturers use pictures or drawings of sections of the vehicle with the component highlighted in some manner.

A drawback to locators and wiring diagrams is their publication date. The manuals should be in the dealerships when their corresponding vehicle model is available for public sale. This means that last-minute changes may not be included and either the diagram or locator may be in error. This is not a real issue until the manufacturers update their product during the model year instead of waiting for the next model. Even then, most changes are small and may only be a change in wire color. Experience is the best method of learning how to read wiring diagrams and component locators.

BUZZERS, RELAYS, AND TIMERS LOCATION

Component	Location
A/C Compressor Clutch Relay	In power distribution center.
Automatic Shut Down Relay	In power distribution center.
Combination Flasher	On left side of dash, right of junction block.
Door Lock Relay	On junction block.
Door Unlock Relay	On junction block.
Driver Door Unlock Relay	On junction block.
Engine Started Motor Relay	In power distribution center.
Fog Lamp Relay	On junction block.
Front Wiper High/Low Relay	In power distribution center.
Front Wiper On/Off Relay	In power distribution center.
Fuel Pump Relay	In power distribution center.
High Beam Headlamp Relay	On junction block.
High Speed Radiator Fan Relay	In power distribution center.
Horn Relay	On junction block.
Low Beam Headlamp Relay	On junction block.
Low Speed Radiator Fan Relay	In power distribution center.
Park Lamp Relay	On junction block.
Rear Window Defogger Relay	In left side of truck, behind carpet.
Spare Relay	On junction block.
Transmission Control Relay	In power distribution center.

Figure 7-36 Component locators may be separated into component groups. This figure shows the location of buzzers, relays, and timers.

Electricity plays a large part in today's vehicles. It powers the engine controls, accessories, and many options. It can be expected to play a much larger role in the years to come. Electricity made the use of electronics possible on cars built during the last two decades of the twentieth century. The twenty-first century will see electronics expand greatly in the automotive industry.

Hydraulic Theory

Shop Manual
pages 157–162

Hydraulics is the use of a liquid to transfer force. Liquids used in hydraulic systems are selected based on the work to be done and the chemical properties of the liquid. Brake fluid will not work well in steering systems.

Blaise Pascal formulated his law on the use of **hydraulics** in 1647. Known as Pascal's Law, it states that pressure in a closed hydraulic circuit will be the same everywhere in the circuit. The law is based on the fact that liquid in a closed or sealed circuit cannot be compressed.

A hydraulic circuit is similar to an electrical circuit. There are hoses and lines (conductors), pumps (power source), load, and controls (Figure 7-37). The pump pressurizes the fluid and the control valve directs it into the hoses and lines. At the output end of the line or hose is a working device. It may be a brake component, an electronic fuel injector, or a piston that helps the operator apply force. Remember the inflated balloon example from earlier in this chapter? The atmospheric pressure against the outside of the balloon forced the inside gas outward through the balloon's opening. The exhausting gas created a force to move the balloon.

As for electricity, there are three measures normally used in hydraulic circuits: *force, pressure,* and *area.* The force is applied to the contained fluid or delivered by the fluid. Area is the surface on which the fluid acts. Pascal's Law uses a mathematical formula to calculate hydraulic values. It is written as $F = PA$, where F is force (ft.-lb), P is pressure per square inch (psi), and A

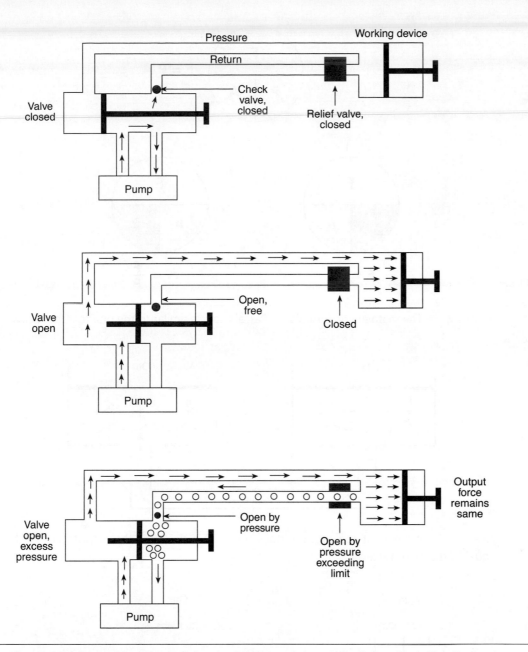

Figure 7-37 Note the actions of the check valve and the relief valve in this hydraulic circuit. Valves of this type are common in hydraulic circuits.

is area in square inches (sq. in.) (Figure 7-38). A hydraulic system can deliver force and increase or reduce it. An increase is called a *mechanical advantage*.

In Figure 7-39, the input and output pistons have the same area. Trapped between them is a liquid. Applying force to the input piston will create an equal force being delivered by the output piston because of the equal areas. However, any reaction against the output piston will deliver the same force backward through the circuit to the input piston and anything that is pushing on it.

In Figure 7-40, the output piston has twice the area as the input. In theory, this means the amount of force applied at the input is doubled at the output. We can prove this by using the formula for Pascal's Law. Since the pressurized fluid between the two pistons is the *transfer median*, the amount of pressure generated must first be determined. The figure shows an area of 1 square inch for the input piston and a force of 10 pounds. Using the known values (1 and 10) and the adjusted formula P = F/A (F-PA), we can calculate that the pressure is 10 psi (P = 10/1). Moving

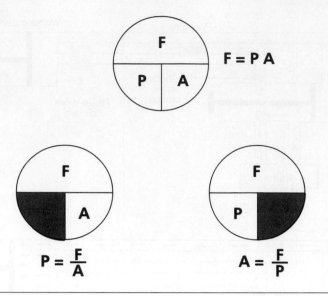

F = P A

$$P = \frac{F}{A}$$

$$A = \frac{F}{P}$$

Figure 7-38 The mathematic formula for Pascal's Law is similar to the one for Ohm's Law.

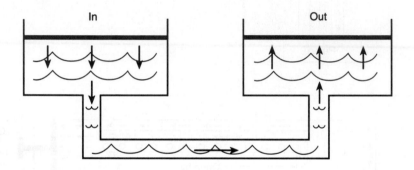

In

Out

Figure 7-39 If the input and output pistons are equal in size, then the output force will be equal to the input force.

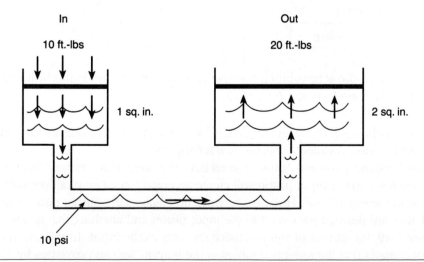

In

10 ft.-lbs

Out

20 ft.-lbs

1 sq. in.

2 sq. in.

10 psi

Figure 7-40 Mechanical advantage can be gained or reduced by having input and output pistons of different sizes.

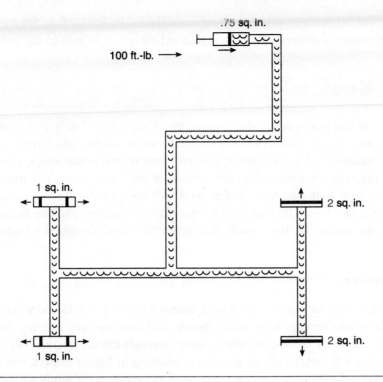

.75 sq. in.

100 ft.-lb. →

1 sq. in.

2 sq. in.

1 sq. in.

2 sq. in.

Figure 7-41 This system has 8 square inches of output area versus $^3/_4$ square inches of input area.

to the output piston and using F = PA, we can calculate the force that can be delivered by the output piston at 20 ft.-lbs (F = 10 × 2). Remember that pressure is per square inch. The input is 1 square inch and the output is 2 square inches. Since there are 2 square inches on the output and 10 psi on each inch, the delivered force is 20 ft.-lbs. This would give a mechanical advantage of 2:1 (20/10). Calculate the total output force in the following simple braking system.

Refer to Figure 7-41 for the known values for this problem. The figure shows a very simple hydraulic braking system. Notice that the input piston is .75 square inches. The rear brakes have 1-square-inch pistons at each wheel while the front has 2-square-inch pistons at each wheel. First, the system pressure must be calculated. The input piston is being applied with 100 ft.-lbs. The known values, force and area, are used in P = F/A or P = 100/.75, resulting in a pressure of 133 psi rounded off. Since the pressure is the same throughout the circuit, the pressure against each square inch of the output pistons is 133 psi. Compute the rear pistons' force using F = PA (F = 1 × 133) or 133 ft.-lbs at each wheel or a total of 266 ft.-lbs for the rear. The front is calculated in the same manner. F = 133 × 2 for 266 ft.-lbs per wheel for a total of 532 ft.-lbs in the front. The operator is applying 100 ft.-lbs at the brake pedal and the vehicle is being stopped with a total of 798 ft.-lbs (front plus rear) of braking effort. This is a 7.98:1 mechanical advantage.

Hydraulic fluids are selected on the basis of the type of hydraulic system in use and the type of work the system must perform. In theory, you could use tap water as the fluid. It is noncompressible like all liquids and it can transmit force as long as the system is sealed. But when a liquid is compressed its temperature goes up in direct proportion to the pressure applied. After about three or four braking efforts using water as brake fluid, the water would boil, thus creating gases (vapors) in the system and resulting in very poor braking from that point on because gases can be compressed. Brake fluid is formulated for a high boiling point and is corrosion resistant. Other hydraulic fluids are designed and selected with the same overall specifications in mind. Hydraulic fluids used in heavy equipment work under pressures as high as 2,200 psi while maintaining a relatively low outside temperature. The fluid would definitely be hot enough to cause injury if it escaped pass a seal or gasket, but most of the cylinders and lines/hoses could be touched safety with regular gloves. In modern vehicles, there is a different hydraulic fluid for the brakes, power steering, and automatic transmission. Each is selected based on its role in the vehicle.

Shop Manual
pages 157–165

Vehicle Design

Without going into too much detail, the engineering of an automobile covers not only the theories discussed earlier, but also how different materials can be shaped and formed and their actions under different conditions. The local library has information on how the shape, composition, treatment, and the reactions of metal affect the actions of the component. For instance, aluminum warps at a lower heat than cast iron. In this book, we will not discuss vehicle design and construction at length, but be aware that the material choice, shaping, and the manufacturing treatment of each component on the vehicle was researched and designed to satisfy the laws of physics.

Aerodynamics

Until quite recently, the study and use of **aerodynamics** was not considered when building everyday vehicles. Even smaller cars were bulky, heavy, and had the aerodynamic design of a brick! Aerodynamics is the study of how an object moves through the air (Figure 7-42). A vehicle with poor aerodynamics will create high air resistance, resulting in higher noise levels and fuel usage. Aircraft and racing vehicles were the first major users of aerodynamic study and application.

Fuel mileage is required to meet standards in the federal Corporate Average Fuel Economy (CAFE) regulation. A manufacturer's vehicle line must be tested and the average of all models offered must meet CAFE.

Federal regulations and the increased price of fuel required vehicle manufacturers to increase **fuel mileage.** Aerodynamically designed vehicles incorporated a rounded outside body shape, but all of the different components, including the passenger, had to fit inside this rounded body. Weight also had to be reduced drastically. Newer headlights were redesigned and configured with aerodynamics in mind. The result is the slick body contours of today's vehicles.

One major drawback to smaller-size vehicles with increased accessories is the lack of working area for technicians. Older vehicles provided room under the hood and dash to reach components. On some new vehicles, it is almost impossible to locate or reach some components. Another drawback associated with the lower body profile and front-wheel-drive (FWD) systems

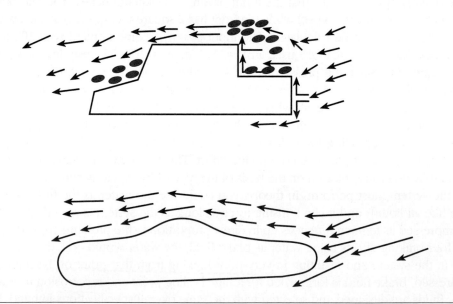

Figure 7-42 On the top vehicle, note how the air is blocked and makes a turbulence area as different airflows meet. The slicker, rounded vehicle at the bottom allows the air to flow freely over and around the vehicle.

is the lack of airflow around and over the engine. Rear-wheel drive generally allows the air from the radiator and cooling fan to encircle the engine, flow past the transmission, and dissipate the collected heat as it moves under the vehicle. With FWD the air leaves the radiator and fan and slams into the side of the engine and transaxle. The firewall side of the engine gets very little airflow. To reduce this problem more efficient water pumps and radiators coupled with precise engine controls are being used. The major advantages to smaller, more aerodynamic vehicles are greatly increased fuel mileage and driver protection.

Passenger Protection

Today, almost every car on the road has seat belts (Figure 7-43) and most have air bags (Figure 7-44). Though these devices are needed to keep the passenger in his or her seat during an accident, they are minor parts of overall vehicle crash protection.

While older vehicles had stout bodies and frames, they usually delivered most of the shock of an accident to their passenger compartments. The automotive and insurance industries, along with various government agencies, have destroyed many vehicles to study the effect of accidents on passengers and vehicles. Newton's Laws of Motion and the dissipation of kinetic energy are key points in designing a safe vehicle.

Automotive studies have resulted in vehicles designed and built with material that will crumple on impact and absorb energy. Such areas are called **crumple zones.** More energy absorbed by the vehicle means less shock and damage delivered to the passenger compartment. The first point or crumple zone is energy-absorbing bumpers that incorporate devices between the bumper and frame. Under federal law, newer cars must withstand a frontal impact of 30 miles per hour before the passenger compartment begins to absorb kinetic energy from the impact. Entire vehicles have been redesigned to protect passengers and, to some extent, the vehicles themselves.

Crumple zones extend from bumper to bumper. The sections of the body and frame interact to absorb as much energy as possIble before it reaches the passenger compartment.

Figure 7-43 Seat belts are required to be worn in most communities. Children must be in an approved seat secured in the vehicle with seat belts.

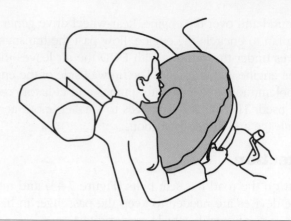

Figure 7-44 Since model year 1997, air bags have been redesigned to accommodate persons of different sizes. In most pickup trucks, the passenger-side bag can be switched off so a child seat can be safely utilized.

Materials

Like other things in nature, materials respond to different situations in different ways. The materials used in vehicles are selected to meet specific properties. In the engine, materials must be capable of enduring high pressure, twisting, and heavy loads, and must still function properly. The most prominent materials in engines are cast iron for the block, aluminum for the cylinder heads, and treated steel for the internal components. Using different materials together presents another problem. The engine produces a great deal of heat and different materials react differently to heat. Aluminum expands much quicker than cast iron. Since they are mated to each other in the engine, engineers had to design some method to compensate for the differences. A gasket was made with materials that would react with the cast iron on one side and aluminum on the other. Gaskets are covered in Chapter 6, "Automotive Bearings and Sealants."

Other systems of the vehicle require different materials to perform their purpose. The material may be high strength and flexible such as suspension components, thin sheet metal for the body, or plastics for trim and some no-load components. All of the materials are selected, shaped, treated, and used based on their physical properties and the operational theories involved.

Horsepower versus Torque

Finally we get to the section that most drivers want to know: horsepower versus **torque.** One is speed; the other is pulling. Both can be obtained by designing and tuning the engine for a particular purpose. In construction work, a high-torque pickup truck is desirable, but fuel mileage will suffer. In a drive-to-work/home truck, fuel mileage is important so great torque is not needed. Horsepower is the key in such instances.

Torque, as applied to a vehicle, is the total rotational force applied to the crankshaft by the pistons during the power stroke and is usually measured as foot-pounds at the flywheel. Horsepower is work done in a straight line. But when the work is not in a straight line, horsepower is calculated as torque times rpm divided by 5,252 (torque*rpm/5252). Typically maximum torque is produced at a much lower engine speed than horsepower (Figure 7-45).

At this point, we have only discussed how to calculate the torque and/or horsepower of the engine. Though this provides a good reference point to determine which vehicle we are going to purchase, it does not necessarily tell us the amount of power delivered to the drive wheels. Before determining how fast or how strong the vehicle is, the driveline components, mainly the transmission, must be calculated. The driveline changes the engine's torque/horsepower to the final

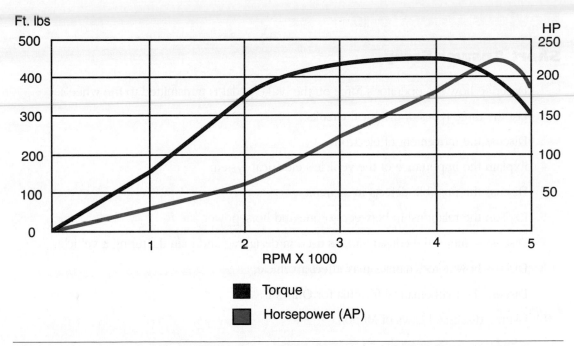

Figure 7-45 A torque/horsepower curve. Note the high torque curve with the sudden drop after 4,000 rpm. The horsepower continues to increase slightly after 4,000 rpm.

actual force moving the vehicle. The selection of a particular gear may allow the vehicle to pull a heavy load at low road speed and high engine speed or cruise at high road speed and low engine speed. Generally speaking, the more torque the less speed, and vice-versus. Drivelines are discussed in Chapter 8.

Summary

❏ Thermodynamics is the study of using rapidly expanding gases to produce energy.

❏ The theory of thermodynamics is used to achieve mechanical power from a liquid fuel.

❏ The motion of an object is called work. Work is the use of energy.

❏ Pressure and vacuum are used to perform work on a vehicle.

❏ Vacuum is pressure below atmospheric.

❏ Newton's Laws of Motion apply to almost every operation on a typical vehicle.

❏ Electricity is made when negatively charged electrons move from atom to atom in large numbers.

❏ Volts, ohms, and amperes are basic measurements of electricity.

❏ Hydraulic systems use liquids to transfer force.

❏ The measurements of hydraulics are force, area, and pressure.

❏ Vehicle design and manufacturing is based on the capabilities of materials used and their functions within the vehicle.

❏ Aerodynamics is the study of objects moving through air.

❏ Horsepower provides capability of high speed.

❏ Torque provides pulling or hauling power.

Terms to Know
Aerodynamics
Ampere
Atom
British Thermal Unit (BTU)
Circuit
Circuit breaker
Coil
Combustion chamber
Component locator
Conductor
Conventional Theory of Electricity
Corrosion
Crumple zones
Current
Electricity
Electromotive force (EMF)
Electron Theory of Electricity
Explosion
Foot-Pounds (ft.-lbs)

Review Questions

Short Answer Essay

1. Describe how the operator's force on the brake pedal is transmitted to the wheels.

2. List the three types of electrical circuits.

3. Discuss the movement of electrons.

4. Explain the importance of the vehicle's electrical system.

5. Discuss the possible defects in a circuit.

6. Explain the relationship between torque and horsepower.

7. Discuss some of the considerations used in designing and manufacturing a vehicle.

8. Discuss how aerodynamics may affect a vehicle.

9. Discuss the mathematical formula for Ohm's Law.

10. List the discussed Laws of Motion.

Fill-in-the-Blanks

1. Basic electrical measurements are in _____, _____, and _____.

2. Basic hydraulic measurements are _____, _____, and _____.

3. An electrical system with 12 volts and 5 amperes will have a resistance of _____ _____.

4. Voltage is a measurement of _____.

5. Electrons flowing through a conductor will create a(n) _____ _____.

6. A hydraulic circuit consists of _____, _____, _____, and _____.

7. A(n) _____ is the most common electrical measuring instrument in automotive shops.

8. The tendency of an object to stay at rest is called _____ _____.

9. The study of using expanding gases to perform work is called _____.

10. Trucks will generally have less _____ and more _____ than a car with a comparable engine.

ASE-Style Review Questions

1. Electrical systems are being discussed.
   Technician A says the theory is based on one volt pushing 1 ampere of current through one ohm of resistance.
   Technician B says that based on the conventional theory, electrons flow from positive to negative.
   Who is correct?
   A. A only
   B. B only
   C. Both A and B
   D. Neither A nor B

2. Hydraulic systems are being discussed.
   Technician A says the force applied against a piston creates the pressure of the transferring liquid.
   Technician B says the pressure of the transferring liquid creates the force on the piston.
   Who is correct?
   A. A only
   B. B only
   C. Both A and B
   D. Neither A nor B

3. Electrical systems are being discussed.

   Technician A says a circuit is comprised of a load, current, a conductor, and controls.

   Technician B says a system may have more than one circuit.

   Who is correct?

   A. A only   **C.** Both A and B

   B. B only   **D.** Neither A nor B

4. Leaks are being discussed.

   Technician A says a leak could cause problems within the system if hydraulic theory cannot be maintained.

   Technician B says leaks may cause problems because the proper application of pressure is not maintained.

   Who is correct?

   A. A only   **C.** Both A and B

   B. B only   **D.** Neither A nor B

5. Thermodynamics are being discussed.

   Technician A says a vehicle uses thermodynamics to produce horsepower.

   Technician B says expanding gases is the study of thermodynamics.

   Who is correct?

   A. A only   **C.** Both A and B

   B. B only   **D.** Neither A nor B

6. All of the following conditions will affect the hydraulic pressure in a system EXCEPT:

   A. External leaks   **C.** Weak pump operation

   B. Internal leaks   **D.** Jammed actuator

7. Application of electrical theory will be affected by each of the following EXCEPT:

   A. incorrect conductors.

   B. damaged controls.

   C. overcharged battery.

   D. undercharged battery.

8. Pascal's Law is being discussed.

   Technician A says a liquid is used to transfer force.

   Technician B says hydraulics is the study of using liquid to transfer force.

   Who is correct?

   A. A only   **C.** Both A and B

   B. B only   **D.** Neither A nor B

9. *Technician A* says current will flow through insulators.

   Technician B says current is pushed by voltage.

   Who is correct?

   A. A only   **C.** Both A and B

   B. B only   **D.** Neither A nor B

10. Electrical circuits are being discussed.

    Technician A says a series circuit provides multiple paths for the current.

    Technician B says series circuits and parallel circuits can be combined into one circuit.

    Who is correct?

    A. A only   **C.** Both A and B

    B. B only   **D.** Neither A nor B

Power System Operation and Subsystems

Upon completion and review of this chapter, you should be able to:

❑ Describe the design of an internal combustion engine and its general operations.

❑ Describe the operation of a four-cycle internal combustion engine.

❑ Discuss the primary differences between gasoline- and diesel-fueled engines.

❑ Explain the function and operation of the starting and charging system.

❑ Discuss the lubrication system and engine oil.

❑ Discuss the engine coolant system and coolant.

❑ Discuss the intake and exhaust system.

❑ Explain the operation of the ignition system.

❑ Discuss the design of a hybrid-powered vehicle and its general operation.

❑ List and discuss engine and emission control systems and devices.

Mechanical linkages are the piston, connecting rod, and crankshaft.

Block configuration is the shape of the block.

Displacement is the volume of air the engine can exchange in one complete engine cycle.

Valvetrain refers to the placement of the engine valves.

A **two-stroke** engine has a power stroke with each crankshaft revolution, whereas a **four-stroke** engine has one every two revolutions.

Rotary engines have no reciprocating pistons, but use a rotor to move the air/fuel mixture and produce power.

The machined portions of a basic **block** are the cylinders, mounting deck for the cylinder head, lifter bores, and the saddles or bores for the bearings.

Introduction

The engine or power system has become the most complex system in the vehicle. Its original design and purpose is still intact. However, the internal combustion engine operations have been refined internally and electronically. The engine itself has more computerized sensors and actuators than the entire vehicle had in the mid-1980s. Modern diesel-fueled engines have just as many electronic devices as gasoline-fueled engines. Working on today's light vehicle engine is no longer a Saturday morning job in the backyard.

Engine Design

The basic engine design from the 1890s remains intact. It has been engineered, made of better material, and made to more exacting manufacturing standards, but it is still basically an air pump. When the starter is engaged, air is sucked into the cylinders. Power is possible when fuel is mixed into the airflow and lit with an electrical spark. The expanding gases (thermodynamic theory) supply the necessary power that pushes on the **mechanical linkages** in the engine (Figure 8-1).

An engine is classified using three criteria: **block configuration, displacement,** and **valvetrain.** Its operation is classified as **two-stroke, four-stroke,** or **rotary.** When an engine is referred to as gasoline or diesel, it is a reference to the fuel used, not the type of engine.

Engine Block Design

The engine **block** starts as a large block of treated cast iron or aluminum. It is bored, drilled, machined, and polished to form the basic support for all the engine's operating systems, internally and externally.

The block may be shaped in various configurations. An **in-line engine** has all of the cylinders aligned in a series of vertical bores. The most common in-line blocks are a four- and six-cylinder design (Figure 8-2). More cylinders involve a longer and heavier block. Four-cylinder engines are available in almost all models of cars and light trucks. Six-cylinder, in-line, gasoline engines are used less frequently in modern cars because of the block's weight and size. They are

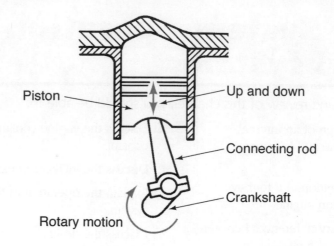

Figure 8-1 The ignition spark lights the air/fuel and the piston is forced downward, thereby turning the crankshaft.

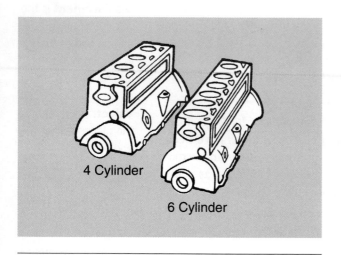

Figure 8-2 In-line engines have all cylinders in a straight line.

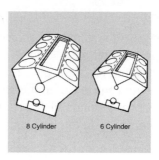

Figure 8-3 V-blocks have the cylinders in two equal rows.

most commonly used in light trucks. Many over-the-road trucks still use the in-line six and in some cases, in-line eight or ten cylinders.

Another block is known as the **V-block.** The number of cylinders are divided into two rows set at an angle to each other (Figure 8-3). The angle between the rows of cylinders is 90 degrees or 60 degrees. The V-block allows for a lower hood line and for more cylinders to be added in a shorter, linear space. Then, the engine could be more powerful and contained in a smaller block. V-block engines also tend to create less vibration than a comparably sized in-line.

Volkswagen made the **horizontally opposed engine** famous during its use in the original VW Beetle. The rows of cylinders are laid out 180 degrees apart (Figure 8-4). This provided a very low profile and air cooling. The engine was mounted in the rear of the vehicle. Subaru still uses a front-mounted, horizontally opposed engine in some of its models. Another thing about VWs and Subarus is the location of their spare tires. Both are mounted in the front of the car. The VW Beetle has storage space under the front hood along with the tire stored there. The spare tire in a Subaru is laid across the top of the engine.

Another engine design was used almost exclusively by DaimlerChrysler. It is an in-line six-cylinder that leans or slants to the right. It became known as the **slant 6.** This engine is still operating in many types of industrial equipment.

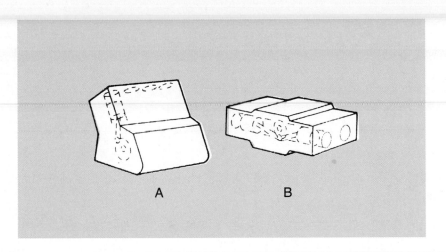

Figure 8-4 The slant 6 (A) and horizontally opposed (B) engines were only used in certain makes.

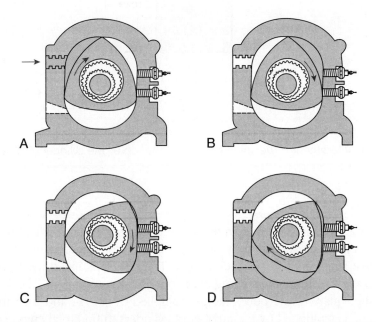

Figure 8-5 The rotary's oval chamber is shaped to create the same internal combustion conditions found in a piston-driven engine.

Mazda uses another engine type for its RX7 model that does not use pistons or a cylinder to generate and transfer the combustion power to the crankshaft. Instead, two triangular rotors within matching oblong chambers intake, compress, and exhaust the air/fuel mixture (Figure 8-5). The rotors are attached to the crankshaft through a **planetary gear set,** which increases the force delivered by the rotors (Figure 8-6). This design is known as a rotary engine.

Fuel economy has caused some types of engines to be discontinued in many car models. The most common type put into today's car is either a four-cylinder in-line or the V-6. During the 1998 and 1999 model years, some V-8 engines were reintroduced in high-performance and luxury model vehicles. The V-8 engine produces more horsepower, and new electronics and better engineering keep the fuel mileage and emissions within mandated limits. Some engines are built using aluminum for the blocks and cylinder heads. A few research engines have been made of ceramic in an attempt to reduce weight and durability. Ceramic material is expensive, but the research provides extensive data on how to use other materials.

A simple **planetary gear set** has a sun gear that fits inside a planetary carrier. The carrier, in turn, fits inside a ring gear.

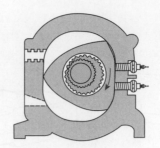

Figure 8-6 Planetary gear sets are also found in transmissions and transaxles. They are used to multiply torque.

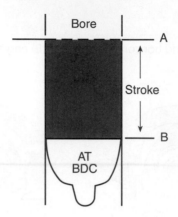

Figure 8-7 Displacement is the volume between the top of the cylinder (A) and the piston at BDC (B).

Displacement

Engine **displacement** may be measured in cubic inches displacement (cid), liters (L), or cubic centimeters (cc).

The amount of air an engine can move in one cycle is its displacement. **Displacement** is the total volume of the cylinders calculated by using the bore, stroke, and number of cylinders (Figure 8-7). The **bore** is the diameter of the cylinder. The **stroke** is measured from the top of the cylinder to the top of a piston that is resting at bottom dead center (BDC). The volume of a cylinder is equal to the square of the bore's radius multiplied by 3.1414 and then multiplied by the stroke. The volume is multiplied by the number of cylinders to find total engine displacement. A typical displacement of late 1990's engines is 2.0, 3.0, 3.8, 4.0, and 5.0 liters (L). There are others that vary in displacement, but they generally fall within the ranges mentioned here. As a point of reference, an engine with 318 cubic inch displacement (cid) is listed as a 5.2 L while a 360 is a 5.9 L.

A Comparison—Old versus New

During the 1950s and 1960s, high-performance, high-horsepower engines ranged from 350 cid (5.7 L) to 440 cid (7.2 L). Stock engines were capable of producing up to 400 horsepower and high vehicle speeds, but they were fuel-hungry and produced enormous amounts of harmful emissions. Late 1990s engines can produce 300 horsepower with high fuel economy, high vehicle speeds, and very low levels of harmful emissions, all with a smaller displacement.

Valvetrains

The **valvetrain** consists of valves, valve springs, a camshaft, valve retaining devices, and may have valve **lifters,** pushrods, and **rocker arms.**

There are two basic valvetrains in use today. The older version, **overhead valve (OHV),** has been used in all types of engine block designs. The OHV system uses a crankshaft-driven camshaft in the block to move the valve mechanism. The camshaft turns at half the crankshaft speed and must be timed to open and close valves in time with piston position (Figure 8-8). The camshaft pushes upward on a solid or hydraulic **lifter,** moving a push rod. The push rod applies upward force to one end of a **rocker arm.** The rocker arm is pivot-mounted and the opposite end applies downward force to the valve (Figure 8-9). With the valve open, air and fuel can enter the cylinder or exhaust can exit. As the camshaft rotates, force is removed from the valvetrain. A spring mounted around the valve stem forces the valve upward and closed. The valve spring also applies force throughout the mechanism to maintain constant contact between the various components (Figure 8-10). There is usually one intake and one exhaust valve per cylinder. Newer, more powerful

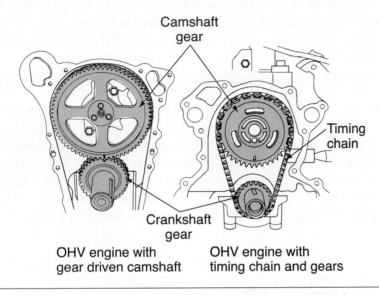

Camshaft gear

Timing chain

Crankshaft gear

OHV engine with gear driven camshaft

OHV engine with timing chain and gears

Figure 8-8 The camshaft and lifters are located in the block for an OHV engine.

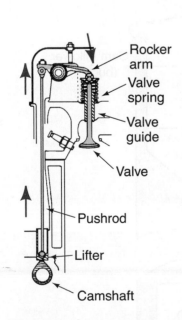

Rocker arm

Valve spring

Valve guide

Valve

Pushrod

Lifter

Camshaft

Figure 8-9 The rocker arm makes the connection between the pushrod and the valve stem.

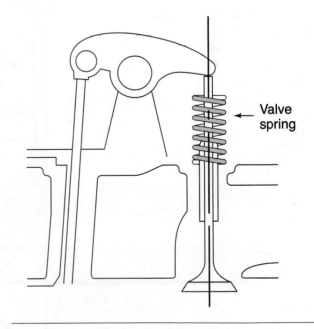

Valve spring

Figure 8-10 The valve spring applies continuous force on all components to prevent sloppy or noisy movement and returns the valve to the closed position.

and fuel-efficient engines may have two intake and two exhaust valves. This provides better flow characteristics for the air/fuel mixture and exhaust gases.

> **AUTHOR'S NOTE:** NASCAR allows only pushrod or OHV engines in its CUP series. Some vehicle manufacturers are returning to OHV engines in their performance models.

The system being used on many newer engines has the camshaft installed in the cylinder head with the valves. It is known as an **overhead cam (OHC)** engine. The camshaft is mounted directly over the valve and exerts downward force directly onto the valve stem (Figure 8-11). This system eliminates the lifter, pushrods, and rocker arms providing better durability, less maintenance, and less noise. Newer versions of the OHC system may have small, hydraulic lifter-type mechanisms that absorb the vibration and shock of the valve opening and closing. Other versions act directly on the rocker arms to operate the valves. There may be a roller bearing between the camshaft and rocker arm (Figure 8-12).

More powerful OHC engines use two camshafts per cylinder head. Each camshaft operates either the intake or the exhaust valves. The **dual overhead cam (DOHC)** reduces stress on the camshaft and allows for multiple valves per cylinder (Figure 8-13). Each camshaft is timed to another and the crankshaft.

The camshafts on an OHV engine are driven through meshed gears or a chain connecting the crankshaft and camshaft gears. When worn, both systems must be lubricated and replaced as a set. The gears or chain are not readily visible, and noise or a breakdown is the first sign of a trouble. A chain and gear may drive the camshafts on an OHC system, but many use a flexible belt. The belt does not require lubrication, is quieter, and is replaced based on mileage or age

An **overhead cam (OHC)** engine is used on many newer vehicles. The camshaft is installed in the cylinder head with the valves.

The term **dual overhead cam** does not represent a different valvetrain. It's a version of the OHC system.

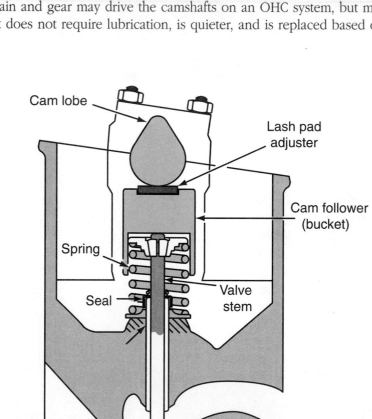

Figure 8-11 Overhead camshafts eliminated the lifter and pushrods.

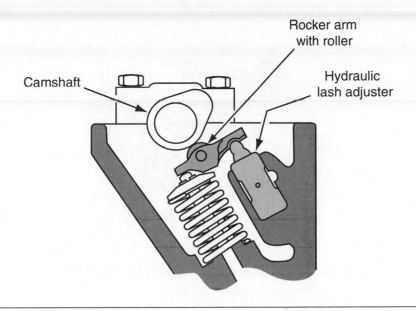

Figure 8-12 The roller bearings reduce camshaft wear and provide quicker operation.

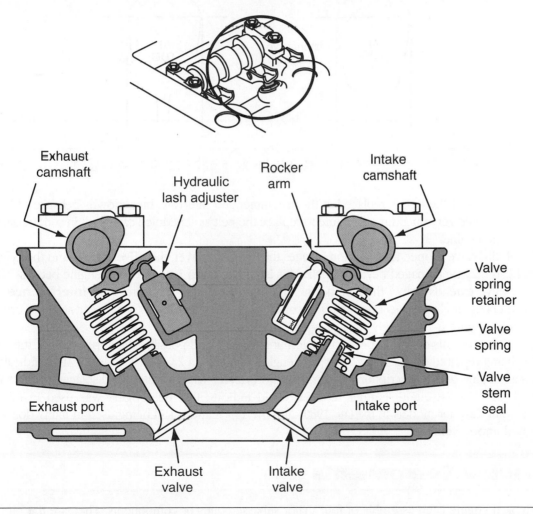

Figure 8-13 Dual overhead camshafts allow multiple valves per cylinder, providing better fuel economy and more performance.

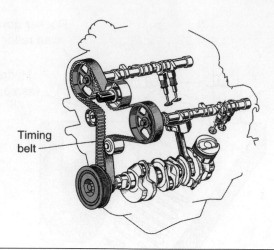

Timing belt

Figure 8-14 Timing belts are quicker and easier to maintain than chains.

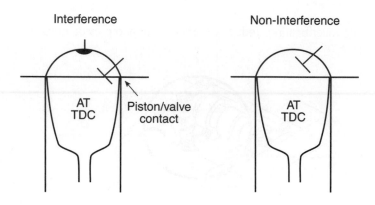

Interference   Non-Interference

AT TDC   Piston/valve contact   AT TDC

Figure 8-15 Non-interference engines keep the valve within the combustion chamber.

(Figure 8-14). If the belt is replaced at the recommended intervals, breakdowns because of belt failure become very uncommon. Failure to replace the belt as scheduled can result in serious damage to the engine.

If the valve is open as the piston reaches the top of its travel, the valve, piston, and the cylinder head may be damaged beyond repair. This is known as an **interference** engine because the open valve extends below the cylinder head and into the cylinder bore. **Non-interference** or freewheeling engine valves open into the combustion chamber but do not extend into the cylinder bore (Figure 8-15).

Based on this discussion, we can determine the size and possibly the application of the engine being repaired or researched. A typical engine would be a 2.0-L OHC in-line 4. This indicates that an engine of this description is an in-line four-cylinder with two liters of displacement and an overhead cam valvetrain. Further research could indicate that this engine was used in many Chevrolet Cavalier models in the late 1980s and the 1990s and is still being used in many domestic and import vehicles.

Engine Components

A typical engine is an assembly of four major subassemblies or components. They are the block assembly, cylinder head, intake manifold, and exhaust manifold (Figure 8-16). The block assembly houses the piston assemblies, crankshaft, and the camshaft and valve lifters on an OHV-type

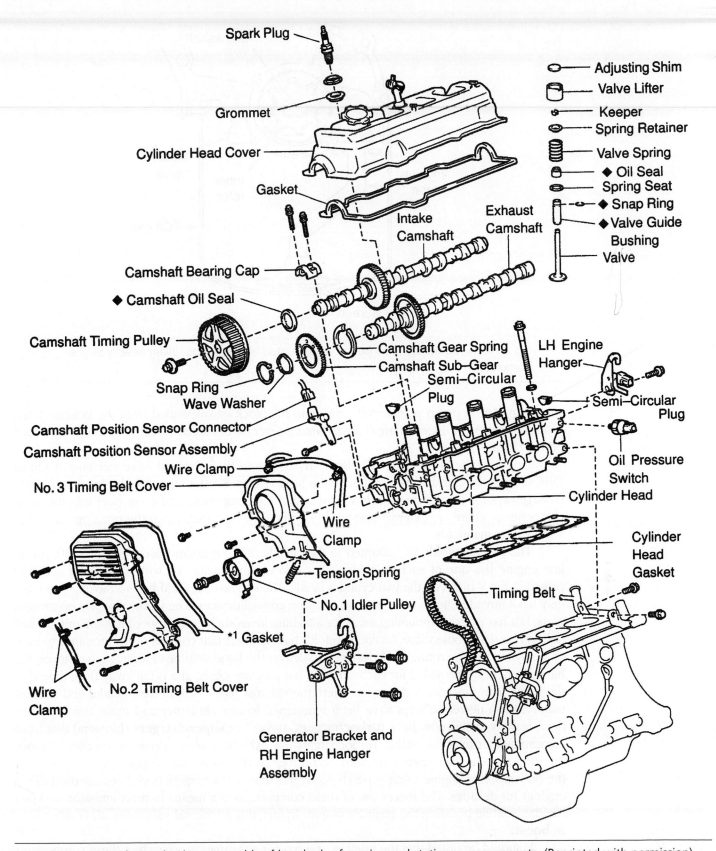

Figure 8-16 The engine is an assembly of hundreds of moving and stationary components. (Reprinted with permission)

Spark Plug

Grommet

Cylinder Head Cover

Gasket

Adjusting Shim

Valve Lifter

Keeper

Spring Retainer

Valve Spring

◆ Oil Seal

Spring Seat

◆ Snap Ring

◆ Valve Guide

Bushing

Valve

Intake Camshaft

Exhaust Camshaft

Camshaft Bearing Cap

◆ Camshaft Oil Seal

Camshaft Timing Pulley

Snap Ring

Wave Washer

Camshaft Position Sensor Connector

Camshaft Position Sensor Assembly

Wire Clamp

No. 3 Timing Belt Cover

Wire Clamp

Tension Spring

No.1 Idler Pulley

*1 Gasket

Wire Clamp

No.2 Timing Belt Cover

Generator Bracket and RH Engine Hanger Assembly

Camshaft Gear Spring

Camshaft Sub-Gear

Semi-Circular Plug

LH Engine Hanger

Semi-Circular Plug

Oil Pressure Switch

Cylinder Head

Cylinder Head Gasket

Timing Belt

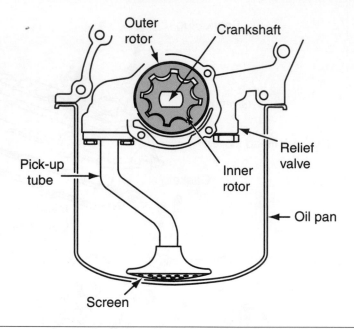

Figure 8-17 Some newer oil pumps are mounted internally and driven directly by the crankshaft.

valvetrain. An oil pump is mounted either into the block or suspended from the bottom of the block (Figure 8-17). External areas for mounting brackets of external components are machined on the block.

On top of the block is the cylinder head that houses the valve and valve mechanism. On an OHC engine, the camshaft is mounted in the head (Figure 8-18). In the cylinder head and block, passageways for oil and coolant flow exist within the component and cross over into the other components. There are external ports in the cylinder head(s) that match the ports on the intake and exhaust manifolds.

The two manifolds are mounted to the outside of the cylinder heads (Figure 8-19). An in-line engine has one of each. A V-block has two exhaust manifolds with one intake manifold placed in the V between the two cylinder heads. The exhaust manifold has no moving parts, but may have threaded holes for mounting emission control devices. The intake also has no moving parts, but has external mounting areas for attaching items such as an upper intake manifold, fuel delivery, and a throttle valve. Intake manifolds have runners that connect the air entrance to each of the cylinders. The runners are designed to deliver the same volume of air to each cylinder. Exhaust runners are provided for each cylinder, but they are joined at a point away from the cylinder head. This junction is the point where the exhaust pipe is connected. Intake and exhaust manifolds on newer vehicles have been redesigned to provide better and more consistent airflow. Many vehicles now have **turbochargers** (turbos) or **superchargers** (blowers) that force compressed air into the intake. Turbochargers use exhaust gases to drive an impeller to move the extra air, whereas superchargers are belt-driven by the crankshaft. At one time racecars were the only gasoline engine using superchargers, whereas turbochargers have been used on diesel engines for decades. The recent use of these components is a means to meet emission and fuel standards while boosting the engine's output. In fact, this additional volume of air is referred to as **boost.**

It should be noted that the number and type of attachments on each of these components depends on the engine design and passenger comfort available on a particular vehicle model. It could be a very simple system with the engine assembly, a water pump, and alternator or a system that is so loaded with extras that the engine assembly is not visible at first glance under the hood.

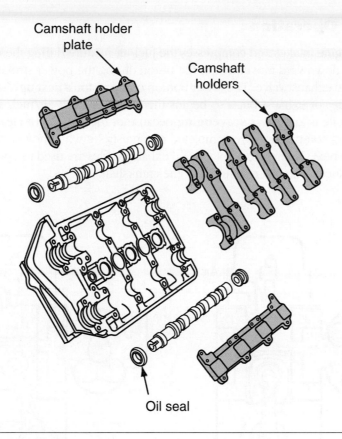

Figure 8-18 Overhead camshafts are mounted over or next to the valves.

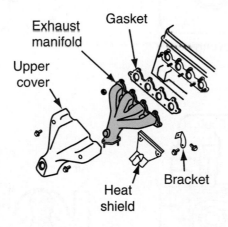

Figure 8-19 Exhaust manifolds are mounted directly to each cylinder head. They may share one exhaust pipe or have dual exhaust pipes.

Engine Operation

A typical gasoline-fueled engine uses a four-stroke cycle operation. Other types include the two-stroke and the rotary. Each type uses similar systems for fueling, lubrication, cooling, and the other components needed for operation. Two-stroke engines are used primarily in light equipment. However, they are generally less efficient than four-strokes and some require the lubrication to be mixed with the fuel. They also tend to emit excessive smoke and pollutants if not properly tuned.

Two-Stroke Operation

The two-stroke engine intakes and compresses the fuel/air mixture during the upward movement of the piston. The downward movement of the piston during the power stroke rotates the crankshaft and opens the exhaust valve. The lower portion of the piston's next upward movement starts clearing the cylinder of exhaust but also begins the intake of fresh air/fuel. The operation performed on this stroke is known as **scavenging** because of the method of clearing and filling the cylinder. Intake and compression are completed on one stroke while power and exhaust are accomplished on the next (Figure 8-20). Intake and exhaust valves are used to open or seal the cylinder. Power is produced on each revolution of the crankshaft.

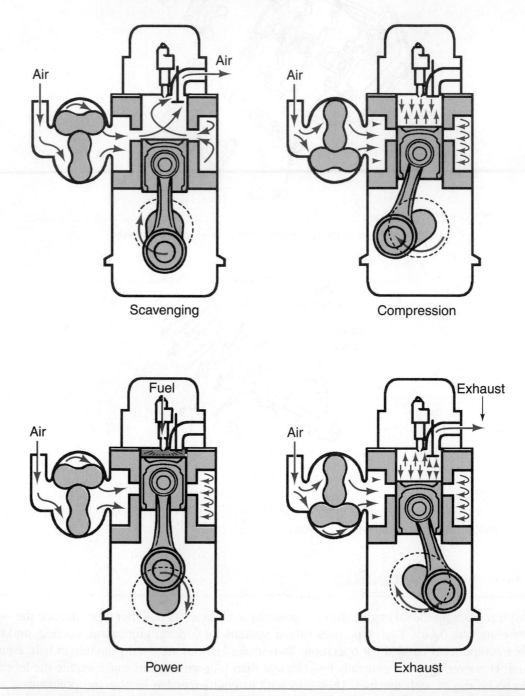

Scavenging

Compression

Power

Exhaust

Figure 8-20 Some two-stroke engines use rotary valves to control the flow of the air and fuel.

Rotary Engines

In a **rotary engine,** the air/fuel mixture is compressed using a triangular rotor inside an oblong chamber (Figure 8-21). At each apex (corner) of the rotor, a seal keeps contact with the chamber wall and pushes the air/fuel ahead of it or is pushed by expanding gases (Figure 8-22).

As the rotor turns, a seal forces the air/fuel mixture into a small area of the chamber. This reduced chamber volume compresses the mixture until ignition occurs. The expanding gases are directed against the seal ahead of the compression seal. This power is transmitted through the rotor to the crankshaft. As the rotor continues to turn, the gases enter a larger area where the spent gases are exhausted. As this area widens, a low-pressure area is formed under the intake valve. A new mixture is drawn in and the cycle is repeated. It should be noted that each side of the rotor and each seal act as the power, intake, compression, and exhaust component during each rotation of the rotor. Since the rotor always turns in one direction without the stop/start motion of a piston, the rotary engine could produce more horsepower with less waste energy than a conventional gasoline engine. However, the rotary engine does not produce the same torque as a piston engine of the same displacement and is considered less efficient. Rotary engines have the same problem with harmful emissions as piston engines.

Rotary engines usually are not rebuilt by a shop, but are purchased from Mazda. This is because of the type of machining required.

Four-Stroke Operation

The four-stroke engine is used in nearly every car and truck currently being produced. A power stroke is produced every two revolutions of the crankshaft. This engine requires four movements of the piston to complete a cycle.

The first stroke is the intake stroke. As the piston moves from top dead center (TDC) downward, the camshaft opens the intake valve. Reduced pressure in the cylinder, combined with the atmospheric pressure in the intake manifold, forces the air into the bore (Figure 8-23). Depending

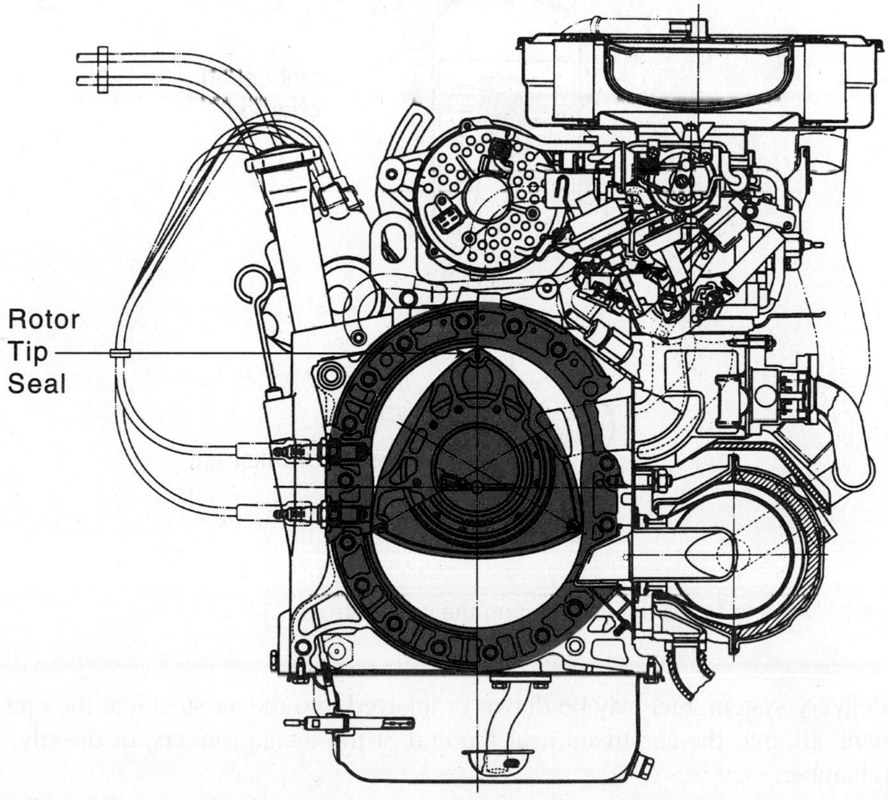

Rotor
Tip
Seal

Figure 8-21 There are two chambers and two rotors in a Mazda RX7 engine. (Courtesy of Mazda Motor of America, Inc.)

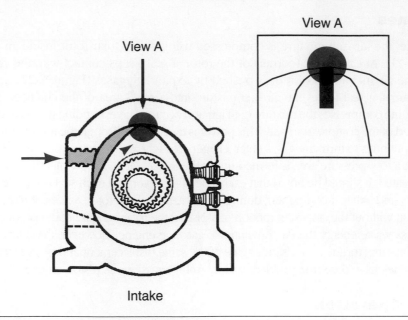

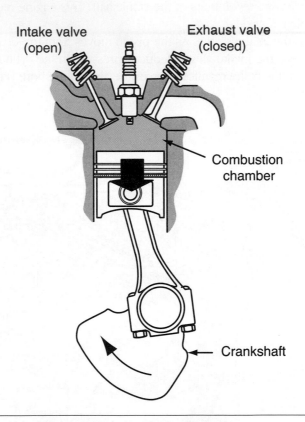

Figure 8-22 The seal slides along the inside of the chamber to capture and move the air/fuel mixture.

Figure 8-23 Air and fuel is drawn in during the intake stroke.

on the fuel delivery system, fuel may be drawn or injected into the air stream at the entrance of the intake manifold, into the air stream near the end of the intake runners, or directly into the combustion chamber.

The intake valve remains open until the piston nears bottom dead center (BDC). During the upper movement or compression stroke, the air/fuel mixture is compressed into the combustion chamber (Figure 8-24). Both valves are closed to seal the chamber during this stroke (Figure 8-25).

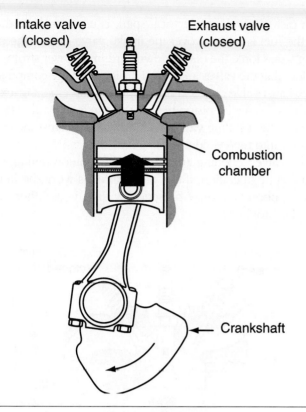

Intake valve (closed)

Exhaust valve (closed)

Combustion chamber

Crankshaft

Figure 8-24 With the air/fuel trapped in the combustion chamber, the piston moves upward on the compression stroke.

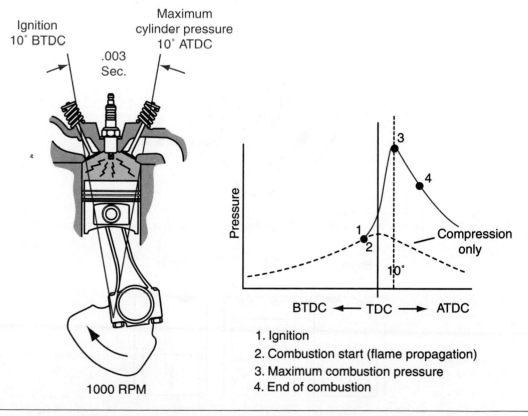

Ignition 10° BTDC

.003 Sec.

Maximum cylinder pressure 10° ATDC

Pressure

3

4

1

2

Compression only

10°

BTDC ← TDC → ATDC

1000 RPM

1. Ignition
2. Combustion start (flame propagation)
3. Maximum combustion pressure
4. End of combustion

Figure 8-25 The spark plug fires about 10 degrees before the piston reaches TDC on the compression stroke.

At a specific point in piston travel, an electrical spark is used to ignite the mixture. On multiple fuel injection (MFI), the fuel is injected over the intake valve or into the chamber just before the spark. The expanding gases force the piston downward in a power stroke, which turns the crankshaft (Figure 8-26). Note that the valves are closed during both the compression and power stroke.

Near BDC the exhaust valve opens and the upward movement or exhaust stroke of the piston exhausts the spent gases. As the piston nears TDC, the intake valve begins to open just as the exhaust begins to close. The resulting **valve overlap** helps exhaust the spent gases and assists in starting the new mixture into the cylinder (Figure 8-27).

The timing of the valves' opening and closing, the ignition timing, and the air/fuel mixture are critical to smooth engine operation. If any of the three is worn, broken, or out of adjustment, the engine will perform poorly and may be seriously damaged. Other operating systems of the engine can cause similar problems.

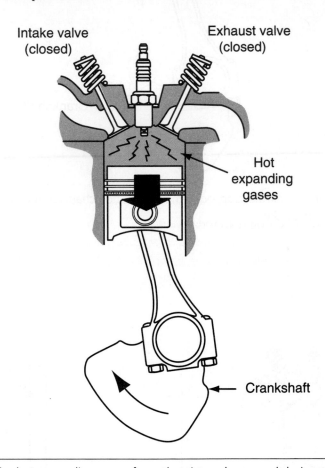

Figure 8-26 The hot expanding gases force the piston downward during the power stroke.

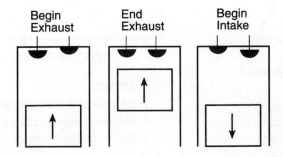

Figure 8-27 Valve overlap occurs at the end of the exhaust stroke and the beginning of the intake.

Newer technologies allow the engineers to design an engine that can change valve timing and valve opening. The PCM uses sensor data to change intake and exhaust valve timing and amount of opening. This helps reduce harmful emissions, increases fuel mileage, and increases power. One system eliminates the entire valve drive mechanism with a solenoid, which opens and closes the valve on command from the PCM.

The four-stroke engine operation is fairly simple if only one piston is being considered and the engine speed is low. However, when multiple pistons, piston weight, piston speed, and other stress factors are considered, the operation is no longer simple. A four-cylinder engine running at 2,000 rpms will have each of the four pistons stop and start about 4,000 times in one minute. Remember Newton's Laws of Motion on inertia.

Lubrication

All components that come in contact with each other need some type of lubrication. The engine with its close-fitting parts, high speed, and high temperatures is no exception. Poor or no lubrication will be apparent within several minutes of operation. Most damage from poor lubrication requires expensive repair or replacement of the complete engine.

The lubrication used today is a high-tech combination of lubricants and cleaning agents designed to reduce friction and improve fuel mileage. There are many different brands and combinations on the market today. Engine oil is rated by its thickness and resistance to breakdown by the Society of Automotive Engineers (SAE) and the American Petroleum Institute (API). The oil is rated by numbers that refer to its thickness or weight (Figure 8-28). Common engine oil used in modern engines is rated as 5W30. The multiple designations, 5 and 30, indicates that this oil can be used in cold or hot weather. The number 5 indicates thin oil while 30 is thicker oil. The newest engines have lower tolerances between moving parts and require thinner oil for lubrication. Older engines typically used thicker oil than today's engine.

Also available on the market are synthetic engine oils (Figure 8-29). Synthetic oil is made from chemicals, does not use natural petroleum products, and is rated by thickness in the same manner as natural oils. In some cases, synthetic oils are better than natural oils; however, they are usually much more expensive then regular petroleum products. There are also mixtures of

Shop Manual
pages 199–206

Figure 8-28 Engine oil comes in several ratings, but only use the one specified by the manufacturer.

Figure 8-29 Synthetic oils are not made from natural products but are rated the same as natural petroleum oil.

synthetics and natural oils available known as **synthetic blends.** The blends are a little cheaper than full synthetics, but are more expensive than naturals.

One of the problems with lubricating an engine is the time during shutdown. The oil tends to run or drain back from engine components, resulting in a dry or nearly dry startup where the moving components have little, if any, lubrication. This causes metal-to-metal contact and wear. Some synthetic brands, Castrol for one, offer a lubricant with magnetic tendencies. The oil has synthetic ester molecules that attach to the components and prevent drain back. In this manner, the moving components are lubed sufficiently even after an overnight shutdown.

The main point to remember is to check the service manual for the type of oil to be used in the various systems of the vehicle. In some engines, the use of synthetic or blends may be a total waste of time and money. Other engines can have longer life, better fuel economy, and a quieter run with synthetics or blended lubricants. Today's vehicles usually have to pass a federally mandated emissions test. Engine oils have become so critical to engine operation that sometimes a simple oil and filter change will cause the vehicle to pass the test. There are synthetic and blend lubricants available for automatic transmission/transaxles as well. Always consult the service manual.

The engine's oil is stored in an oil sump attached to the bottom of the block. A crankshaft or camshaft-driven oil pump is mounted directly over the oil with a pickup tube extending into the oil. The tube has a coarse screen pickup to filter out larger trash particles. The pump draws oil from the sump, pressurizes it, and forces it into the oil filter before it enters the engine's oil galleries. A **check valve** is installed in most pumps to prevent oil drainback during engine shut down.

An oil filter removes most of the harmful particles floating in the oil. The particles are metal and carbon that result during normal engine operation. The filter adapter has a **bypass valve** that opens when oil pressure within the filter becomes too high (Figure 8-30). This allows oil to flow past the filter and into the engine. The valve is primarily designed to allow oil flow when the engine and oil temperatures are cold. However, if the filter is clogged or the oil is extremely dirty, the valve will open. In both cases, unfiltered oil is pumped into the engine. Damage could result.

The oil moves under pressure into the block's oil galleries. Some engine components are *pressure lubricated* whereas others are *splash lubricated*. Pressure-lubricated components include the crankshaft bearings, connecting rod bearings, camshaft bearings, and some rocker arms (Figure 8-31). All of the bearings operate under extreme conditions and the pressurized oil is required to maintain a **clearance** between the bearing and the supported component. Splash lubrication

The **clearance** between a bearing and its journal is called "oil clearance" and is usually about 0.001 inch to 0.003 inch.

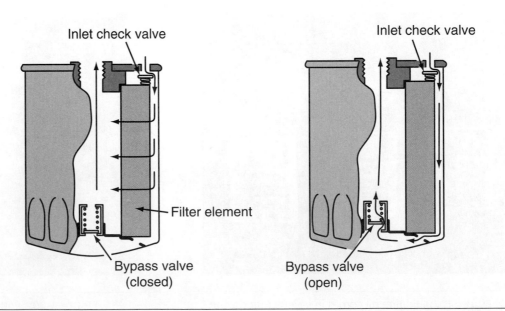

Figure 8-30 The bypass valve allows cold, unfiltered oil to flow to engine components during startup and cold driveaway.

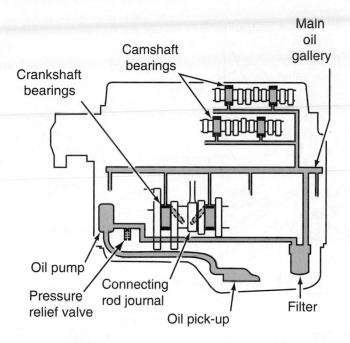

Figure 8-31 Pressurized oil is required for such mated components as the crankshaft and its bearings.

is accomplished when the pump supplies a large pool of oil for the component to move through. The oil is splashed over the components before returning to the sump through large passageways or galleries. Some components using splash lubrication are camshaft drive gears and chains, valvetrain mechanisms, and cylinder walls (Figure 8-32).

Some people believe that the crankshaft rotates through the oil in sump. If this did occur, **cavitation** or foaming would cause the oil pump to suck air and the engine would be without lubrication. Too much oil can do as much damage as too little oil. The type and amount of lubricant specified by the manufacturer must be followed.

Starting and Charging Systems

Different circuits control the starting and charging systems. The starter controls are the ignition switch, a park/neutral or clutch switch, and the relay. Antitheft circuits may also be in series with or controlling power to the starter relay. The **park/neutral switch** is in series with the ignition switch, starter relay, and the PCM. If the automatic transmission is not in park or neutral, the switch is opened and current cannot flow to the starter relay or PCM. A **clutch switch** performs the same function on a manual transmission vehicle. If the clutch pedal is not fully depressed, the starter and PCM will not function. Some vehicles also have an antitheft system that will disable the starter and/or ignition system if not disarmed by the correct security code.

The charging system consists of the generator, battery, and associate conductors. Some technicians state that the battery is part of the starter system. However, this makes no difference to the vehicle. The generator, as noted in Chapter 1, uses magnets (stator) around an electrically charged magnet (windings) on a rotor. The rotor is driven by a belt and the crankshaft. The amount of electricity allowed into the rotor windings determines the magnetic force of the rotor. The more current flow the stronger the magnetic field. This rotating magnetic field moves through the magnetic field(s) of the stator, and each time a magnetic field is "cut" by another magnetic field an electrical current is produced. The strength of the rotor field is controlled by the alternator's regulator and the rotational speed of the rotor.

The regulator senses the electrical load being used by the vehicle's electrical systems and adjusts the input voltage to the rotor so the generator's output from the stator is sufficient for the

Cavitation is the introduction of air into a fluid, in this case, air forced into the oil by the rotating crankshaft and connecting rod.

The **park/neutral switch** is known as the manual valve position sensor and sends a signal to the PCM. It is sometimes referred to as the PRNDL (pronounced pindil) switch.

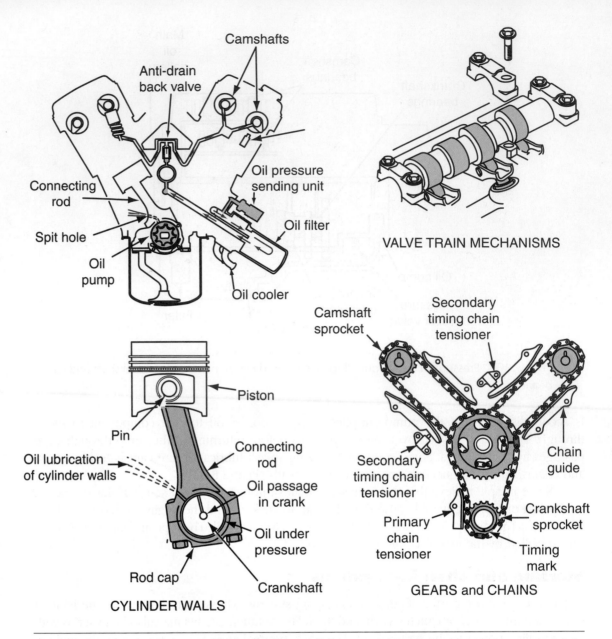

VALVE TRAIN MECHANISMS

CYLINDER WALLS

GEARS and CHAINS

Figure 8-32 Splash lubrication takes place in those areas where the components are not tightly mated. The oil shown here is sprayed from around the camshaft and its bearings.

load (Figure 8-33). In this manner, there is enough voltage and current to allow for full operation of all systems and charge the battery. However, the addition of more and more electrical devices and computers is overloading the typical alternator's capacity.

To meet this electrical demand manufacturers are trying to develop a 42-volt or 48-volt power supply. In general terms, plans call for a 42- or 48-volt battery with the appropriate sized generator to recharge it. There are two ways of wiring the remainder of the vehicle's electrical components. One is to eventually switch all components including the conductors to match the 42/48-voltage generator and battery voltage. This would be very expensive and would require massive reorganization of parts suppliers and all associated technical fields. A second way would be to retain the current 12-volt systems but add a transformer of sorts between the battery and the vehicle. This would allow a steady 12-volt supply while reducing the cost of the changeover drastically. It would also be possible to supply a specific amount of current to each device, which in turn means more accurate sensors and actuators.

Figure 8-33 The brushes are attached to the regulator and are in constant physical contact with the slip rings on the rotor.

Hybrid-powered vehicles may be the answer to this increasing electrical load. In general terms, a hybrid vehicle uses an internal combustion engine to drive a high-output generator that charges the large drive battery pack. Electrical power for the vehicle systems including the drive motors is drawn from the drive battery pack. A second type of hybrid vehicle uses the engine to drive the generator and power the vehicle drive wheels. The battery pack provides the power for the vehicle's electrical systems and when commanded by the on-board computer can assist in powering the vehicle. Both types have the ability to provide a constant electric current for all electric devices on the vehicle.

Fuel Systems

Shop Manual
pages 207–211

Fuel systems deliver fuel from the tank to the mixing device, usually a carburetor or fuel injector. The fuel volume must be delivered at a specified volume and pressure. A mechanical pump usually produces between 3 and 6 psi with a volume of 1/2 pint in about 30 seconds. An electric pump usually produces up to 59 psi and a similar volume. Diesel engines use an injector pump that produces up to 1,800 psi and sometimes use a mechanical pump to provide a constant volume to the injector pump.

Mechanical Fuel Delivery Systems

Fuel is delivered to the mixing unit by a fuel pump. A mechanical pump driven by the camshaft is used on most carburetor systems. The pump operates from a camshaft-driven pushrod. Its fuel delivery pulses on and off as the camshaft turns. The carburetor is the mixing unit, which has a reservoir or bowl to hold sufficient fuel. The bowl is required due to pump pulses and the pump's inconstant pressure and volume. The mechanical pump delivers fuel to the carburetor on each pump stroke.

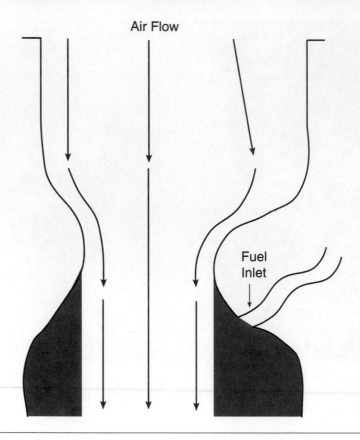

Air Flow

Fuel Inlet

Figure 8-34 Low pressure is created by air moving rapidly past the recessed area.

The carburetor has various internal passageways to deliver fuel into the airflow moving through its **venturi** (Figure 8-34). The venturi is a large, vertical air passage near the center of the carburetor body. It opens directly into the top of the intake manifold. As the air flows past a narrow portion of the venturi, a low-pressure area is established at the point where the venturi widens again. In the bowl, atmospheric pressure pushes the fuel into the low-pressure area to mix with the air. The fuel passages are sized to deliver a certain amount of fuel depending on the volume of air moving through the carburetor.

Mechanical fuel systems are almost nonexistent in modern vehicles. However, there are still vehicles and equipment in use today that rely on this type of fuel system.

Electronic Fuel Injection (EFI)

The electronic fuel injection (EFI) system requires an electric fuel pump. This type of pump delivers fuel at a constant pressure and volume. The fuel may be delivered to injectors in a throttle body unit similar to the carburetor or individual injectors at each cylinder. Throttle body injection has some of the same problems as the carburetor system. However, the pressure to the injectors can be regulated with an engine load-sensing fuel regulator to ensure a nearly constant air/fuel ratio at all engine speeds and loads.

Some vehicles, most notably Fords, use a low-pressure supply pump in the tank and a high-pressure pump mounted on the frame rail. Most vehicles have only one electric pump mounted inside the tank. There are two critical concerns with electronic fuel delivery maintenance. This type of system depends on a constant pressure and the volume of fuel. A clogged or restricted filter can cause numerous problems with engine operations. Sometimes, the filter is forgotten because some vehicle manufacturers do not require a typical service for 100,000 miles. The technician should keep this in mind when the vehicle comes in for an oil change. If the fuel fil-

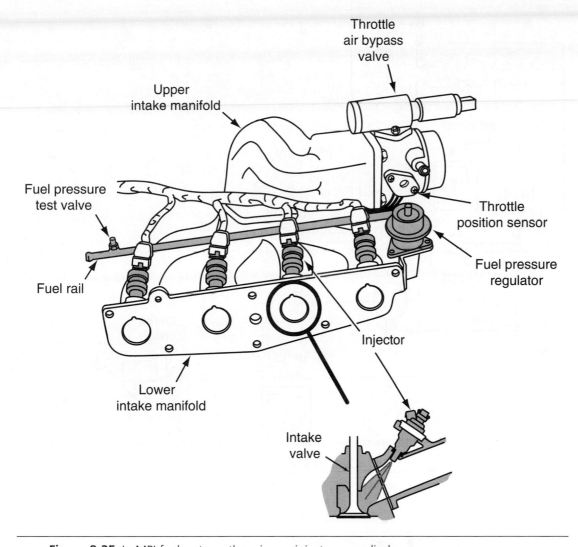

Figure 8-35 In MPI fuel systems there is one injector per cylinder.

ter is factory-installed and the vehicle mileage is over 30,000 miles, it may be best to question the owner on the history of the fuel filter.

A second and potentially more expensive problem is running the fuel tank dry. An in-tank electric pump is cooled and lubricated by the fuel passing through it. A pump running without fuel in the tank can be damaged, but this may not become obvious until a later date when it fails. If a fuel pressure problem develops and the filter is good, ask the owner if he/she knows whether or not the vehicle ran out of gas recently. Sometimes, a simple question and answer can resolve a complex problem, and the owner is usually the best source of vehicle history.

Most modern EFI systems have individual injectors for each cylinder. The injectors are mounted in the intake runners at the intake valve in the cylinder head (Figure 8-35). The injectors are opened on command from the PCM for a specific period of time depending on engine speed, load, temperature, and other factors. The constant fuel pressure and volume ensure that the air/fuel ratio meets the operating condition of the engine.

The fuel is injected over the open intake valve on most vehicles. However, a new gasoline system called **direct injection** has the fuel sprayed (injected) directly into the combustion chamber near the intake valve. This provides a better burn and cleaner emissions because the fuel is injected into the air being compressed in the cylinder. Some direct injection systems use an engine-mounted injector pump to pressurize the fuel. Direct injection with an injector pump is the fuel system used on diesel-fueled engines.

Diesel-fueled engines use the heat of the compressed air to ignite the fuel.

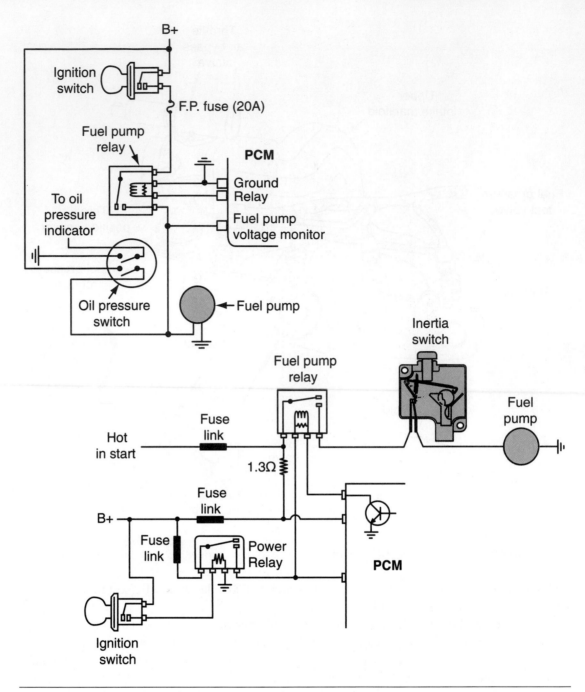

Figure 8-36 A typical starter control circuit.

Controls for the electrical fuel pump are the ignition switch, pump relay, and the PCM (Figure 8-36). Some vehicles use a separate switch in series with the relay and pump. This switch may be referred to as a rollover or inertia switch. Its purpose is to cut power to the pump if the vehicle is involved in an accident. Other systems use an oil pressure switch or engine speed sensor to signal the PCM that the engine is not running or oil pressure has been lost. General Motors uses an oil pressure sensor to stop the pump if oil pressure is lost. A bypass circuit in the fuel pump relay will supply power to the pump during starting. Like the starter circuit, an antitheft circuit may be incorporated into the fuel pump circuit.

Air for an EFI system is drawn into the engine in a manner similar to the carburetor. However, dirty air and fuel filters affect EFI systems more than a carburetor system. The best air/fuel mixture with current engine technology is 14.7 parts air to 1 part fuel, or a 14.7:1 ratio. A cold engine will need a rich mixture of about 13:1 for starting. A warm engine can operate on a leaner mixture of about 15:1 ratio or higher. The volume or amount of intake air depends on engine displacement, the fuel system, and the mechanical condition of the engine. A dirty air filter can enrich the fuel mixture to the point of flooding the engine by restricting the air intake.

Ignition Systems

The ignition system provides the spark to ignite the air/fuel mixture. The system consists of controls, an ignition coil, spark plugs, and spark plug wires. The heavy current circuit for the spark plugs is called the **secondary circuit.** A mechanical **distributor** or the PCM controls the system. The distributor is geared to the camshaft. The distributor shaft turns a rotor inside the cap (Figure 8-37). The cap has an internal ignition coil terminal and a terminal for each spark plug. The rotor points at each spark plug terminal once in each revolution. If the coil is discharging, the current moves through the cap's coil terminal, travels along the rotor's conductor, and bridges the gap between the rotor tip and the nearest spark plug terminal. The distributor has an advanced mechanism that adjusts timing of the spark based on engine speed and load.

Distributor systems have gone the way of carburetors. They have been replaced with electronic systems that are faster and more accurate in controlling ignition spark. But, like the carburetor fuel system, many vehicles and equipment with the distributor ignition system are still in use.

The newer systems employ electronics to control the ignition spark. The distributor has been replaced completely with an electronic package that determines crankshaft and camshaft position, engine load, engine temperature, and engine speed to determine **ignition timing** within each individual cylinder (Figure 8-38). This same data is also used to fire the fuel injectors and vary the time of injector opening.

Shop Manual
pages 216–220

The **ignition timing** must be adjusted because the speed of the pistons changes as the engine speed changes.

Figure 8-37 An electronic distributor used on some modern vehicles.

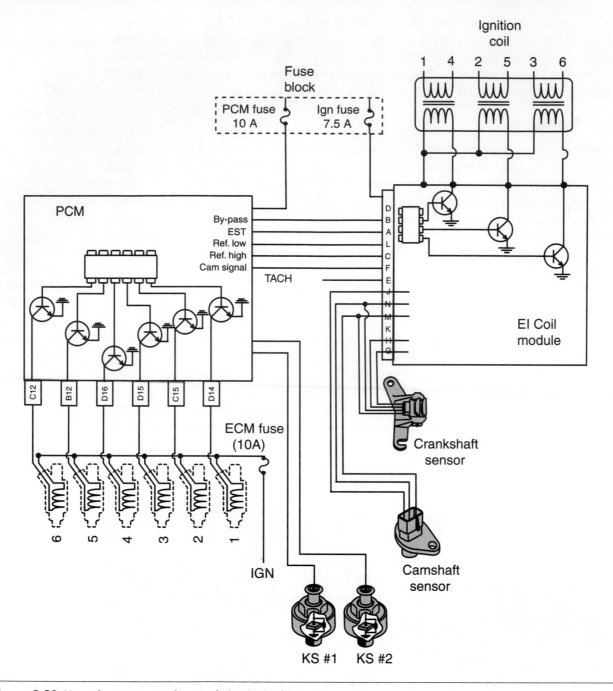

Figure 8-38 Note the sensors and control circuits in this ignition system.

A **coil pack** consists of a primary circuit control package and an ignition coil mounted over the package. The coil is in the secondary circuit.

The PCM performs all the calculations necessary based on engine, transmission, and other sensor input. This provides maximum fuel mileage, cleaner emissions, and more available power for any given engine application. Some of the vehicle sensors and actuators are shown in Figure 8-39. Some of the terms and explanations listed there may be unfamiliar, but they are only offered at this point as an indication of the electronics used on an engine. Also, this figure does not list all of the electronic devices, and some manufacturers use different terms for the same type device. Some of the devices are not used on some vehicles. Detailed information is available in the *Today's Technician* series, *Automotive Engine Performance.*

Delivering current to the spark plug is similar to the mechanical ignition system. The major mechanical difference is a separate **coil pack** for each pair of spark plugs or a coil for each

Sensor/Actuator	Information	Determines
Throttle position (TPS)	Throttle opening	Driver's speed intention
Oxygen sensor (O_2)	Oxygen in exhaust gases	Air/fuel mixture ratio
Manifold absolute pressure (MAP)	Pressure in intake manifold	Engine load
Mass air flow (MAF)	Air flow in grams per minute	Engine load (may be used with or in place of MAP)
Engine coolant temperature (ECT)	Engine temperature	Engine temperature
Air charge temperature (ACT)	Intake air temperature	Intake air temperature
Vehicle speed	Vehicle speed	Vehicle load (may be used to send data to speedometer)
Overdrive switch	Engages/disengages overdrive	Provide better fuel mileage
Torque converter clutch (TCC)	Engages/disengages lockup clutch	Locks/unlocks engine output directly to driveline
P/N; clutch switch	Gear or clutch position	Allows engine to be started
Brake light switch	Brake on or off	Turns off cruise control, activates antilock brakes, vehicle is to be slowed
Injector	Fuel delivery	Injects fuel into intake or cylinder
Idle air bypass (IAC)	Air control during idle speed	Controls engine idle speed by adjusting air flow
Cooling fan	Cools coolant	Draws air over radiator fins
Air condition clutch	Drives air-conditioning compressor	Cool air to passenger compartment

Figure 8-39 The sensors listed here are standard in most vehicles, but they constitute only a small portion of the electronic devices used on today's vehicles.

spark plug (Figure 8-40). On the paired system, each time a coil pack fires, it ignites two separate spark plugs. One of the spark plugs will fire a compression stroke and generate the controlled burn to power the piston. The second plug is on the exhaust stroke and acts as a waste spark. There may be very little fuel left in that cylinder, but, primarily, the spark is used to discharge the coil pack. The very newest systems use a coil per spark plug. This is known as **coil-on-plug** or **coil-near-plug.** The coil pack is mounted where the spark plug wire would normally be attached to the plug or near the plug with a short spark plug wire (Figure 8-41). This system eliminates the distributor and spark plug wires. It also allows very precise timing of the ignition spark.

Diagnosing an electronic ignition or fuel system can be done accurately as long as the technician understands how the system works. With the systems produced since 2000 and later, electronic diagnostic equipment can tell the technician almost everything that is happening in the two systems by connecting one cable between the test equipment and the vehicle. However, if the technician does not understand the operation of the systems, all of that data is worthless.

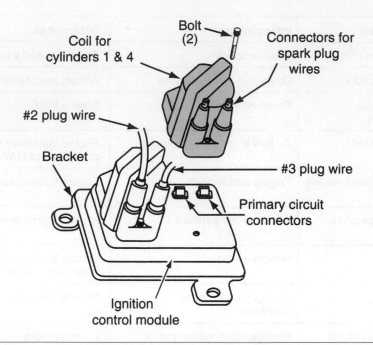

Figure 8-40 Two coil packs supply secondary voltage to this 4-cylinder engine. Note the companion plugs, #1 and #4.

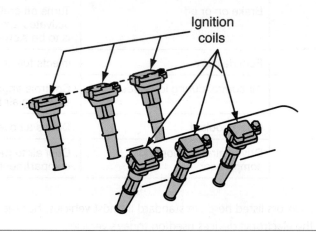

Figure 8-41 Typical coil-on-plug layout.

Shop Manual
pages 220–222

The water pump may be mounted behind the timing belt cover, on the timing chain cover, or to the side of the block on a bracket.

Cooling Systems

Horsepower is produced from the expanding gases in the combustion chamber. High temperature is created and must be dissipated. Some of the heat is absorbed by the engine's lubricant and carried to the oil sump. However, most of the engine heat is absorbed by the cooling system.

A crankshaft-driven water pump assists in moving cool coolant into the engine block. The coolant circulates through passageways in the block and cylinder head. Heat from combustion and friction is transferred to the coolant. The coolant leaves through a thermostat housing near the top of the engine. The thermostat is a temperature-sensitive valve mounted at or near the highest and hottest part of the engine (Figure 8-42). Some vehicles have the thermostat mounted at the top entrance into the radiator. During cold and warm-up conditions, the thermostat remains

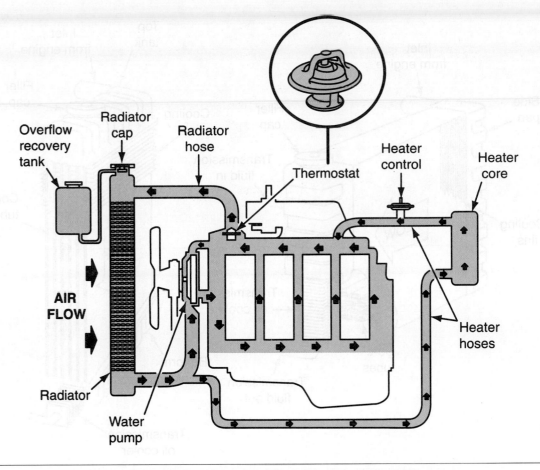

Figure 8-42 This thermostat is mounted slightly below the top of the engine. Some thermostats are mounted in the lower (return) hose near the bottom of the engine.

closed, forcing the coolant to circulate back into the engine block. By not cooling the coolant, the engine heats quicker. Most thermostats are fully open at about 190 degrees Fahrenheit (88 degrees centigrade). As the temperature rises, coolant is allowed to flow through the upper hose into the top of the radiator.

The radiator is either a cross flow or a vertical flow. The two types are very similar except for the direction of coolant flow (Figure 8-43). The coolant moves through small tubes down or across to the radiator's exit. The tubes are finned to expose as much surface area as possible to the passing airflow. A fan draws air from the front of the vehicle over the radiator fins and exhausts the air over the engine block. As the coolant exits the radiator through the lower hose, the water pump picks it up and pushes it into the engine block.

Most cooling fans are driven by electric motors controlled by a cooling fan switch mounted in a coolant passage or the PCM (Figure 8-44). When overheating is sensed, the switch or PCM closes the fan relay. This control of the fan's operation is based directly on engine temperature and reduces the load placed on the engine and saves fuel by using the fans only when necessary. Some vehicles use two fans, which may be referred to as cooling fans 1 and 2 or as a cooling fan and an air-conditioning fan. Since operating the air conditioning creates more heat in the engine, switching on the air conditioner also switches on a cooling fan. If only one fan is present, it will operate, whereas a two-fan system will have one of the two working. It should be noted that under extreme, overheating conditions, the PCM could switch on both fans while switching off the air conditioning system.

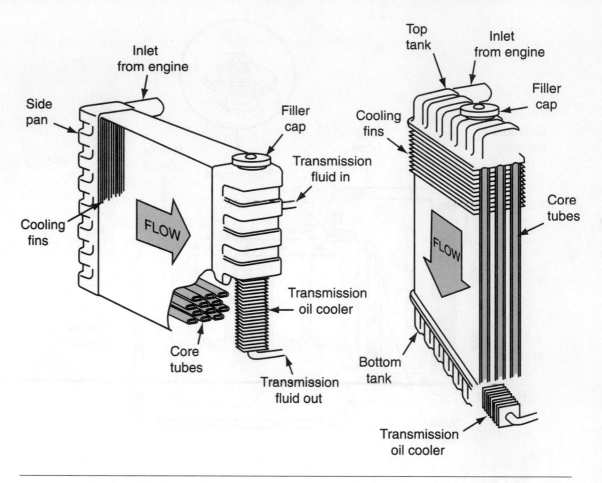

Figure 8-43 A typical down-flow radiator.

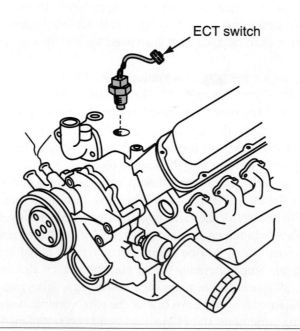

Figure 8-44 This engine coolant temperature sensor is mounted at about mid-block. Most are mounted near the thermostat housing.

Exhaust Systems

The original purpose of the exhaust was to clear the combustion chamber and quiet the noise of the engine. That same purpose remains but some components of the exhaust system are now used to clean the exhaust gases of harmful emissions.

The exhaust manifold is attached to the cylinder head with a separate opening or port for each cylinder (Figure 8-45). The manifold runners run together at some point shortly after leaving the cylinder head. The exhaust pipe extends from the manifold to a muffler (pre-1974) or to a catalytic converter (post-1973). **Mufflers** and **resonators** are still used to quiet the exhaust noises but are positioned after the converter. Between the converter, muffler, and resonators is a tail pipe made of stainless steel. A section of the tail pipe extends from the last muffler to the rear of the vehicle.

AUTHOR'S NOTE: A few years back, several children of the same family died while riding in the camper shell on their family pickup. They were returning from a family outing less than a hundred miles from home at night. The children were placed in the camper so they could sleep. During the trip home, the children succumbed to carbon monoxide poisoning. The cause was a broken tail pipe under the truck that allowed exhaust gases to enter and be trapped in the camper shell. During the twenty-one-point safety inspection, inspect the exhaust system thoroughly.

For the best power and mileage, the exhaust must be tuned for the engine. Failure to exhaust gases quickly causes poor performance, particularly at higher engine speeds. Restricted or too-small exhaust pipes will not clear the combustion chamber for fresh air and fuel. A fully opened system with no noise or pollution devices is not necessarily the best means to achieve the

Figure 8-45 An exhaust manifold of this design is more efficient at gas flow. It is commonly referred to as a "lake pipe" design.

most performance *unless* the system is designed or tuned to the vehicle and engine. In most areas, removal of noise-reducing devices is illegal. It is illegal everywhere in the United States to remove or disable the catalytic converter or other factory-installed emission devices.

Emission Control Systems

Shop Manual
pages 224–227

The burning of fossil fuel produces harmful emissions. Hidden within an engine's exhaust gases are three major chemicals or particles that harm the environment and living things. **Hydrocarbons (HC)** are particles of carbon left after ignition. HC is measured in parts per million of air. **Carbon monoxide (CO)** is a colorless, odorless, tasteless gas that can kill oxygen-breathing animals. **Nitrogen oxides** (NO_x) are emitted in the exhaust when combustion chamber temperatures exceed 2,500°F.

At one time, six basic emission control systems were used on cars and light trucks. Electronics and engine designs have eliminated some of them even though allowable emission limits have been lowered. The first four are still in use today while the last two have been replaced with better technology. In addition to the operating systems, a great deal of emission control is accomplished during the vehicle's engineering and design stage.

Design and Construction

The design of the engine and its internal components and operation help determine the amount of harmful emissions produced. Even though no design alone can fully eliminate pollutants, the shape of the combustion chamber, valve timing, ignition timing, fuel and air delivery, and a host of other engineering details can significantly reduce emissions. These details involve items such as the size of the air filter, placement of the spark plug, and even the type of oil filter.

Catalytic Converters

Catalyst action is a chemical reaction between two or more elements. One element (catalyst) causes a chemical change in the other element(s) without changing itself.

The two-way converter uses two metals, platinum and palladium, to reduce carbon monoxide and hydrocarbons through **catalyst action.** During the catalyst stage, the exhaust gases are heated to about 1,600°F (878° centigrade) and the unburned fuel is burned off. The hydrocarbon (HC) molecules break apart, forming hydrogen (H) and carbon (C) atoms. This burning or oxidation of the exhaust also causes chemical changes in the carbon monoxide molecule by breaking it into oxygen (O) and carbon dioxide atoms (CO_2). Some of the oxygen forms with the released hydrogen to make H_2O, or water. The best exhaust gas mixture at the tail pipe is H_2O and CO. A three-way converter does the same as the two-way but uses rhodium to help reduce nitrogen oxides (NO_x) (Figure 8-46).

Exhaust Gas Recirculating (EGR) Systems

The sole function of the **EGR system** is to lower nitrogen oxide emissions. NO_x is formed in the combustion chamber when temperatures exceed 2,500°F. The EGR valve allows some inert exhaust gases to enter the intake manifold and mix with the air/fuel (Figure 8-47). Since there is very little, if any, fuel in the exhaust gases, the recirculated exhaust gases occupy space in the combustion chamber, thereby preventing a full air/fuel intake. The lower amount of air/fuel available reduces the heat of the combustion and holds the chamber temperature low enough to block the formation of NO_x.

The EGR valve is controlled by the PCM, which controls the amount of vacuum allowed to the valve's diagram. Some vehicles have a sensor that signals the PCM on the movement of the valve. This provides feedback so the valve can be used more effectively. Other vehicles use a valve operated by a PCM-controlled electrical motor.

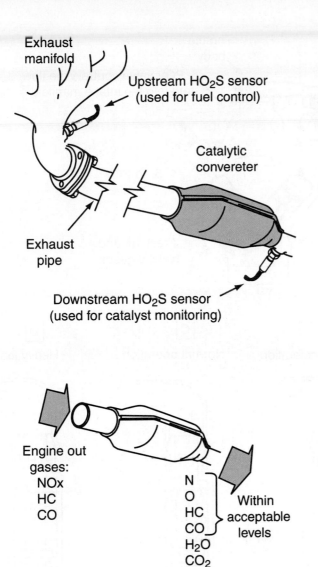

Figure 8-46 A three-way converter helps reduce HC, CO, and NO$_x$ emissions.

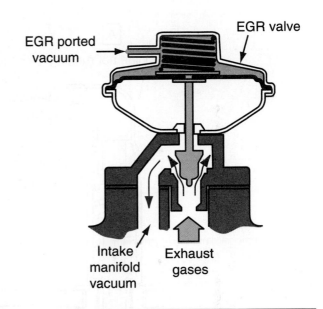

Figure 8-47 The exhaust gas recirculation valve helps lower combustion chamber temperature and reduces the formation of NO$_x$ gases.

Positive Crankcase Ventilation (PCV) Systems

The PCV system captures gases in the oil sump and routes them back into the intake system (Figure 8-48). The PCV valve is connected to a port on the intake manifold. The controlled vacuum draws gases from the top of the oil sump and feeds them into the intake for burning. As the gases are drawn out, a low pressure develops in the sump. Clean, fresh air is drawn down from the air filter into the sump, where the air and sump gases are mixed before entering the PCV valve. The most common harmful pollutants found in the sump are unburned fuel (HC) and acids. The oil sump must have some type of pressure relief or the various engine seals and gaskets will leak oil. The PCV relieves the pressure without releasing the pollutants to the atmosphere.

Evaporative (EVAP) Systems

With fuel in the fuel tank, vapors are released that could escape into the atmosphere. To prevent this, the **EVAP system** contains and routes the fuel vapors to the intake manifold. The first component is

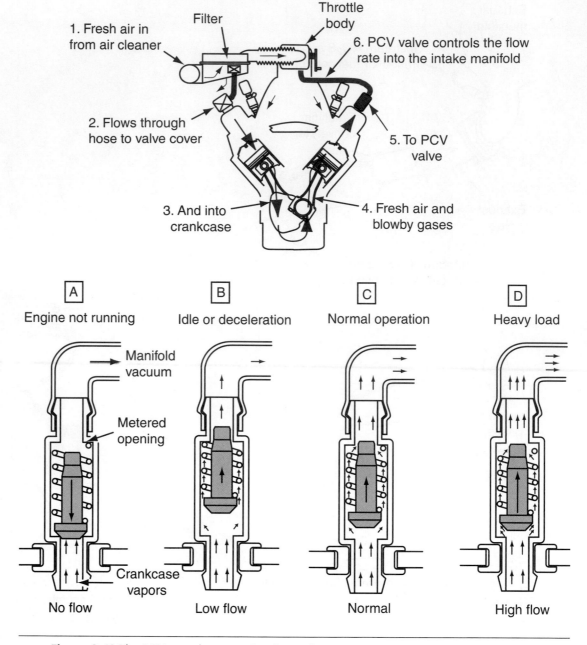

Figure 8-48 The PCV must be operational to reduce pressure in the oil sump and route harmful emissions back through the intake.

the fuel tank filler cap. The cap is designed to allow air into the tank but prevents any vapor from escaping (Figure 8-49).

Vapor pickup tubes in the top of the tank are arranged so liquids cannot enter the system, possibly flooding the engine. The tubes route the vapor to a charcoal-filled canister located in or near the engine compartment. The canister acts as a reservoir to store the vapors until the PCM or other control device opens the purge valve. Intake vacuum is applied to the purge valve any time the engine is operating. If the PCM opens the purge valve, fuel vapors are drawn from the canister and mixed with the air/fuel in the intake manifold.

The newest EVAP systems are sealed tighter than the older ones. Sometimes, they cause a problem with engine operation. Loosening the fuel filler cap relieves the pressure and lets the engine operate. This is not a very common problem, but it is a point to remember if a vehicle comes in on a wrecker with a no-start problem.

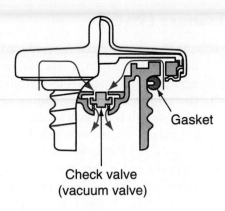

Gasket

Check valve
(vacuum valve)

Figure 8-49 This is the only type of fuel cap allowed with the newest EVAP systems.

Secondary Air Injection Systems

The **secondary air injection** works with the catalytic converter. Since the converter needs heat to work correctly, secondary air injection supplies the necessary oxygen to promote burning of any fuel in the exhaust.

A crankshaft-driven, high-volume, low-pressure air pump supplies the air (Figure 8-50). The air is directed by a valve mechanism. The air may be pumped into the exhaust manifold or the exhaust pipe before the converter, into the front of the converter, or diverted to the atmosphere. The direction of flow depends on the type of secondary air-injection system, the type/make/model of the engine, and engine load. The only time air should be diverted to the atmosphere is during engine deceleration. During deceleration, the air/fuel mixture in the exhaust can become very rich. Adding oxygen to the rich exhaust could cause a backfire within the exhaust system and may damage exhaust components. The secondary air-injection system is one of the systems that has been replaced with better engine design on many vehicles.

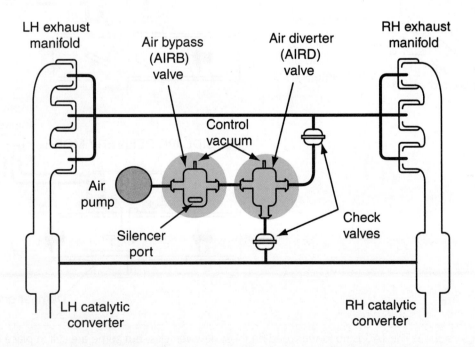

Figure 8-50 The secondary air injection system has been replaced with better engine controls, but is still around on many older cars and trucks.

Thermostatic Air Cleaners (TACs)

The TAC is used to quickly warm the incoming air during cold start and cold driveaway. It was used on carburetors and earlier fuel-injection systems. The air snorkel for the air filter has a vacuum-controlled valve (Figure 8-51). The vacuum is drawn from the intake manifold through a temperature-sensitive switch. The switch is generally mounted on or near the thermostat housing. When the engine is cold, the switch is opened to allow full vacuum to the TAC valve. The valve closes a small door to any outside air. Incoming air has to pass over an exhaust manifold stove, through a flexible tube, and into the air snorkel.

The exhaust manifold heats quickly and warms the air. When the engine reaches about 100°F (38° centigrade), the switch closes off the vacuum and a return spring forces the door open to outside air. The TAC is another system that has been deleted because of new air/fuel mixture control technology.

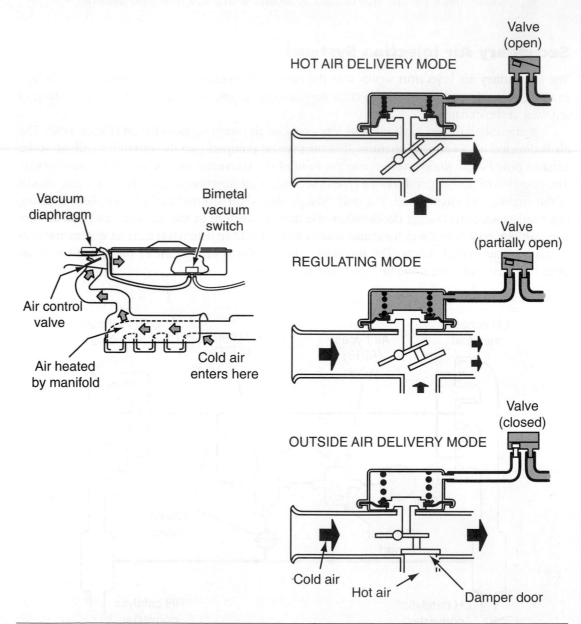

Figure 8-51 The TAC is no longer used on most new vehicles, but some are still in place on older road-worthy vehicles.

Computer Systems

Computer systems on modern vehicles probably represent the biggest and most comprehensive change to the transportation industry since the introduction of a credible internal combustion engine. Computer systems have provided more efficient vehicles to the public and can be a boon to technicians and repair shops. The only real drawback to vehicle computer systems is the fact that many technicians-to-be and in-field technicians either don't understand electronics systems or are afraid of them. If a technician can diagnose a headlight system, then she/he can diagnose a computer system. Different testers, procedures, and different values are present, but it is still electricity and it will follow certain rules. The *Shop Manual* highlights some tests that new technicians can do without causing damage or injuries.

Shop Manual
page 227

AUTHOR'S NOTE: Eight of the ten weeks in our program's first quarter cover automotive electricity and electronics, followed by another 225 hours of advanced electronics in the last two quarters. When the students turn pale at the mention of micro-amps, nano-amps, and corresponding voltage values with extremely high resistance levels, we present a simple water solution. If you are measuring the volume of a flood, you use a large device capable of measuring cubic feet (or gallons) per minute. If you are measuring the drops per minute from a coffee maker, you use a cup. The electrical test principles are the same whether testing computer circuits or headlights.

Sensors

Before we start this discussion, let's define electrical and electronic circuits. First, *electrical systems* or circuits on a vehicle are those that use 12 or more volts with up to 60 amperes of current, such as the exterior lights and fans. *Electronics* refers to electrical devices that are controlled or regulated by other electrical devices. Both use the movement of electrons, voltage, and resistance to make a usable circuit. The television remote control is one such example of an electronic device. One other thing to remember about "fully automatic" electronic devices is that there is no true completely automatic device. At the bare minimum, someone has to plug it in or install the batteries and turn it on the first time. Even if a robot builds a car "without" human assistance, someone turned on the power before anything happened.

Sensors are electrical devices that measure some mechanical, chemical, or electrical action. There are many varieties, with many different names, on modern vehicles so we will discuss them as types and relate some as examples. Vehicle sensors are basically **transducers** that convert one form of energy to another. **Potentialometers** or potentiometers are one type of sensor.

AUTHOR'S NOTE: The spelling difference between *potentialometer* and *potentiometers* is based on electronic terms. A potentiometer is technically defined as an electrical current-controlling device (e.g., by turning a knob a lamp can be dimmed or brightened). A sensor, as discussed in a following paragraph, does not control, but registers a voltage signal for use by a computer. That sensor type in many automotive circles is considered a potentialometer because it is measuring the potential voltage at a given point on a resistor. I have not heard an explicit definition of the sensor discussed next that precludes the use of the term *potentialometer*. Technically the "tio" spelling is correct; but spelling it either way will get the correct part. I feel more comfortable with the "tialo" spelling, but check with the instructor for her/his opinion.

Figure 8-52 The TPS converts mechanical action into an electrical signal.

Potentialometers on a vehicle usually change mechanical action (energy) into an electrical voltage or a signal. The most prominent is the **throttle position sensor (TPS)** (Figure 8-52). As the driver moves the accelerator pedal and in turn moves the throttle (air) valve at the intake, the TPS mounted to the side of the throttle body measures this mechanical movement, converts it to electricity, and transmits a signal to the PCM. This is done by a wiper arm moving over a resistor wire inside the TPS. At idle speed the signal voltage to the PCM is between .5 volts and 1.1 volts on most vehicles. This signal "tells" the PCM the operator has the throttle valve closed and gives the appropriate commands for a smooth-running engine. There are usually three wires to a potentialometer: one is the reference voltage (REFV or VREF) of either 5 or 7 volts, one is the ground for the sensor, and the last is the signal return to the PCM. As the operator opens the throttle (depresses the pedal), the wiper arm moves, resistance between the VREF and the signal return conductors change, and the voltage signal increases (Figure 8-53). At wide-open throttle the voltage is about 4.5 volts. The PCM sends the proper commands to change the air/fuel mixture, spark timing, and may even cause the transmission to shift gears.

Another type of sensor is the alternating current (AC) generator usually called a permanent magnet (PM) generator, which is nothing more than a very small version of the vehicle's generator. Granted, the signal voltage will be very low, but its operating principles are the same as the vehicle's generator. Magnetic teeth are attached to a mechanical device like a brake rotor or axle shaft. The sensor has an electrical magnet energized by current from the PCM. As stated in Chapter 7, as each tooth passes the sensor magnet an electrical voltage or current is produced and transmitted back to the PCM. One sensor of this type that has been used for many years is the **vehicle speed sensor (VSS),** which does exactly what its name implies—it measures vehicle speed. The second common AC sensor is the wheel speed sensor used in antilock brake systems (ABS). They measure the rotational speed of the wheels so the ABS controller can make braking decisions and issue commands in a panic stop.

A **Hall-Effect switch** is a type of pulse generator signal that either measures the mechanical speed of a device or the position of certain components. Hall-Effect switches have been used for years as crankshaft and camshaft position sensors. These two sensors signal the PCM on how fast the shafts are rotating and when to command the fuel injectors on/off and when to fire the

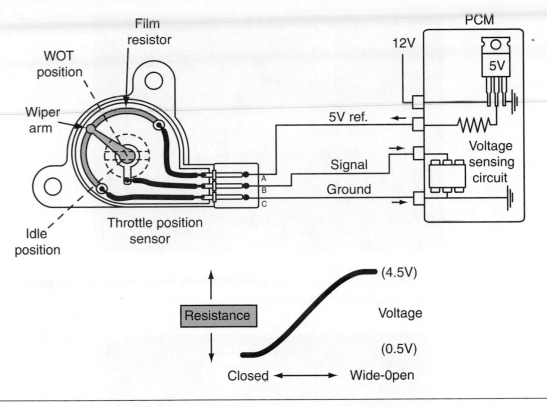

Figure 8-53 The TPS works by measuring the voltage at various points on a resistor wire.

spark plugs. A new use since 2003 models is the replacement of AC or PM sensors at the wheels with Hall-Effect switch sensors. The main reason for this is the on/off type of signal generated by the Hall-Effect switch as opposed to the constantly changing signal generated by AC generators. The Hall-Effect switch is also grounded when it is mounted. Computers respond better to a simple on and off signal. A following section deals with types of signals.

AUTHOR'S NOTE: Cut a soda can in half and recycle the top half. Punch a hole in the center bottom sized to fit snuggly over a shaft. Cut four or more equally spaced slots in the sides of the can. Put some sort of toothed gear on the shaft and slide the can over the shaft so it covers the gear. Mount a magnet near, about 0.015 inch, the outside of the can aligned with the gear and you have the makings of a Hall-Effect switch.

Other sensors change temperature or pressure into electrical signals. The engine coolant temperature (ECT) sensor changes temperature into an electrical signal while the transmission fluid pressure (TFP) sensor transducer changes pressure into an electrical signal. Other sensors measure fuel pressure, refrigerant pressure in the air conditioning, engine oil temperature, and engine oil pressure, and others will be added to future vehicles.

The **manifold absolute pressure (MAP)** sensor measures the vacuum or absolute pressure in the engine's intake and converts that to an electrical signal (Figure 8-54). This signal informs the PCM on how hard the engine is working (engine load). As the throttle is open initially, air rushes into the intake and the pressure increases to near atmospheric. The computer interprets this signal as a high engine load and holds the fuel injectors open longer and retards the spark timing. As the vehicle speeds up and the engine load decreases, the intake pressure drops and the computer changes the injector on-time and may even cause the transmission to upshift. A **mass air flow (MAF)** sensor measures the volume of the air leaving the air filter enroute to the intake (Figure 8-55). This is another indication of engine load. The MAF sensor may also be designed to measure the temperature and density of the incoming air. This makes the MAF more sensitive to engine load and the type of air entering the engine (i.e., the colder the air, the higher

Figure 8-54 The MAP converts pressure measurements in the intake manifold and generates an electrical signal. This signal is a measurement of engine load.

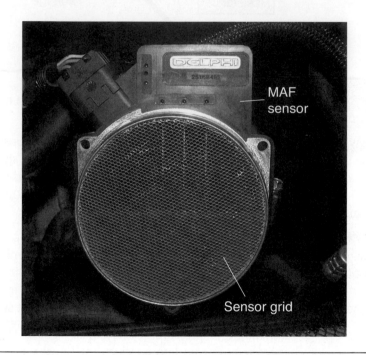

Figure 8-55 The MAF is mounted in the air intake duct between air filter output and the throttle valve.

its density, hence more oxygen is available for thermodynamics). The PCM can now give more precise commands to the fuel and ignition systems.

Switches are probably the most common sensor on a vehicle. Switches may be operated by the driver or electronically controlled. Generally speaking, a switch gives an on or off signal to the PCM and/or other computer modules. In some instances, this signal does nothing but tell the computer what the operator wants. When the air conditioning button is pressed on, a signal is sent to a pin on the PCM labeled "AC Request." The air conditioning will not switch on until the PCM looks at certain other signals, like the ECT. If all meet the required parameters, the compressor is energized. If one of the required signals is missing or out of limits, then there is no air conditioning, no matter how hot the vehicle's interior gets. This circuit can also switch off the air conditioning if engine temperature is too high.

Figure 8-56 A fuel injector.

There are many sensors on today's vehicles. A few were discussed here, but all sensors fall into one of the categories mentioned.

Actuators

Actuators are electrical devices that do the actual work commanded by the PCM or other computer. They range from fuel injectors to cooling fans. Most are **solenoids** or electric motors. A solenoid is an electrical device that changes electrical energy into mechanical energy. Probably the most common is the fuel injector (Figure 8-56). On command from the PCM the injector coil is energized, creating a magnetic field. This field acts on a metal, moveable core that is part of the injector pindle. The core's movement opens the injector, allowing pressurized fuel to enter the airflow. When the injector is switched off, a spring forces the core to move backwards, shutting off the fuel. The second most common actuators on most modern vehicles are the shift solenoids in the transmission. They control fluid flow to and from transmission shift valves. Other solenoids may be used to operate power door locks or the blend doors on the climate control systems.

Electric motors are used everywhere on a vehicle. They can be used to operate the throttle valve, power windows and locks, seats, blower and cooling fans, air conditioning blend doors, and the starter motor. They range from the heavier motors such as those for power seats and fans to small, palm-size ones used as blend door actuators, and operate based on commands from their controlling computers.

Signals and Commands

Signals, in this discussion, are those electrical currents generated by a sensor and transmitted to a computer. The signal may be **analog** or **digital.** Analog signals would be like those generated by an AC (PM) generator where the voltage goes from zero to a positive voltage back to zero then to a negative voltage (Figure 8-57). This is a signal that a computer has a problem interpreting. Usually an analog signal is sent to an AC/DC converter where it is changed into a digital signal. Digital signals are usually interpreted as on or off (Figure 8-58). Based on the computer's programming, the "on" signal may be considered as "high" or "yes." The "off" signal can be considered as "low" or "no." The signal is used for the computer to make decisions and issue commands.

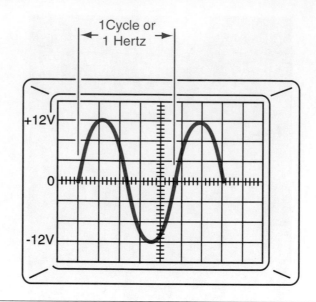

Figure 8-57 Note that an analog signal moves from positive to negative voltage. This is known as a sine waveform.

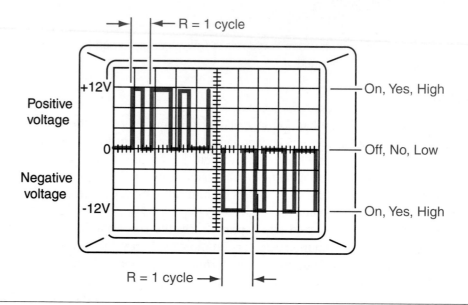

Figure 8-58 Digital signals can be positive or negative voltage. The ones shown here are listed as a rectangular waveform. The differences between the widths indicate the amount of time the signal was in the on (or yes, high) positon.

AUTHOR'S NOTE: Computer decisions are not like those produced by the human brain. A computer makes a decision by comparing input signals to installed programming and issues commands according to the program instructions. If something is missing, then the computer does not even try to figure out what is wrong. It just quits and turns on the check engine light mounted in the instrument panel.

Commands are simply electrical signals sent by a computer to an actuator to do something. A simple command may only ground a relay coil to close a switch that turns on the cooling fan, or it may switch the fuel injector on. Usually the commands constitute an on/off signal to the actuator and can be found on the negative side of the circuit.

Computers

Computers are electrical devices that use electronic circuits to make programmed decisions and issue commands based on some input. Computers can only be as smart as their human programmers. A computer receives a signal, compares it to internal instructions, selects something very specific that needs to be performed, issues an electrical command, and then sits back and sees what happens. One of the first things to check when diagnosing a computer is to first determine if it is getting signals. If all of the appropriate signals are present and correct but there is no output command, then, simply put, the computer is bad. Automotive technicians do not repair computers; they only replace them.

Automotive computer systems are here to stay. They will become more powerful and associated with more mechanical components and systems. Already the mechanical linkages for the throttle and the transmission shift valves have been eliminated on some drive-by-wire vehicles. Some steering systems use electrical motors to drive the steering gear instead of being directly driven by the steering wheel. Many of these computer systems come from military development and research into electric-powered vehicles. When diagnosing a computer system or one of its sensors or actuators, just consider it as a very small headlight and collect the high labor salary top technicians receive.

Summary

❏ The major systems of a vehicle are: electrical, engine, driveline, steering, suspension, climate control, frame, and body.

❏ The two-stroke engine produces a power stroke for each revolution of the crankshaft.

❏ Air and fuel are drawn in during the compression stroke of the two-stroke engine.

❏ The rotary engine does not have the stop-go movement of a piston engine.

❏ The four-stroke engine completes intake, compression, power, and exhaust strokes in one cycle.

❏ Valves are used to open and close ports in the cylinder head.

❏ Valvetrains are either OHV or OHC.

❏ OHC engines may have two camshafts per cylinder head.

❏ The lubrication system cools, lubricates, and cleans the internal components of the engine.

❏ Synthetic lubricants and blends are made from chemicals rather than natural petroleum products.

❏ The starting system provides power to rotate the crankshaft, which draws air and fuel into the cylinders.

❏ The charging system charges the battery and supplies the electrical current to operate the vehicle.

❏ Fuel is delivered to the intake manifold by a mechanical or electrical pump.

❏ EFI systems require an electrical fuel pump.

Terms to Know

Actuator
Analog
Block
Block configuration
Boost
Bore
Bypass valve
Carbon Monoxide (CO)
Catalyst action
Cavitation
Check valve
Clearance
Clutch switch
Coil-near-plug
Coil-on-plug
Coil pack
Commands
Digital
Direct injection
Displacement
Distributor
Dual Overhead Cam (DOHC)
EVAP system

❑ Air is drawn into the intake manifold by engine vacuum and atmospheric pressure.

❑ The ignition system supplies the electrical spark to ignite the compressed air/fuel mixture.

❑ The ignition system must be timed to deliver the spark at the proper time in piston travel.

❑ The cooling system removes the heat from the engine block and cylinder head.

❑ In order to get fresh air/fuel into the combustion chamber, an exhaust system must quickly route the exhaust gases from the chamber.

❑ The catalytic converter reduces CO, HC, and NO$_x$ emissions.

❑ The EGR helps reduce NO$_x$ emissions.

❑ HC emissions are reduced using the PCV and EVAP systems.

❑ The secondary air injection system provides oxygen to assist the catalytic converter in reducing emissions.

❑ TAC systems warm the intake air during cold starts and cold driveaway.

❑ Computer systems monitor and control many mechanical devices on modern vehicles.

❑ Sensors provide input signal to the computer.

❑ Actuators are commanded by a computer to perform a specific action, usually mechanical.

Review Questions

Short Answer Essay

1. List the four strokes of an internal combustion engine, and explain the actions that happen on each stroke.

2. List the major systems of an automobile.

3. Explain how the lubrication system works.

4. List and describe the actions of the cooling system and its components.

5. Explain the purpose of the exhaust system.

6. Explain the elements and action of a catalytic converter.

7. Describe the purpose of the sensors.

8. Discuss the purpose of an actuator.

9. List and explain the difference between the two types of signals used in automotive computer systems.

10. Explain the common purpose of a MAP and a MAF.

Fill-in-the-Blanks

1. The EGR system is designed to help reduce _____ emissions.

2. The ideal exhaust emission would be _____ and _____.

3. On the compression stroke, the valves are _____.

4. The exhaust system quiets engine noise and helps _____ _____.

5. The ignition system may have one coil per spark plug or one coil for _____ _____ _____.

6. A 13:1 air/fuel ratio is a(n) _____ mixture.

7. EFI systems can change the amount of fuel injected by varying the _____ the injector is open.

8. The secondary air injection system supplies air to promote the _____ action of the catalytic converter.

9. The cooling fan is used to draw air over the _____ _____.

10. The _____ stroke increases the pressure on the air/fuel in the cylinder.

ASE-Style Review Questions

1. The ignition system is being discussed.
   Technician A says a distributor is used on all engines.
   Technician B says the PCM calculates the timing of the spark.
   Who is correct?
   A. A only
   B. B only
   C. Both A and B
   D. Neither A nor B

2. *Technician A* says it is illegal to remove the muffler from the exhaust in all localities.
   Technician B says the catalytic converter was installed in 1974 cars.
   Who is correct?
   A. A only
   B. B only
   C. Both A and B
   D. Neither A nor B

3. Emission controls are being discussed.
   Technician A says the TAC captures fuel tank vapors.
   Technician B says the EVAP is used to route oil sump vapors to the intake manifold.
   Who is correct?
   A. A only
   B. B only
   C. Both A and B
   D. Neither A nor B

4. *Technician A* says the secondary air injection uses a pump similar to the one used in the cooling system.
   Technician B says the EGR system helps reduce NO_x pollutants.
   Who is correct?
   A. A only
   B. B only
   C. Both A and B
   D. Neither A nor B

5. The lubrication system is being discussed.
   Technician A says the camshaft may drive the oil pump.
   Technician B says the pressure relief valve is used to bypass oil around the oil filter.
   Who is correct?
   A. A only
   B. B only
   C. Both A and B
   D. Neither A nor B

6. *Technician A* says the battery can operate the vehicle's electrical systems for a short time.
   Technician B says the starter motor draws current from the AC generator.
   Who is correct?
   A. A only
   B. B only
   C. Both A and B
   D. Neither A nor B

7. *Technician A* says every carburetor requires an electrical fuel pump to maintain the fuel level in the carburetor bowl.
   Technician B says the camshaft drives the mechanical fuel pump.
   Who is correct?
   A. A only
   B. B only
   C. Both A and B
   D. Neither A nor B

8. The engine's theory of operation is being discussed.
   Technician A says valve overlap happens when the valves are not timed correctly.
   Technician B says an inference engine may have serious damage if the timing belt breaks.
   Who is correct?
   A. A only
   B. B only
   C. Both A and B
   D. Neither A nor B

9. *Technician A* says the air/fuel mixture enters the cylinder on the compression stroke on a two-stroke engine.
Technician B says oil must be mixed with the fuel on a four-stroke engine.
Who is correct?
A. A only   **C.** Both A and B
B. B only   **D.** Neither A nor B

10. Rotary engines are being discussed.
Technician A says a rotary can produce more power because the piston does not have to start or stop.
Technician B says the combustion chamber is round.
Who is correct?
A. A only   **C.** Both A and B
B. B only   **D.** Neither A nor B

Brakes

Upon completion and review of this chapter, you should be able to:

❏ Discuss the application of Pascal's Law in brake systems.

❏ Explain the purpose and operation of the brake system.

❏ Discuss the use of power boosters.

❏ Explain the purpose and operation of master cylinders.

❏ Discuss and explain the use of control valves.

❏ Explain the operation of drum brakes.

❏ Explain the operation of disc brakes.

❏ Explain the purpose and operation of the parking brake.

❏ Discuss the purpose and general operation of antilock brake systems.

❏ Explain the basic operation of traction control systems.

Introduction

An object in motion tends to stay in motion. Obviously, our vehicles cannot always be in motion. They must be slowed and stopped at times. The brake system supplies the necessary force for the driver to slow or stop the vehicle. The system is built on the application of hydraulics theory.

A BIT OF HISTORY

The earliest automobiles used a mechanical linkage to force a block of wood, leather, or metal against the wheel for brakes. This is still used on some horse-drawn wagons.

Hydraulic Theory and Pascal's Law

The automobiles produced in the early 1900s were stopped by using a mechanical linkage attached to what we now call drum brakes. The linkage required constant adjustment. The most that could be said for those brakes was that the top speed in those days averaged about five miles per hour. With heavier vehicles and higher speeds, mechanical linkage was not sufficient. Hydraulic brakes were first used in the 1920s.

Hydraulic Theory Review

As discussed in Chapter 7, "Automobile Theories of Operation," liquids cannot be compressed and make ideal transmitting agents when contained in sealed systems. In this manner, any force applied at one end would result in an equal force being applied at the other end. The force is transmitted by pressure buildup of the fluid within the system. The fluid's pressure can also be used to increase or decrease the forces at work.

Pascal's Law states that pressure applied to fluid in a sealed system is transmitted equally in all directions and to all parts of the system. Pascal also developed a mathematical formula to calculate the use of pressure to transmit force. The formula is force (F) equals pressure (P) multiplied by area (A) (Figure 9-1). Force is measured in foot-pounds (ft.-lbs) and pressure is measured in pounds per square inch (psi). Area is in square inches (sq. in.).

Shop Manual
pages 237–239

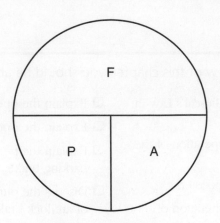

Figure 9-1 Force (ft.-lb) is equal to pressure (psi) times area (sq. in.).

The use of Pascal's formula is similar to the one used when applying Ohm's Law in Chapter 7, "Automobile Theories of Operation." A major mistake during the first usage of Pascal's Law was mixing force and pressure during the calculations. An example of the correct formula is:

F = PA where P is 10 psi and A is 2 sq. in.
F = 20 ft.-lbs

In this case, a 20-ft.-lb force results when 10 psi are pressing a 2-sq.-in. piston. Another method of the same formula is P = F/A.

P = F/A where F is 15 ft.-lbs and A is 3 sq. in.
P = 5 psi

The formula can also be used to calculate **mechanical advantage.** This is similar to gear ratios where a small force can be increased to a higher force within the same system. In Figure 9-2,

A **mechanical advantage** is the amount of force increased at the output compared to the input force.

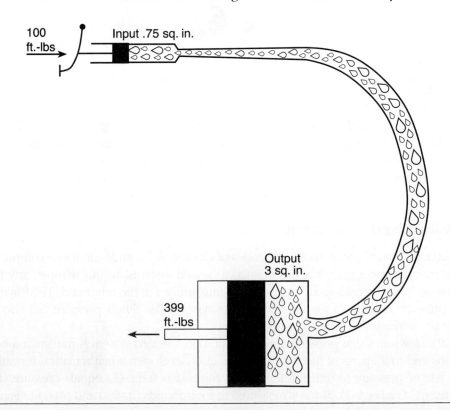

Figure 9-2 An output piston that is larger than the input piston can increase the input force.

the .75-sq.-in. input piston is being applied with a force of 100 ft.-lbs. Using the formula, the pressure within the system is 133 psi (P = 100/.75). Since the pressure is the same everywhere within the system, 133 psi are pressing on each square inch of the 3-sq.-in. output piston. The force of the output piston is calculated by F = 133 by 3 or 399 ft.-lbs. This gives a mechanical advantage of 3.99 to 1 (399/100). Another method of calculating mechanical advantage is to divide the output by the input or 3/.75 = 4:1.

Brake System Operation

Shop Manual
pages 239–246

The automotive brake system uses hydraulic theory to transfer the driver's input force throughout the system to an output force that applies the mechanical brakes. The components of a basic brake system include the pedal and linkage, the **master cylinder,** lines and hoses, controlling valves, and either drum or disc brakes at the wheels. Most systems also use a power boost to assist the driver's efforts in braking. The tires are not considered to be part of the brake system even though they provide the contact between the applied braking force and the road. If the tire is slick with little or no tread, or the road surface is slick, there may be no slowing of the vehicle even when brakes are forcibly applied. An inspection of the tires' tread should be made when the vehicle is receiving brake repair or inspection.

The **master cylinder** converts the driver's mechanical input force to hydraulic pressure for the brake system.

Braking

As the driver presses downward on the brake pedal, the movement of the pedal is transmitted to the piston in the master cylinder (Figure 9-3). The piston pressurizes the brake fluid in the bore and attempts to force it through the brake lines. Since this is a sealed system the pressurized fluid transmits the applied driver's force to the output pistons at the wheels. The pistons may be in a **wheel cylinder** or disc **caliper.** The pistons act against the drum brake shoes or disc brake pads, respectively.

Wheel cylinders work with drum brakes. **Calipers** are used with disc brakes.

Releasing

When the brake pedal is released, return springs at the pedal, inside the master cylinder, and at the drum brake shoes retract each of the brake components to the rest (release) position. Within

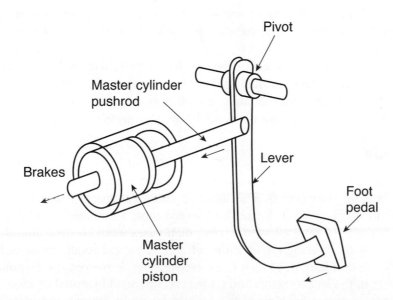

Figure 9-3 The pressure generated in the master cylinder transmits the applied force to the wheel brakes.

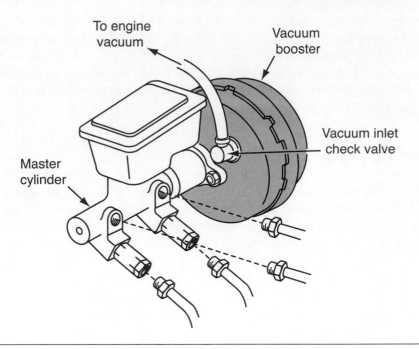

Figure 9-4 The booster check valve is mounted on the front side of the booster where the vacuum line from the engine is connected.

the disc brake caliper, a distorted square-cut O-ring straightens when the brake pedal is released. As the ring straightens or returns to its normal shape, the caliper piston is pulled backward slightly. This reduces the contact of the pads with the rotor.

Power Boost

Most vehicles use some type of booster on the brakes to reduce the driver's effort. The most common type used on passenger cars and light trucks is the vacuum chamber (Figure 9-4). The engine supplies the vacuum. A diesel engine requires the use of a vacuum pump to provide the vacuum. Movement of the brake pedal opens an air valve that allows atmospheric pressure to enter one side of the chamber. This extra force acting on the booster interior **diaphragm** increases the force exerted by the driver. Some vehicles use hydraulic systems to boost the driver's braking input.

A dual brake system is used on almost every vehicle now sold in the United States. The brake system has two hydraulic circuits, including a dual chamber master cylinder. In this manner, a leak or other failure on one system will not affect the other system. It will take additional space and effort to stop the vehicle, but the driver will still be able to maintain control.

> The **diaphragm** is a large, flexible disc stretched across the interior of the vacuum brake booster to divide it into two chambers.

Brake Fluids

Brake fluid acts as the transfer agent to transmit the driver's braking input to mechanical force at the wheels' braking components. The different classifications of brake fluid are generally based on the fluid's boiling point. Brake fluid should have a minimum boiling point of 400°F (204°C). If the brake fluid were allowed to boil, then gases would form within the brake lines and hoses, reducing the braking effect at the wheels. A second requirement of brake fluid is that it should be noncorrosive or at least corrosive-resistant. However, the manufacturers of all light vehicles sold in the United States and Canada recommend biannual or mileage-based service including the complete flushing of the brake's hydraulic system and installation of new fluid. Even the best brake fluids will deteriorate over a period of time or usage and this will affect the sealants and metal components of the brake system. A third problem with older fluids

TABLE 9-1 MINIMUM BOILING POINTS FOR THE SIX BRAKE FLUID GRADES

Boiling Point	DOT 3	DOT 4	DOT 5	DOT 3/4	DOT 5.1
Dry	401°F (205°C)	446°F (230°C)	500°F (260°C)	500°F (260°C)	585°F (307°C)
Wet	284°F (140°C)	311°F (155°C)	356°F (180°C)	347°F (175°C)	365°F (185°C)
DOT 5.1 long-life has the same dry and wet boiling point of 424°F (218°C).					

is the fact that they are **hygroscopic,** which means they absorb moisture from the air very easily. Newer synthetic fluids have almost eliminated the moisture-absorbency problem, but they are not commonly used in new vehicles or as replacement for old fluids. Other qualities of brake fluids include:

Free flowing at high and low temperature
Low freezing point
Lubricate metal and flexible parts

The two most commonly used classifications of brake fluids are DOT 3 and DOT 4. Both are polyglycol compounds, but DOT 4 has a higher boiling point than DOT 3, as outlined in Table 9-1. The two fluids can be mixed and can replace each other within a system. It is suggested that if a DOT 3 brake system needs to be flushed, that is a good time to install DOT 4 fluid rather than just mixing them together. However, do not replace a higher-rated brake fluid with a lower one (e.g., DOT 3 in place of DOT 4). Three synthetic fluids are available. Each has a boiling point over 500°F (260°C) and can be mixed or can replace DOT 3 or DOT 4. They are classified as DOT 3/4, DOT 5.1, and DOT 5.1 long-life. The synthetics last longer, absorb less moisture, and are in theory better for braking. DOT 5.1 has a life of about 60 months and DOT 5.1 long-life is supposed to be good for about 10 years. DOT 3/4 has a life of about 24 months.

> **DOT** is the United States Department of Transportation.

One classification of fluid, DOT 5, is a silicone-based fluid and is used in very few vehicles. It should never be mixed with any other classification of brake fluids nor should it be used as a replacement for any of them. DOT 5 will damage the metal and flexible components of a typical brake system. Antilock brake systems will not work with DOT 5 fluids.

Brake System Apply Components and Operation

Each component of the brake is essential to the proper operation of the complete system. A failure of one component will affect the entire system in some manner, ranging from a substantially reduced braking action to complete system failure.

Shop Manual
pages 244–250

Pedals and Linkages

The pedal is connected to the master cylinder's piston with a **pushrod** (Figure 9-5). If the system uses a power booster, the pushrod is connected to the booster's diaphragm or valve to operate the assist. On most power boost systems, the pedal's pushrod does not act directly on the master cylinder piston.

On some vehicles, the pushrod must be adjusted for length when the master cylinder, booster, or pedal is replaced. Better machining techniques have eliminated this adjustment requirement on many vehicles; however, some manufacturers still require an inspection of this mechanical connection when new components are installed.

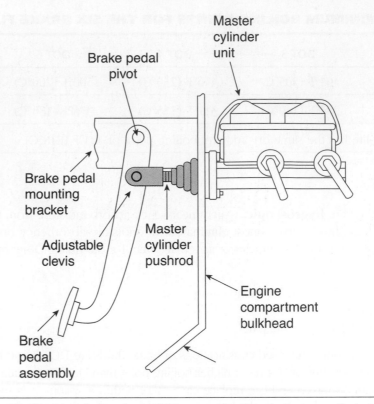

Figure 9-5 The force applied at the pedal is transmitted to the master cylinder's primary piston by the pushrod.

The mounting for the pedal is generally a shaft from which the pedal hangs downward. A return spring is used to return the pedal to the fully up or off position after the brakes are released. Also mounted on or near the pedal shaft is the brake (stop) light switch (Figure 9-6). The switch is normally open. When the brake is applied, the pedal's shank or the pushrod presses (releases) a plunger on the switch and closes the brake light circuit. At times, the position of the switch may have to be adjusted to work properly.

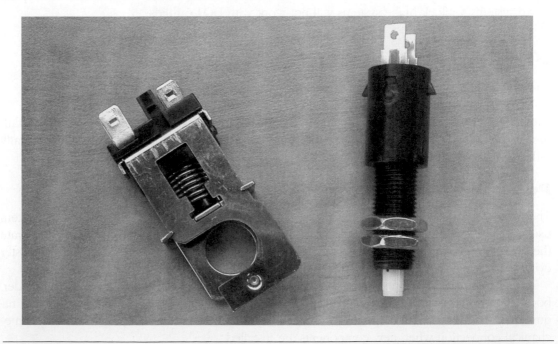

Figure 9-6 Typical switches for the stop light only. Newer versions may have additional connections.

Modern vehicles have a multiple function brake light switch. The switch may be used to activate the **antilock brakes (ABS),** switch off the cruise control, and signal the PCM that the brakes are applied.

Power Boosters and Master Cylinders

The master cylinder is the point at which mechanical force is changed to hydraulic pressure. The booster is placed between the brake pedal and master cylinder. It increases the driver's ability to apply force to the master cylinder piston.

The brake's power booster may be vacuum or hydraulic operated. The **vacuum brake booster** is connected to the engine's intake manifold or vacuum pump. The operating engine draws air from the vacuum chamber and maintains a vacuum in both sides of the chamber. A check valve, normally located at the connection of the hose and chamber, maintains the vacuum during engine shutdown or hard acceleration. A large diaphragm extends across the interior of the vacuum chamber. The brake's pushrod is connected to the driver's side or rear of the diaphragm. A power piston is mounted to the front side of the diaphragm and extends forward to contact the master cylinder's primary piston (Figure 9-7). The power piston houses vacuum and air valves. During brake release, vacuum is supplied through the vacuum valve to the rear chamber of the diaphragm. This helps keep the diaphragm in the release position.

As the brake pedal is depressed (brake apply), the air valve is opened, thereby allowing atmospheric pressure to enter the rear chamber. At the same time, the vacuum valve closes the vacuum port to the rear chamber. The controlled opening of the air valve regulates the amount of pressure that can actually be applied to the diaphragm. This prevents full boost and brake lockup during light braking. Since a vacuum is still applied to the front chamber, the diaphragm and

Antilock brake systems (ABS) help increase driver control by reducing wheel lockup during panic stops.

A diesel engine requires a vacuum pump for the **vacuum brake booster** and other vacuum-operated devices.

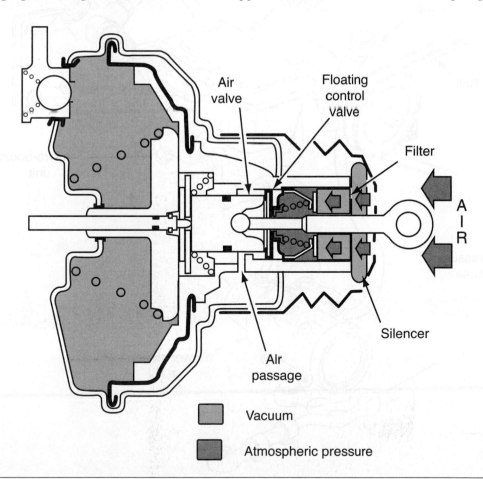

Figure 9-7 The volume of air entering the booster is controlled by the linear movement of the pedal's pushrod.

power piston move forward, thereby increasing the driver's input force. The master cylinder piston is moved forward within its bore and a hydraulic pressure is created in the brake system. When the brakes are released, the rear vacuum valve is opened, the air valve is closed, and the applied vacuum and a return spring force the diaphragm to the center or off position.

Hydraulic boost is accomplished by directing a fluid under high pressure to the driver's side of a master cylinder. A hydro-boost system uses the power steering system to supply the pressurized fluid. Fluid from the power steering pump is routed through a valve system fitted between the brake pedal's pushrod and master cylinder (Figure 9-8). The valve assembly is part of the master cylinder, but may be replaced individually on most units. As the pedal is depressed, the pushrod opens a valve that directs pressurized fluid to a power piston on the driver's side of the master cylinder. The valve's movement is proportional to the movement of the pedal. The further the pedal moves downward, the more the valve opens. The pressurized fluid forces a power piston and pushrod to move the master cylinder piston. This, in turn, pressurizes the brake fluid and the brakes are applied.

When the brakes are released, a release valve is opened to allow the pressurized fluid to flow directly back into the power steering system. A small accumulator is filled with pressurized fluid in the event of a power steering failure. If there is not enough pressure within the boost system when the brakes are applied, the accumulator fluid is directed to the power piston and offers enough brake boost for two or three stops.

The General Motors **PowerMaster** system uses a special master cylinder. A small, electric-motor-driven pump supplies pressurized fluid to the power piston. The motor and pump are

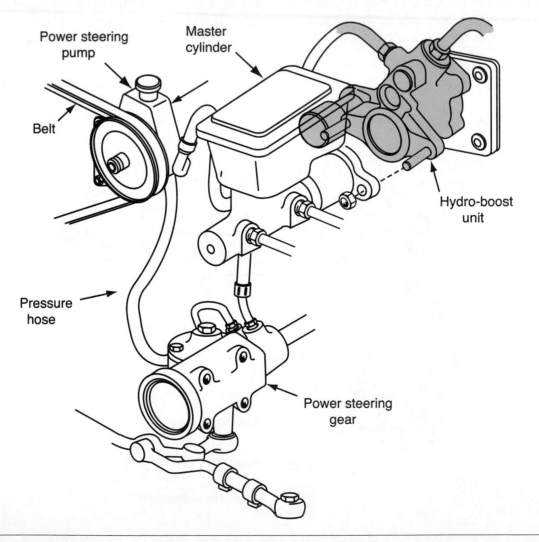

Figure 9-8 Layout of a typical hydro-boost power brake booster.

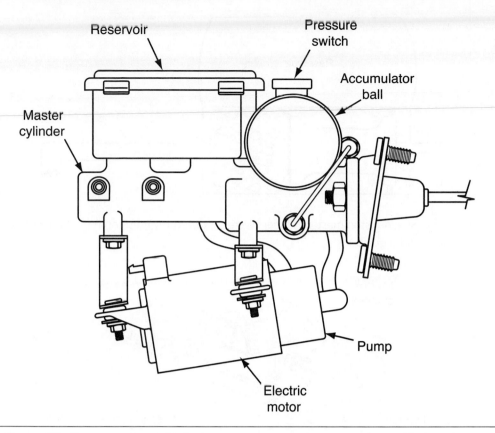

Figure 9-9 The PowerMaster brake booster's motor and pump are located below the master cylinder.

mounted under the master cylinder (Figure 9-9). Brake fluid is routed from the cylinder's reservoir to the pump. The fluid is pumped into an accumulator. Pressurized nitrogen gas on one side of the accumulator's diaphragm pressurizes the fluid. When the brakes are applied, a valve is opened between the accumulator and power piston bore at the driver's end of the master cylinder. The movement of the valve is proportional to the downward movement of the pedal. The fluid acts against the power piston and assists the driver in applying the brake. When the brakes are released, a release valve is opened to allow the boost fluid to flow back to the reservoir. A pressure switch on the accumulator closes when the pressure drops below a set limit and turns the pump on. In this manner, pressurized fluid is always available for brake operation. As with the hydro-boost system, the accumulator stores enough fluid for two or three stops in case of pump or motor failure.

Master Cylinders

Single-piston master cylinders were installed on the original hydraulic brake systems (Figure 9-10). A single-piston master cylinder fed a single brake system. If there was a failure in any component, the complete system failed. The operation of the single-piston master cylinder works in the same manner as a dual-piston except for the extra components used in a dual system.

Dual or split brake systems have been required since 1967. This setup allows half of the brake system to be available to control the vehicle if there is a leak or some other failure. The most common system is a front/rear split where one master cylinder piston feeds the brakes on one axle. A **diagonally split system** has a front and a rear wheel fed by one master cylinder piston (Figure 9-11).

The split system requires a dual-piston master cylinder. The master cylinder is mounted on the forward side of the brake booster or directly on the firewall. The basic components of a dual master are the primary and secondary pistons, return springs, body, and reservoir.

Some master cylinders and reservoirs are molded into one unit, but most reservoirs are plastic and can be dismounted from the master cylinder. The reservoir is mounted to the top of the

The **diagonally split system** allows the driver to have braking power at each axle if a leak develops in the brake system.

241

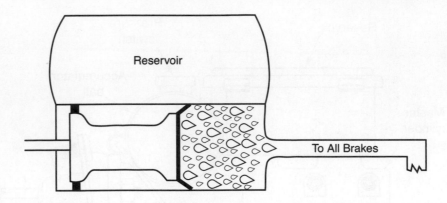

Figure 9-10 Single-piston master cylinders were outlawed from use on motor vehicles by federal laws and regulations.

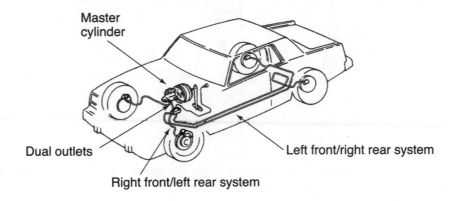

Figure 9-11 The diagonally split system requires a separate brake line from the master cylinder or combination valve to each wheel.

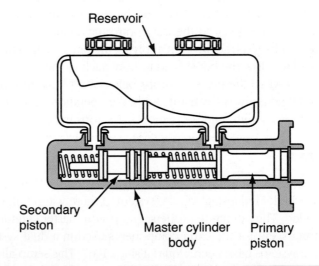

Figure 9-12 The primary piston operates the rear and applies hydraulic or mechanical force against the secondary piston.

master cylinder. There are two chambers in the reservoir that separately feed the two bores of the master cylinder (Figure 9-12). The reservoir cover can be quickly removed to add fluid. Most reservoirs are translucent so the fluid level can be checked without opening the system. Some reservoirs have a fluid level sensor that activates a dash light to warn the driver of low fluid levels.

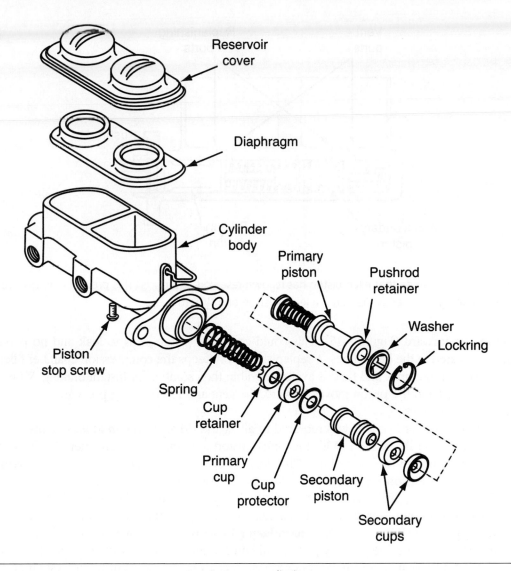

Figure 9-13 Internal components of a master cylinder.

The master cylinder body houses the internal components and is drilled to accept the reservoir and brake lines. The threaded ports on the side of the body are connections for the brake lines and may have a metering orifice installed. There are also vent and replenishing ports drilled into the body that open into the bore and the reservoir. This allows fluid to flow to and from the bore as the pistons move.

The open end of the body's main bore is installed toward the firewall. Pistons and springs are slid into the bore and a locking ring is installed for retention (Figure 9-13). Some older master cylinders may have bleeder ports at the forward or closed end of the master cylinder, but most master cylinders are bled using the line ports.

The two pistons are the primary and secondary pistons (Figure 9-14). The primary piston applies the rear brakes and the secondary piston. As the brake pedal is depressed, the primary piston cup forces the fluid in the compression chamber to move into the brake line and against the secondary piston. As the pressure builds in the chamber and brake lines, the secondary piston moves, starts to pressurize the fluid in its compression chamber, and begins to apply the front brakes. If the vehicle is disc on front and drum on the rear, the initial movement of the primary piston moves the shoes out toward the drums. Return springs push the two pistons back to their release positions when the driver releases the pedal.

As the pistons move to their release positions, fluid in the spool area behind the piston head vents back into the reservoir. Fluid behind the head is needed to remove the pressure drop as the

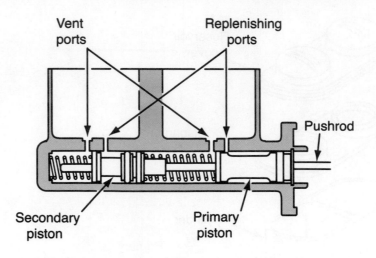

Figure 9-14 Note that each piston has its own reservoir and ports. This prevents the reservoir from being completely drained if a leak exists in one system.

piston moves forward (Figure 9-15). This could cause the piston seals to leak and no pressure would be created in the brake lines. A replenishing port keeps the compression chamber filled so there is no air in the lines and fluid is available within the chamber for instant braking. When the brake is applied and the piston moves forward, the vent and replenishing ports are closed to the compression chamber.

The dual master cylinder was required to allow the driver to retain at least some braking if there is a leak in the system. With the single-piston system, a leak anywhere in the system may cause a complete loss of braking. The dual system reduces that possibility. The **primary piston** has an extension screw that extends into the compression chamber that can apply the secondary piston. If a leak occurs in the rear brake system and the brake pedal is depressed, the primary piston moves in its normal manner. However, the fluid ahead of that piston leaks out and no pressure is created. The **secondary piston** will not move because of the lack of pressure. The screw on the primary piston will contact and push the secondary piston forward (Figure 9-16). In this manner, the secondary piston can charge the front brake system and the driver will have some braking action available. If a leak occurs in the front system and the

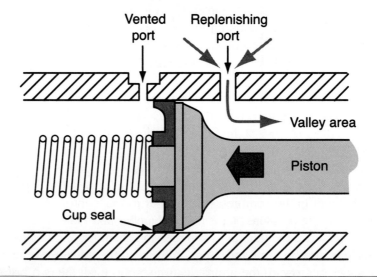

Figure 9-15 The replenishing port keeps the chamber behind the piston head full as the piston moves forward.

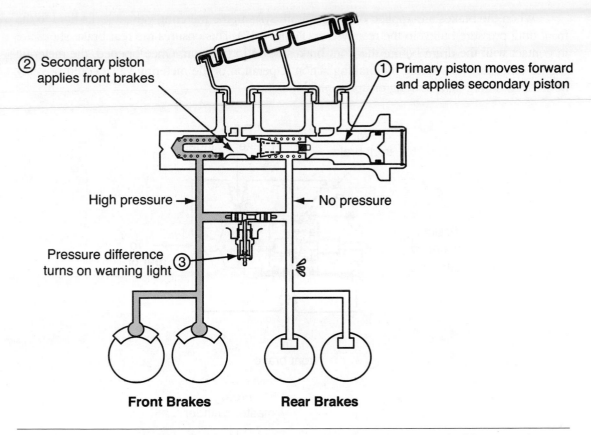

② Secondary piston applies front brakes

① Primary piston moves forward and applies secondary piston

High pressure →   ← No pressure

Pressure difference turns on warning light ③

Front Brakes   **Rear Brakes**

Figure 9-16 With a leak in the primary circuit, the primary piston cannot build pressure but is pushed forward by the pedal until it mechanically connects and applies force to the secondary piston.

brakes are applied, the primary piston charges the rear brake system, but since the front circuit is leaking, the secondary piston must be pushed completely forward to its stop before pressure can build in the rear brakes. Both conditions will cause the brake pedal to drop drastically as the brakes are being applied.

Control Valves, Lines and Hoses, and Warning Lights

There are not too many valves in the automotive brake system, but a failure of one may cause a complete loss of brakes. Steel lines carry the brake fluid to the wheels and usually do not require any maintenance. Flexible hoses connect the steel lines to the individual wheels. This allows the wheels to move but does not bend or break solid lines. The hoses have to be inspected and replaced at intervals.

Metering Valves

Vehicles with front disc and rear drum brakes have to hold the front brakes off until the rear brakes move enough to contact the drums. The **metering valve** accomplishes this task. As the brakes are applied, the disc brake reacts instantly, but the rear brake must move far enough to contact the drum. In addition, enough pressure must be created to overrun the return spring on the shoes. If the front brakes were applied at the instance of brake pedal movement, they would probably lock down before the rear ones would begin to brake.

Shop Manual
pages 252–256, 269

When the brakes are applied on a disc/drum system, the metering valve blocks fluid to the front until pressure builds in the rear system (Figure 9-17). This ensures the rear brake shoes are in contact with the drum before the front brakes begin to function. Once opened, the metering has no other function during the braking action. Operation of the metering valve could possibly be noticed during light braking or upon initial braking.

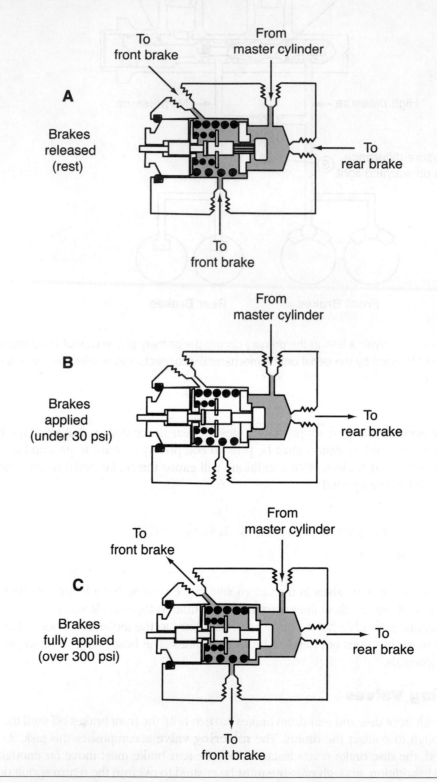

Figure 9-17 Under 30 psi (B) the valve to the front brakes is closed while allowing fluid to the rear drum brake. Over 30 psi, the larger valve is forced open, letting fluid to the front brakes.

Proportioning Valves

In order to control the vehicle during hard braking, it is necessary to apply more brake action at the leading axle than the trailing axle. Many brake systems may direct up to 90 percent of the braking action to the front wheels. During heavy braking, this could still cause the rear brakes to lock up. The proportioning valve divides the braking pressure between the front and rear axle to reduce or eliminate rear-wheel lockup.

The **proportioning valve** is basically a block of metal drilled to accept line fittings, a valve, a piston, and a spring. The lines connected to the valve are from the master cylinder and rear brakes. The piston has a small and a large end with the small end at the valve. A valve stem connects the valve and piston (Figure 9-18). During normal braking, the spring holds the valve open and fluid is directed equally to each axle. As the vehicle's weight shifts forward during braking, the front wheel's road friction increases while the rear contact decreases. High pressure from the master cylinder during a hard stop enters the proportioning valve and presses on the large end of the piston, thereby forcing it to move. The valve stem moves with the piston and moves the valve to restrict the inlet port. This reduces the amount of fluid and pressure that is applied to the rear wheels.

A **combination valve** is used on most vehicles. This combines the metering valve and proportioning valve into one unit (Figure 9-19). The operation of the two valves is the same as stand-alone units. The brake warning light switch may be mounted on the combination valve.

Load-Sensing Proportioning Valves

A truck is extremely light in the rear compared to the front. This is particularly important when the truck is being heavily braked. During braking, the truck's weight shifts to the front and the lightened rear wheels lose traction with the road. If the wheels lock up and skid, the truck can go out of control. A **load-sensing proportioning valve** helps prevent the lockup by reducing the rear brake pressure during braking.

One type load-sensing valve uses the movement of the frame and rear axle to regulate pressure to the rear brakes. The valve bracket is mounted to the frame with a lever attached to the axle (Figure 9-20). As the rear of the frame rises during braking, the axle pulls down the lever. The valve attached to the lever restricts the amount of fluid and pressure being directed to the rear wheels.

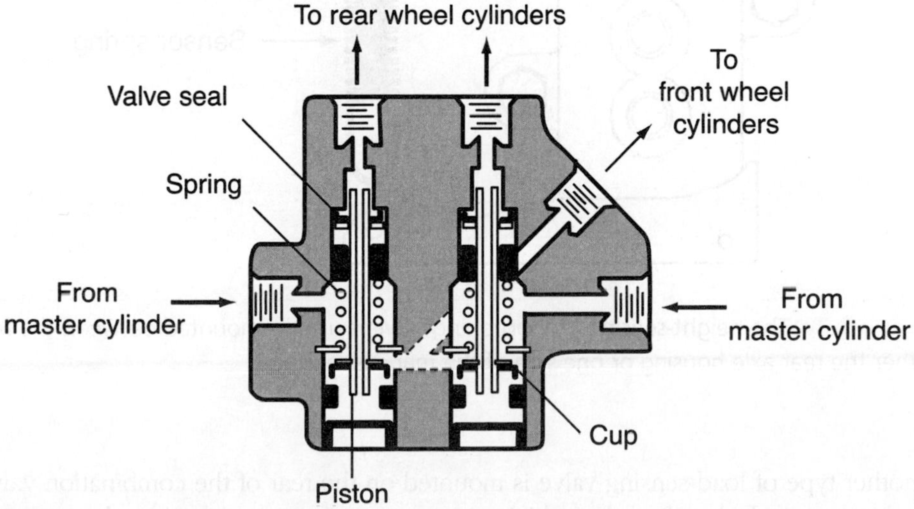

As the pressure from the master cylinder changes, the **proportioning valve** operation also changes, thereby increasing or reducing the pressure to the rear wheels.

Figure 9-18 A dual proportioning valve mounted on the frame away from the master cylinder.

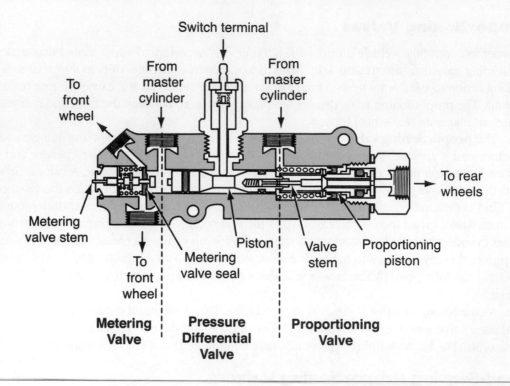

Figure 9-19 A combination or three-function valve.

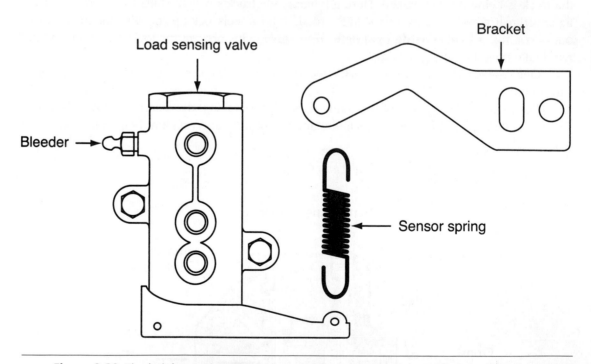

Figure 9-20 The height-sensing proportioning valve is usually mounted between the frame and either the rear axle housing or one side of the rear suspension.

Another type of load-sensing valve is mounted on the rear of the combination valve. It has a ball that moves with the tilt of the vehicle. As the rear of the vehicle rises during braking, the ball moves forward against a ball valve. The ball valve reduces the amount of fluid and pressure to the rear wheels.

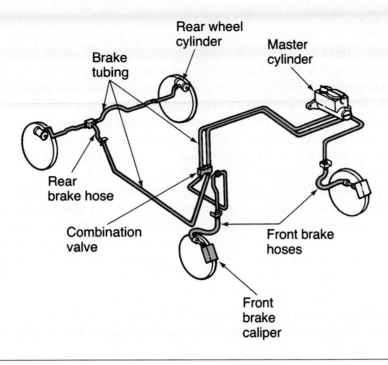

Figure 9-21 Brake fluid is contained in a sealed system consisting of metal lines and flexible hoses.

Lines and Hoses

Brake lines are made of steel with flared fittings at each end. The steel lines connect the master cylinder to the valves and the valves to the wheels. But the steel lines do not extend completely to the wheels. Flexible hoses connect the steel lines at the frame to the brake system at the wheels (Figure 9-21). This is required because of the almost constant movement of the wheel assembly during driving. A steel line would quickly break if it were continuously flexing. While steel lines almost never require maintenance, the flex hoses should be inspected each time the wheels are serviced. They will deteriorate and bulge due to chemical actions and age. They can also have interval damage that is not visible.

Brake lines have two types of fittings. The older type is known as a **double flare** fitting. The newer type is an **ISO**-type flare (Figure 9-22). Each requires a matching fitting and cannot be interchanged. The ISO-type provides a better seal.

ISO are the initials for the International Standards Organization.

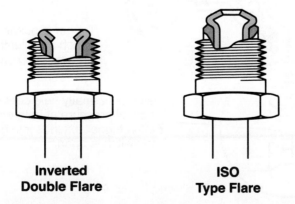

Inverted Double Flare

ISO Type Flare

Figure 9-22 The double flare fitting is being replaced by the ISO fitting. One cannot replace the other unless both sides of the connection are changed to one type of fitting.

Warning Lights

The **brake warning light** is used to alert the driver to a problem in one of the systems. There is usually one red warning light located in the instrument panel. It is turned on by any one of three possible switches or sensors. The first switch is on the parking brake and reminds the driver that the parking brake is applied. The second, a brake fluid level sensor, is not present on all vehicles, but if used, alerts the driver to low brake fluid level in the master cylinder reservoir. The third is used on dual brake systems to warn the driver of a brake failure in one of the two brake circuits. The switch is usually mounted on the proportioning or combination valve. There are two types of switches that may be used in this warning circuit.

On one type, the switch plunger sets in a slot at the center of a long piston in the proportioning valve. The piston is held in the center by equal hydraulic pressures at each end of the piston. When a leak occurs and pressure drops on one end of the piston, the pressure at the other end moves the piston out of the center. The shoulder of the piston slot pushes up on the switch plunger to close the light circuit (Figure 9-23). At the same time, the piston closes the outlet ports of the system that is leaking.

Another type of warning light switch is the hourglass-shaped type. The principle of operation is the same but the switch's ground contact extends into the slot. As the piston moves off center, the shoulder of the slot touches the contact and grounds the switch, closing the circuit.

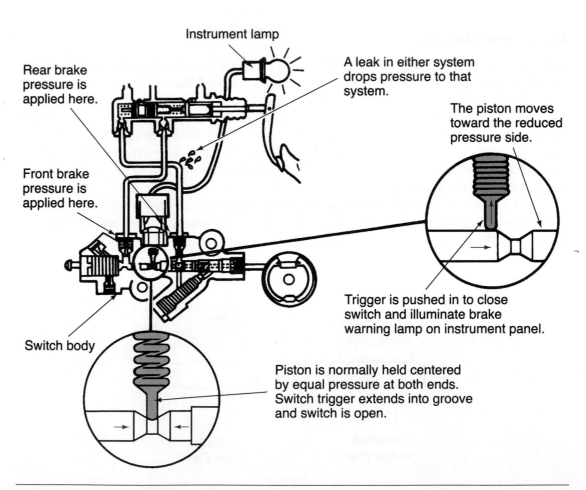

Instrument lamp

Rear brake pressure is applied here.

A leak in either system drops pressure to that system.

The piston moves toward the reduced pressure side.

Front brake pressure is applied here.

Trigger is pushed in to close switch and illuminate brake warning lamp on instrument panel.

Switch body

Piston is normally held centered by equal pressure at both ends. Switch trigger extends into groove and switch is open.

Figure 9-23 The warning light switch is closed by unequal pressures at the outer ends of the piston.

Drum Brake Components and Operation

Shop Manual
pages 263–267

The brakes on all vehicles consist of a **friction material** being clamped against a rotating device by hydraulic pressure. There are two basic braking systems based on how the friction material and rotating device interact with each other. The oldest system is drum brakes.

The wheel components in a drum brake system are the wheel cylinder, drum, shoes, adjuster, and mounting hardware (Figure 9-24). The hydraulic pressure is delivered from the master cylinder via steel and flexible lines. Once inside the wheel cylinder, the pressure is applied against two pistons.

Asbestos was used for many years as brake **friction material.** Because of the health dangers of asbestos, most brake friction is now made of organic materials or compounds.

Wheel Cylinders

The wheel cylinder has two pistons, a spring, sealing cups or seals, two links, and the cylinder body. The body has a machined bore for the pistons. A threaded hole in the bore provides a connection point for the brake line while a second threaded hole drilled into the bore is for bleeding air and fluid from the cylinder (Figure 9-25). The bleeder is always at the highest point of the bore when the cylinder is mounted on the vehicle. Within the bore are pistons, cups, and a spring. The spring is placed between the two pistons and cups to keep all internal components in place and in contact with each other during brake release. This spring does not return the pistons or cups to their starting positions when the brakes are released.

The two cups act as seals when hydraulic pressure is applied to the bore. The interiors of the two cups face each other and the center of the bore. The outside faces of the cups are flat against the inner face of its mating piston (Figure 9-26). As the pressure builds during brake application, the outer edges of the cup are forced outward against the bore to provide a more effective seal.

The pistons are placed at each end of the bore with the spring, cups, and hydraulic fluid trapped between them. The outward face of the piston is usually cupped to provide a mounting point for one end of a link. The links are the connecting point between the pistons and the brake shoes (Figure 9-27). Around each end of the cylinder is a boot, which serves as a dust cover to protect the bore. As hydraulic pressure is increased within the bore, the cups, pistons, and links move outward to apply force against the two brake shoes.

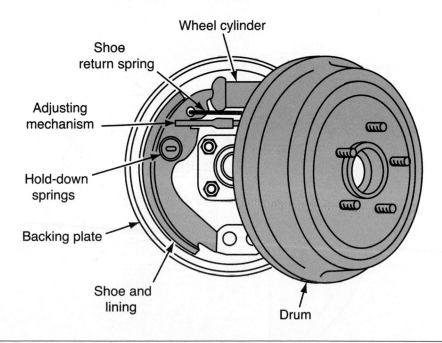

Figure 9-24 All of the drum brake components except the drum are mounted to the backing plate.

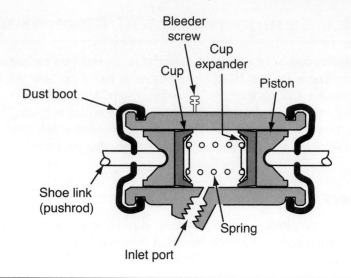

Figure 9-25 Most wheel cylinders have two operating pistons moved by hydraulic pressure trapped between them.

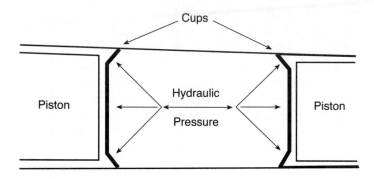

Figure 9-26 The hydraulic pressure applied against the cups seals the wheel cylinder's pressure chamber.

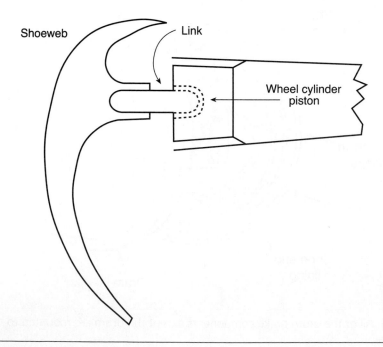

Figure 9-27 The link makes the connection between the wheel cylinder piston and the brake shoe.

Brake Shoes

The brake shoes provide a mounting area for the friction material. The metal portion of the shoe is normally referred to as the **shoe web.** The friction material may be bonded (glued) or riveted to the web. Brake shoes are curved to match the inside wall of the drum (Figure 9-28). Return springs hold the upper ends of the shoes tight against the wheel cylinder links. Retainer springs attach the shoes to the backing plate, but allow the shoes to move out or in as the brakes are applied and released. The backing is bolted to the hub and acts as a rear cover and a mounting platform for the drum brake components. There are several methods to attach the shoes at the bottom, and the type of attachment at that point can help increase braking power.

The brake **shoe web** is the metal skeleton used to mount the friction material and attach the shoe to its mounting.

Brake Linings

The lining thickness, position on web, and shape may be different on the brake shoe set for one wheel. The secondary shoe may have a shorter and thinner liner than the primary shoe. The shoes must be properly placed during installation or brake failure or excessive wear could occur.

Brake linings are made of heat-resistant material and are the part of the brake system that actually contacts the rotating drum. The linings may also have bits of copper, brass, zinc, and other materials to increase friction and prolong the life of the lining. This type of compound has replaced asbestos as a lining material.

Servo Action

Some systems use a stationary **anchor** to hold the lower end of the brake shoes in place (Figure 9-29). In this setup, the only force applied to the shoe for stopping is the hydraulic pressure within the wheel cylinder. This increases the stopping distance and requires more effort from the driver. Servo drum brakes are used on most vehicles.

A servo action uses the force from one operating member to increase the force applied to another operating member. In drum brake systems, each brake shoe receives equal force from the wheel cylinder to push the shoe out against the drum. If the vehicle is moving forward, the

The pressurized fluid trapped between the pistons and shoes acts as one **anchor** for the shoes.

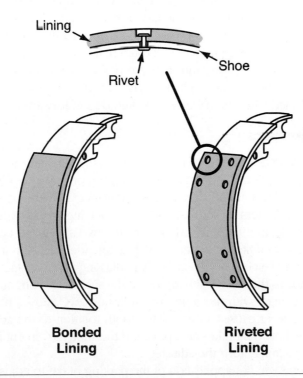

Bonded Lining

Riveted Lining

Figure 9-28 The bonded shoe is more prevalent on light vehicles, and the riveted shoe is most often, but not always, found on heavier vehicles.

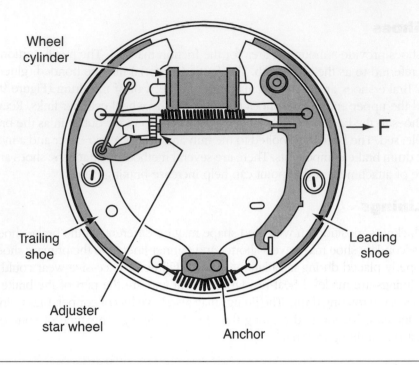

Wheel cylinder

F

Trailing shoe

Leading shoe

Adjuster star wheel

Anchor

Figure 9-29 A typical setup for a leading-trailing drum brake.

front shoe contacts the drum and attempts to move with the rotating drum. An anchor would stop the shoe from moving, but with a servo system, this shoe is connected to the rear shoe. As the front shoe attempts to rotate, it transmits force through the linkage to the rear shoe. The rear shoe is now pressed against the drum by the wheel cylinder and the servo action. The upper end of the rear shoe cannot move inward because of the pressurized brake fluid trapped between the two wheel cylinder pistons. In this manner, the driver seats the shoes against the drum and the force of the rotating drum is also applied against the rear shoe. Stopping power is increased and the driver's effort is decreased. Note that the secondary or rear shoe performs most of the braking effort; however, the brakes may be set so the front shoe is actually the secondary shoe.

The vast majority of vehicles equipped with drum brakes usually have duo-servo action. The servo effect is possible in both forward and rearward braking with duo-servo systems. The technician who works on antique or classic vehicles may encounter a servo system where the servo action is only accomplished when the vehicle is braked during forward travel.

Brake Self-Adjusters

Drum brake systems require some type of mechanism to adjust the shoes outward as they wear. Older vehicles and some larger trucks require the technician or driver to adjust the brakes at regular intervals, normally during routine preventive maintenance. However, the majority of light vehicles use automatic self-adjusters. The adjuster is the connecting link between the upper or lower ends of the brake shoes (Figure 9-30). A spring holds the two shoes and adjuster in contact with each other. The adjuster has a starred wheel, which is threaded onto an adjusting screw and is normally turned by the movement of the shoes. An adjusting lever contacts the adjusting screw and is moved by the shoe as the brakes are applied. If the shoes are worn and have to travel more than a set distance for contact with the drum, the adjusting level moves further than usual. The additional motion of the lever causes a one or two tooth rotation of the adjuster. Additional adjustments are made each time the brakes are applied. If the shoe movement is short, the level cannot move far enough to actually turn the adjuster.

Some newer systems use the application of the parking brake to adjust drum brakes. While this system works well, many drivers do not realize that the parking brake must be used regularly in order to adjust the brakes. The adjustment works the same way except for the operating mem-

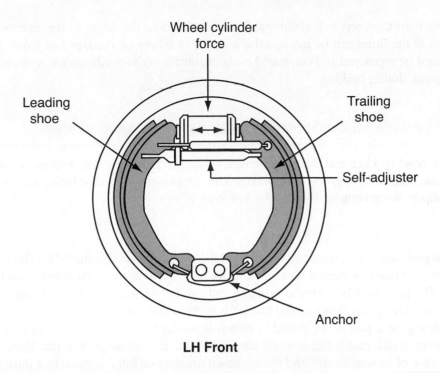

Wheel cylinder force

Leading shoe

Trailing shoe

Self-adjuster

Anchor

LH Front

Figure 9-30 The self-adjuster for a leading-trailing drum brake is usually connected between the tops of the two shoes. On a duo-servo system, the self-adjuster replaces the anchor shown in the illustration.

ber that performs the rotation of the adjusting screw. Most non-servo systems and many other vehicles have the adjuster at the top end of the shoes under the wheel cylinder. Most vehicles have a slot either in the drum or backing plate to allow a brake adjuster tool to be inserted for manual adjustment. This is primarily done to adjust the brakes after repairs have been completed,

Drums

The **drum** covers the brake mechanism and is attached to the axle or hub (Figure 9-31). The drum is drilled to fit over the lug studs. The interior friction area of the drum is machined to provide a smooth braking surface. The installation of the wheel assembly and lug nuts hold the drum in

The **drum** and hub may be manufactured as a one-piece component.

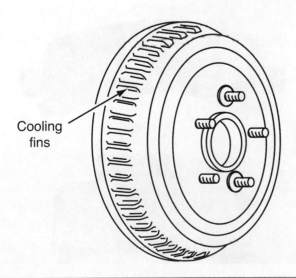

Cooling fins

Figure 9-31 Many larger brake drums have fins molded to the outside to enhance cooling of the brakes.

place. Some manufacturers use additional fasteners that hold the drum to the axle or hub. The friction area of the drum can be machined if grooved or otherwise damaged or worn. The inside diameter must be measured and compared to the manufacturer's specifications. A drum that is too thin may break during braking.

Disc Brake Components and Operation

Disc brakes react quicker and operate differently than drum brakes. Disc brakes are pressurized and controlled the same way as drum brakes. The components of a disc brake are the pads, rotor, and caliper. Many vehicles use disc brakes at all wheels.

Pads

Disc **brake pads** are much smaller than drum brake shoes in area (Figure 9-32). The friction material is either bonded or riveted to a small metal pad. There are two pads per wheel, an inner and outer. The pads may be identical in shape and size but most have some differences between the inner and outer pads. Between the two pads is the brake disc or rotor.

The lining on a pad is composed of materials similar to the ones used on drum brake linings. However, a disc pad lining is much harder than the drum linings. The pad lining has to be harder because of its smaller size and the increased pressure or force applied to it during braking.

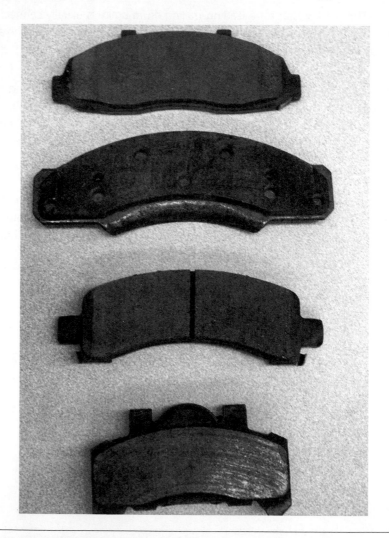

Figure 9-32 Typical disc brake pads.

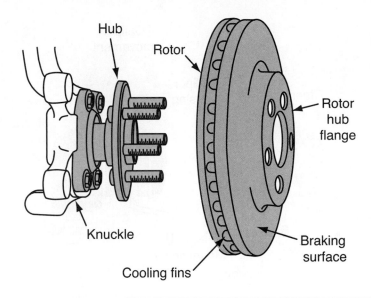

Figure 9-33 labels: Hub, Rotor, Rotor hub flange, Knuckle, Braking surface, Cooling fins

Figure 9-33 Shown is a typical floating rotor or a two-piece rotor and hub assembly. Note the cooling fins between the braking surfaces.

Rotors

The **rotor** is attached to the hub or axle and rotates with the wheel. Each side of the rotor is machined for smooth braking (Figure 9-33). As the brakes are applied, the two pads clamp against the rotor and slow it. When the brakes are released, the pads are moved slightly backward but they should never lose complete contact with the rotor. The rotor is subject to great heat during braking. Excessive and hard braking will cause the rotor to warp. The rotor can be machined if the warpage is not too severe and it is not too thin.

Calipers

The caliper is mounted at the wheel and most extend over the rotor in a U-shaped mounting. Most calipers are mounted to the steering knuckle using an adapter. The caliper may have one, two, or four pistons. The pistons in most calipers are larger than those found in a wheel cylinder. The caliper may be sliding or fixed. It is drilled to accept the line fitting and a bleeder valve.

The operation of the **sliding caliper** is similar to the actions of a wheel cylinder except the caliper can move as a unit. As pressurized fluid enters the piston bore, the piston moves out, pushing the inner pad against the rotor (Figure 9-34). The outer pad is on the opposite side of the rotor and held in place by the caliper portion that extends over the rotor. Since any action creates an equal and opposite reaction, the caliper moves away from the rotor. This action draws the outer pad against the outer edge of the rotor. More force applied against the piston and inner pad results in the same amount of force being applied to the outer pad. A sliding caliper may have one or two pistons mounted in the caliper.

A **fixed caliper** usually has four pistons, two on the inner side and two on the outer side of the rotor. It requires a much larger caliper and the additional pumping to fill, equalize, and bleed the additional cavities for the two outer pistons. As the fluid enters the cavities, the pistons are forced outward, thereby clamping the rotor. This is an older system and is not seen often on light vehicles.

There is no manual adjustment on a disc brake system and no return springs. Each bore is fitted with a **square-cut O-ring** near the outer end of each piston. The ring acts as a seal and adjuster. As the piston moves outward on each brake application, the O-ring distorts slightly (Figure 9-35). When the brakes are released, the square ring returns to its original condition, pulling the piston and pad back from the rotor (Figure 9-36). As the pads wear, the piston moves further outward and the distorted ring slides along the piston. When it recovers its shape after brake release, the piston

Like the drum brakes, the **rotor** and hub may be manufactured as a one-piece unit.

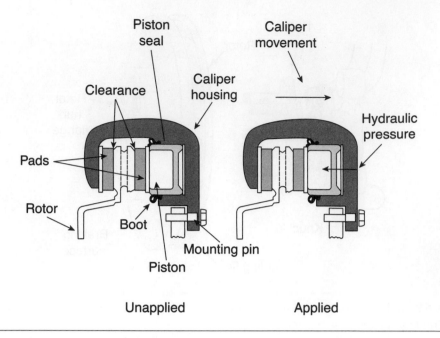

Figure 9-34 The sliding caliper moves on its mounting pins so equal force can be applied by both pads.

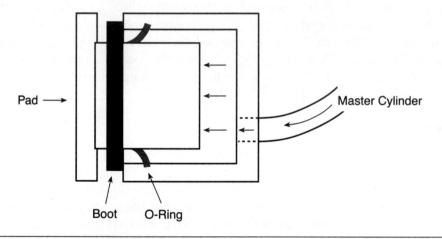

Figure 9-35 The square-cut O-ring distorts as the caliper piston moves outward.

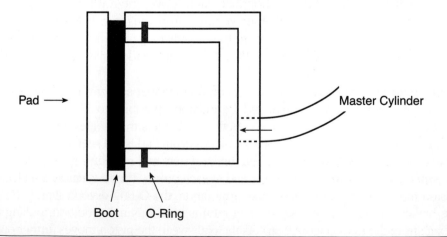

Figure 9-36 The O-ring's normal shape with brakes released.

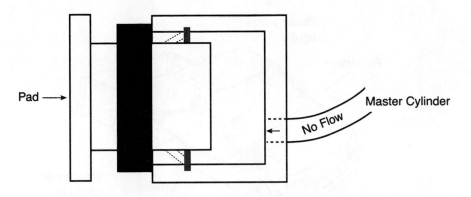

Figure 9-37 The square ring can only return the piston a short distance, which allows the piston to adjust for pad wear.

is further down the bore (Figure 9-37). This action continually keeps the pads flat against the rotor and allows for the adjustment needed as the pads wear.

Parking Brake Operation

Commonly referred to as emergency brakes, parking brakes are designed to hold the vehicle still after it has been stopped. They function on the rear axle only in passenger cars and light trucks. A foot- or hand-operated lever near the driver is used to set the brake. A foot pedal, when used, is located at the left kick panel near the driver's left foot. If a lever is used, it is usually placed between the two front seats. A typical system uses cables to connect the driver's action to a lever or cam in the rear brakes to lock the shoes to the drum. Some rear discs have small drum/shoe mechanisms on the inside of the rotor for parking. Other rear discs have screws that force the pads to clamp the rotor when the parking brake is set.

Shop Manual
pages 267–268

Antilock Brake Systems (ABS)

Contrary to some beliefs, antilock brake systems (ABS) were not designed to reduce the stopping distance of a vehicle. They do prevent the wheels from locking and give the driver better control over the vehicle during hard or panic braking. Stopping distance is reduced to some extent by the fact that the wheels are not sliding and the tire maintains traction with the road. There are several different types of antilock systems. Some work only on rear wheels and are known as **rear-wheel antilock (RWAL)**. RWAL is commonly used on light trucks and vans (Figure 9-38). The typical ABS system on a passenger vehicle is a four-wheel antilock. Four-wheel systems can be three or four channels. Three-channel systems control the braking at each front wheel and the rear axle set. Four-channel systems control the braking at each wheel (Figure 9-39).

Since this unit covers basic theories and operation, we will only discuss the basic components and operation common to all types of antilock systems. The components are the **speed sensors, pressure controllers,** and the **controller.**

Shop Manual
pages 269–271

Ford refers to their RWAL as RABS or Rear Anti-lock Brake Systems.

Speed Sensors

In order to control brake pressure, the ABS controller must know the speed of the wheels. Speed sensors generate an AC signal each time a tooth on a toothed ring or **tone wheel** passes the sensor's magnet (Figure 9-40). The number of signal pulses sent to the module is used to calculate the speed of the wheel or axle.

On RWAL and three-channel, four-wheel ABS, the rear speed sensor is mounted to measure the speed of the rear axle. This is done by placing the sensor at the output shaft of the transmission

---

A **tone wheel** performs the same function as a toothed ring.

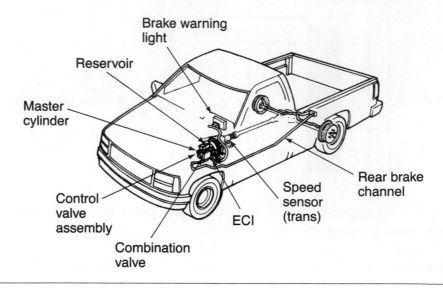

Figure 9-38 A RWAL (RABS) brake system.

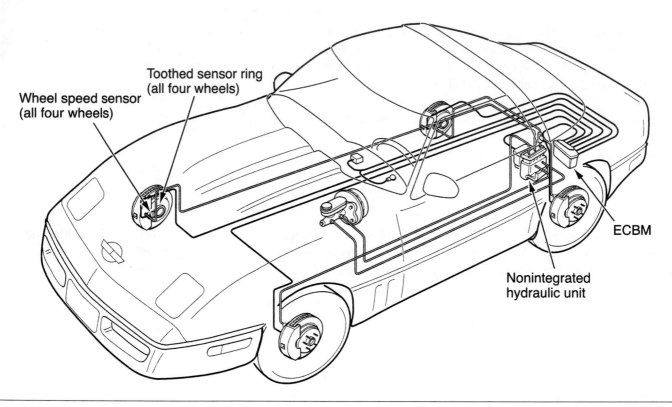

Figure 9-39 Up until model year 2000, this was a typical Bosch ABS-2 system used on GM products. Most new ABS have the components mounted in the trunk (shown here) moved to the engine compartment.

or at the ring gear in the differential (Figure 9-41). The toothed ring is placed on the output shaft or ring gear. The pressure to both rear brakes is controlled as one circuit. Each brake on the rear axle will receive the same amount of fluid and pressure. A three-channel system uses sensors at each front wheel in conjunction with the rear sensor. The front tone rings mounted on the brake rotor or the outer CV joint. The three-channel system can control pressure to each of the front wheels in addition to the rear brakes. It is typed as a four-wheel ABS.

Wheel speed
sensor

Figure 9-40 This wheel speed sensor is mounted to the steering knuckle and the toothed ring on the outer CV joint.

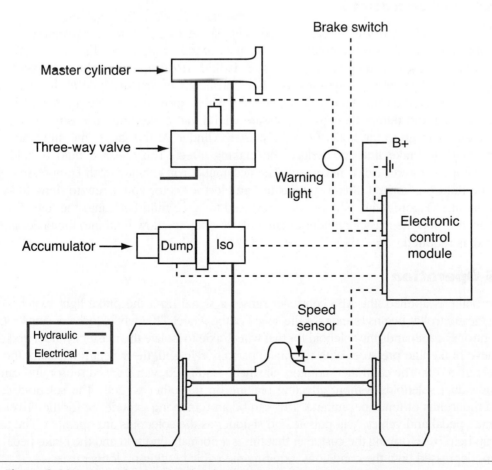

Figure 9-41 Layout of a DaimlerChrysler RWAL system.

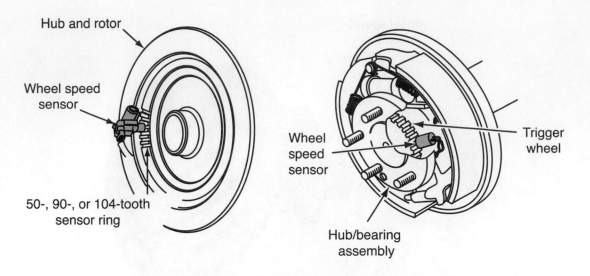

Hub and rotor

Wheel speed sensor

50-, 90-, or 104-tooth sensor ring

Wheel speed sensor

Trigger wheel

Hub/bearing assembly

Figure 9-42 On the left the tooth ring is molded to the inside of the rotor. At the right the ring is around the outside of the axle near the backing plate.

Four-wheel, four-channel ABS controls the pressure to each wheel. This system will have speed sensors and tone wheels at each wheel (Figure 9-42). The pressure and fluid can be regulated independently of each wheel based on that wheel's speed.

Pressure Controllers

The devices actually performing brake pressure regulation are solenoids commanded by the antilock controller. There is a solenoid for each wheel or axle with a speed sensor. A RWAL system has one solenoid commonly known as an **isolation/dump solenoid** or brake actuator (Figure 9-43). While different systems use different means to actually control the brake pressures, they all use the same theory of operation. Each solenoid can isolate the wheel from the master cylinder and dump the existing pressure within that brake line. In this manner, no further pressure can go to the brake from the master cylinder. At the same time, fluid can be released from the brake line and reduce the braking effect. The released fluid is held in an accumulator for future use, if necessary. The accumulator is charged with compressed gas to keep the fluid pressurized. Some systems use an electric motor and screw to draw fluid from the brake line (Figure 9-44). The motor is reversed to force fluid back into the line. The solenoid can also pop the master cylinder isolation valve open to allow fluid into the brake line and solenoid as required.

ABS Operation

As the brake is applied, the ABS controller senses a signal from the brake light switch. At this point, the controller begins to monitor the speed of the wheels. If a wheel locks or stops rotating, the controller commands the solenoid for that wheel/axle to isolate the master cylinder and dump or relieve brake line pressure. With the brake partially released, the wheel rotates and the solenoid is turned off. The controller monitors all wheels equipped with a speed sensor and can control individual solenoids without affecting braking on the other wheels. The solenoid can be cycled thousands of times per minute. This can be an unnerving experience for the driver since the brake pedal and vehicle will pulsate and shudder as the solenoids are operated. The technician can help by informing the customer that this is a normal condition and the brake pedal must be kept depressed until the conditions requiring the panic braking no longer exist.

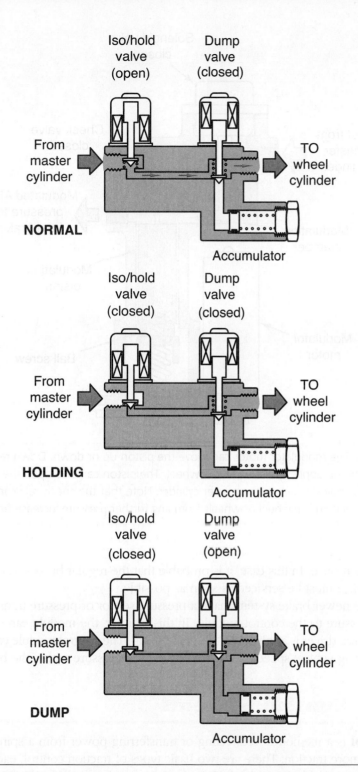

Iso/hold valve (open) Dump valve (closed)

From master cylinder

TO wheel cylinder

NORMAL

Accumulator

Iso/hold valve (closed) Dump valve (closed)

From master cylinder

TO wheel cylinder

HOLDING

Accumulator

Iso/hold valve (closed) Dump valve (open)

From master cylinder

TO wheel cylinder

DUMP

Accumulator

Figure 9-43 The isolator/dump valve for a RWAL system may be referred to as the RWAL actuator.

The antilock brake system has an **amber dash warning light** that can be lit by the controller if a defect is found. The only defect that the controller can detect is an electrical one. The controller cannot detect mechanical brake conditions such as a leaking wheel cylinder or excessively worn shoes or pads unless a mechanical condition directly affects a sensor.

The regular brake system will function normally if the ABS warning light is illuminated, but the ABS will not function. Certain brake failures may illuminate the red brake warning and the

The **amber ABS light** is not to be confused with the red brake light. The red light means the regular brakes may fail.

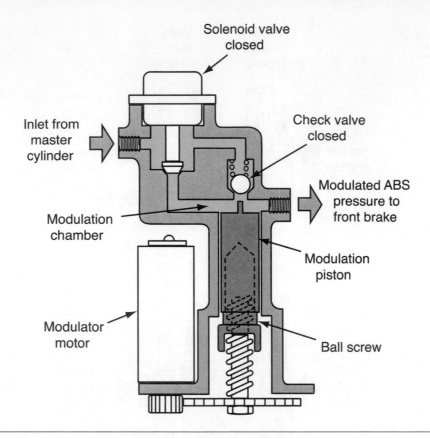

Figure 9-44 The modulator motor can move the piston up or down. Down relieves pressure to the wheel and up applies pressure to the wheel. The piston can also open the check valve so fluid can flow quickly back to the master cylinder. Note that the check valve and solenoid valve are closed, isolating the wheel or wheels from any further pressure increase from the master cylinder.

amber ABS lights at once. In this case, it is probable that the regular brakes have failed or have a fault and the vehicle must be serviced as soon as possible.

Some of the newer brake systems have a pressure sensor or pressure transducer to transmit brake system pressure to the control module. In this manner, the module can determine exactly how much force the driver is applying to the brake pedal. The control module can control lockup better by comparing vehicle speed, wheel speed, and the pressure within the brake system.

Traction Control Systems

Shop Manual
pages 269–271

Traction control is a method of redirecting or transferring power from a spinning drive wheel to a wheel with more traction. There are two basic types of traction control, each relying heavily on electronics. Dependable four-wheel ABS makes traction control a reality on many mid-price vehicles. However, for the driver who likes to accelerate quickly from dead stop, traction control can cause damage to the vehicle's driveline. Traction control is meant to assist on slick road surfaces, not in mud-bogging or hot-rodding.

Antilock Brakes with Traction Control

In this type of traction control the brakes are actually applied to a wheel that has lost traction and the vehicle speed is below 30 mph. When the brake is applied to one drive wheel, the differential or final drive shifts power to the other wheel. Assuming that the second wheel has traction, the vehicle should move. However, extra components and software programming must be added to the ABS.

The basic additions include two pumps driven by an electric motor, a different component, pressure and pressure-reducing solenoids, relays, and the hydraulic and electric circuits. The added sensors are used to track brake fluid levels, brake pedal travel, rotational sensors (motors), pressure switches, and a thermal limiter to calculate brake pad temperature.

The controller antilock brake (CAB) and the ABS perform like all other ABS systems during normal and panic braking. The circuits are monitored for faults in the same manner also. However, when a drive wheel loses traction, the traction control program of the CAB operates the system as a braking device. The CAB monitors wheel spin and actuates the pump and solenoid to apply the brake at that wheel. Driveline power is diverted to the other drive wheel. As the spinning wheel regains traction, the brake is released. The brake is only applied enough to divert the power and slow the spinning wheel. If the terminal limiter calculates the applied brake pad to be overheating, the traction control is turned off until the CAB resets the system.

Traction Control/Acceleration Slip

This system incorporates ABS with the engine and transmission control system to control wheel slippage (Figure 9-45). The system can perform in one or several of four modes at the same time. It should be noted that brake application is only done in conjunction with another mode. The modes are:

- Command the PCM to retard spark timing to reduce engine output.
- Force the throttle blade closed to decrease engine speed and output.
- Upshift the transmission to reduce delivered torque.
- Apply the brake on a spinning wheel.

On older systems without full electronically controlled transmissions, the gear upshift is not possible. Like the previously discussed traction control and other electronic control systems, this system must be tailored, electrically and mechanically, to meet vehicle specifications. A traction

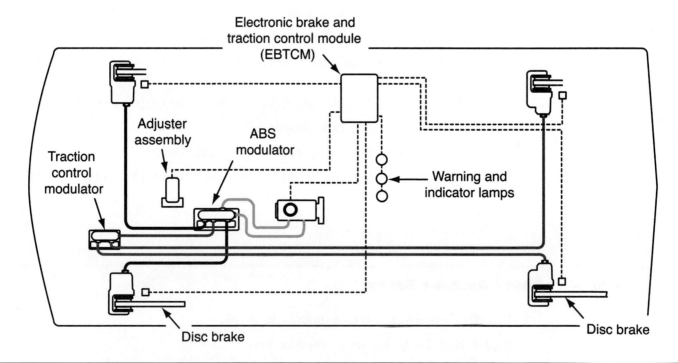

Figure 9-45 Shown is a Delphi Chassis VI ABS with traction control components installed.

control system for a Ford Crown Victoria will not work with a Lincoln Town Car, even though the vehicles are very similar in weight and overall size. Each vehicle is set up to function with electronic differences that must be considered during the design and application of all systems.

Summary

❑ The use of hydraulics in the brake system allows force to be transferred with moving parts.

❑ The operator's force on the brake pedal is mechanically transmitted to the master cylinder's primary piston.

❑ The operator's force can be increased greatly by using a booster between the pedal and the master cylinder.

❑ A single-piston master cylinder may cause all of the brakes to fail if there is leak at any wheel.

❑ A dual-piston master cylinder allows the driver to have braking power on two wheels even if a leak is present on one wheel.

❑ The metering valve holds the front disc brake off until the rear drum brakes have moved to make contact with the drums.

❑ The proportioning valve prevents the rear brakes from locking during heavy braking by reducing the pressure.

❑ A combination valve houses the metering and proportioning valves.

❑ A brake warning light is used to alert the driver to a failure in the system.

❑ Wheel cylinders use two pistons and links to apply the brakes.

❑ Return springs on a drum brake force the shoes and pistons to their release positions.

❑ Most drum brakes use an automatic self-adjuster operated by the movement of the secondary shoe or parking brake.

❑ Disc brakes do not use a mechanical adjuster.

❑ Disc brakes are adjusted by the action of the square cut O-ring.

❑ Flexible hoses are used to connect the wheel cylinder or disc caliper to the steel line located on the frame.

❑ Parking brakes are used to hold the vehicle still but not for stopping the vehicle.

❑ ABS is used to prevent wheel lockup during braking.

❑ The loss of ABS does not affect the operation of the regular brakes.

❑ Traction control uses the ABS or engine controls to control wheel spin during acceleration.

Review Questions

Short Answer Essay

1. Describe duo-servo action on a drum brake system.

2. Explain how disc brakes are adjusted for wear.

3. Discuss the purpose and operation of the master cylinder's primary piston.

4. Explain how brake pedal force is boosted with a hydro-boost system.

5. List and explain the purpose of the wheel-mounted components of a disc brake.

6. Briefly describe the actions taken by an ABS during a panic stop.

7. Explain the purpose of the proportioning valve.

8. Describe the operation of a frame-mounted, load-sensing proportioning valve.

9. Discuss the operation of a metering valve.

10. Discuss the operation of a combination valve-mounted, load-sensing proportioning valve.

Fill-in-the-Blanks

1. The _____ brake shoe performs most of the braking.

2. The _____ system boosts the driver braking effort using an electric motor and pump.

3. The ABS warning light will not detect _____ conditions.

4. The adjustment of disc brakes is done by the _____.

5. The master cylinder secondary piston provides fluid to the _____ brakes.

6. The metering valve controls fluid during _____ braking.

7. The proportioning valve controls fluid pressure during _____ braking.

8. A damaged steel line must be _____.

9. A(n) _____ _____ connects the caliper to the steel brake lines.

10. If the ABS warning light is on, the regular brakes will _____.

ASE-Style Review Questions

1. The combination valve is being discussed.
   Technician A says a leak in the front brake should turn on the warning light.
   Technician B says the brake pedal may be much lower to the floor if the brake warning light is on.
   Who is correct?
   A. A only　　　　**C.** Both A and B
   B. B only　　　　**D.** Neither A nor B

2. The diagonally split brake system is being discussed.
   Technician A says one master cylinder piston will operate the brakes on a front and a rear wheel.
   Technician B says that if a leak occurs, the operator will still have two working brakes on an axle.
   Who is correct?
   A. A only　　　　**C.** Both A and B
   B. B only　　　　**D.** Neither A nor B

3. Disc brake calipers are being discussed.
   Technician A says adjustment for pad wear cannot be performed manually.
   Technician B says the square-cut O-ring automatically adjusts for pad wear.
   Who is correct?
   A. A only　　　　**C.** Both A and B
   B. B only　　　　**D.** Neither A nor B

4. The master cylinder on a front/rear split system is being discussed.
   Technician A says a leak at the front brake will stop the primary piston from building pressure.
   Technician B says the load-sensing proportioning valve is mounted at the rear of all master cylinders.
   Who is correct?
   A. A only　　　　**C.** Both A and B
   B. B only　　　　**D.** Neither A nor B

5. Split brake systems are being discussed.

   Technician A says the diagonally split system is set so a front brake and a rear brake are available for braking.

   Technician B says the front/rear split system has the rear brakes operating from the same master cylinder piston.

   Who is correct?

   A. A only   **C.** Both A and B

   B. B only   **D.** Neither A nor B

6. Disc brakes are being discussed.

   Technician A says the square-cut O-ring adjusts for pad wear.

   Technician B says a fixed caliper uses only one piston.

   Who is correct?

   A. A only   **C.** Both A and B

   B. B only   **D.** Neither A nor B

7. A 2-square-inch input piston is being applied with 10 pounds of force. A 4-square-inch output piston is being used.

   Technician A says mechanical advantage can be calculated using the formula F = PA.

   Technician B says the pressure in this system is 40 psi.

   Who is correct?

   A. A only   **C.** Both A and B

   B. B only   **D.** Neither A nor B

8. The metering valve is being discussed.

   Technician A says the movement of the valve determines rear brake pressure.

   Technician B says the valve adjusts rear brake pressure during initial braking.

   Who is correct?

   A. A only   **C.** Both A and B

   B. B only   **D.** Neither A nor B

9. Drums and rotors are being discussed.

   Technician A says the drum and rotor are mounted between the wheel and axle or hub.

   Technician B says drums are machined to match the brake pads.

   Who is correct?

   A. A only   **C.** Both A and B

   B. B only   **D.** Neither A nor B

10. Proportioning valves are being discussed.

    Technician A says the load-sensing proportioning valve may be mounted on the frame rail.

    Technician B says some proportioning valves use a ball to control the valve.

    Who is correct?

    A. A only   **C.** Both A and B

    B. B only   **D.** Neither A nor B

Suspension and Steering

Upon completion and review of this chapter, you should be able to:

❏ Discuss how the Laws of Motion affect the operation of the suspension and steering systems.

❏ Explain the relationship between vehicle frame, suspension, and steering.

❏ Define the different suspension systems.

❏ Describe the purpose and operation of springs and torsion bars.

❏ Describe the purpose and operation of shock absorbers and struts.

❏ Explain the operation of electronic suspension systems.

❏ Describe the different steering systems and their general operation.

❏ Explain operation of electric steering systems.

❏ Explain operation of four-wheel steering.

❏ Explain tire ratings and markings.

❏ Explain operation of the tire pressure monitor (TPM).

❏ Define alignment angles and wheel alignment.

❏ Describe SAI, setback, and toe-out-on-turn angles.

❏ Describe caster, camber, toe, and wheel alignment.

❏ Explain static and dynamic tire balancing.

Introduction

With power delivered to the wheels, the vehicle must be suspended and steered. The steering system allows the operator to steer the vehicle within designed limits. The suspension system supports the vehicle, provides a better ride, and assists in vehicle control.

Laws of Motion

In Chapter 7, "Automobile Theories of Operation," we discussed some theories of physics that apply to vehicle design and operation. One theory was Newton's Laws of Motion. The laws are based on an object, in the absence of outside forces, performing in a given manner, such as moving at the same speed in a straight line. As in most systems on a vehicle, there are almost always some outside forces at work.

Obviously, the portion of the Laws of Motion discussed do not consider the outside forces that move, stop, or turn an object. For example, friction tries to stop or slow an object while a force greater than the mass (weight) of the object is required to move or accelerate it. When designing a vehicle that must work within the Laws of Motion and other physics, related items like mass, force, torque, friction, and speed must be considered. The suspension and steering systems have components that are almost constantly moving, stopping, or changing direction during vehicle operation. The suspension and steering systems must be able to overcome the straight physics of motion so the vehicle can be controlled.

Frame, Suspension, and Steering Relationship

Shop Manual
page 291

Proper operation of the suspension and steering systems depends on a straight and true frame. The frame, full or unibody, must be strong enough to support the suspension and steering as they operate over the road and under different load conditions. A bent or warped frame may affect the steering to the point that the driver may have problems keeping the vehicle under control. A weak frame can flex as it is stressed when changing steering and suspension angles. A weak frame may also allow the suspension system to transmit the shock to the body, break mounting points, and create steering problems. Frames that are misaligned may not be readily apparent until some other component fails or begins to show excessive wear.

 AUTHOR'S NOTE: A small 1994 "compact" truck on which we had installed four new tires was experiencing a vibration problem at cruising speed. The truck had not been abused, was primarily transportation to and from work, and had no vibration problem until the installation of the new tires. Rebalancing and finally installing another set of new tires did not solve the problem. After consultation with the local dealer a TSB was found that indicated a twisting in the frame that could occur even during normal driving conditions. This problem was most prominent when hard, long-mileage tires were installed. Though we were unable to correct the frame condition, after installing four new "softer" tires (more like the factory originals) the vibration was no longer noticeable.

Suspension System Operation

Shop Manual pages 291–297

The suspension supports most of the vehicle's weight and absorbs most of the various shocks from the bumps and holes in the road. But a second and just as important purpose is the control a properly functioning suspension system offers a driver. A suspension system allows the wheel assembly to move upward and backward or down and forward as it maneuvers over the road. This angle of movement absorbs some of the shock and reduces stress on the suspension and steering components.

The mass of the vehicle is divided into two weights. The **sprung weight** is the mass supported by the springs and usually includes everything mounted above the axles (Figure 10-1). It is important to remember that only the springs support the sprung weight. **Unsprung weight** or *mass* includes components such as the axles, brakes, tires, shock absorbers or struts, and the springs themselves. Vehicle engineers try to reduce unsprung mass by reducing the weight of the wheels and suspension components. Aluminum wheels are now being used on many vehicles instead of steel ones. Another way to reduce unsprung mass is to use an independent suspension system.

Generally, the lighter, **unsprung weight** or *mass* has to steer, support, and control itself and the heavier **sprung weight.**

Figure 10-1 An overview of the steering and suspension systems.

When the tire moves up (**jounce**) or down (**rebound**) over the road surface, a shock or jar is transmitted through the suspension system (Figure 10-2). For the most part, the shock is absorbed by the flexing of the tires and springs before it reaches the passenger compartment. However, a spring tends to rebound back and forth until its energy is expended. Since most roads do not have many long, flat surfaces, the vehicle would bounce up and down continuously unless the spring action was controlled.

A hydraulic cylinder known as a **shock absorber** or **strut** is used to control the motion of the springs. This cylinder allows the spring to give or extend enough to absorb most of the road shock. It controls further spring movement by **metering** fluid within the cylinder.

Ideally, the passengers would never feel the irregularities of the road, but the weight of a typical vehicle makes it practically impossible to eliminate all rough conditions. In addition to absorbing road shock, the suspension also helps control body lean during turns to help maintain vehicle control. **Sway bar** or **anti-roll bars** are used to reduce lean (Figure 10-3). As the vehicle is

Metering is the controlled flow of a liquid or vapor.

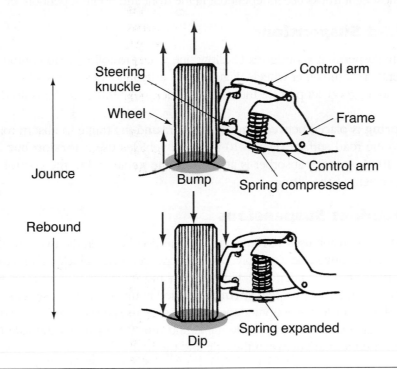

Figure 10-2 Most of the shock generated by the jounce and rebound of the wheel assembly is absorbed by the springs.

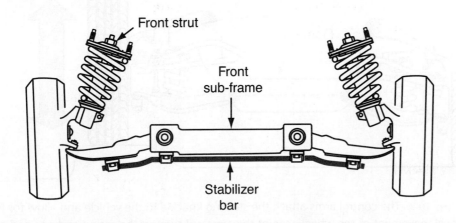

Figure 10-3 The sway or stabilizer bar is mounted at each end to the frame or body and helps control body roll.

turned, its **center-of-gravity** and weight shifts toward the outside of the turn, making the vehicle unstable. The springs and sway bar must be stiff or strong enough for the driver to retain control. Some vehicles have a form of the sway bar mounted at the rear of the vehicle.

The suspension system is comprised of all the factors involved so the driver experiences a smooth ride and controls the vehicle at the same time. Newer and more expensive suspension systems use electronics to control **air springs** for a smoother, more controllable vehicle.

Types of Suspension Systems

Suspension systems can be generally classified as **nonindependent** or **independent.** Today most nonindependent systems are found on heavy trucks, primarily because of their rear-wheel drive and **payload** capacity. The most popular suspension for cars is a four-wheel, independent system while most light trucks use independent in the front and nonindependent on the rear axle.

Independent Suspensions

An independent suspension is designed to allow one wheel to follow road contours without affecting the operation of the other wheels (Figure 10-4). Independent systems allow for a smoother ride while providing good support for the vehicle. Older independent systems used a spring and control arm.

A **coil spring** is placed between one control arm and the frame to absorb road shock and hold the tire to the road surface (Figure 10-5). Some vehicles use a **torsion bar** in lieu of the spring (Figure 10-6). A shock absorber is attached to the frame and to the control arm. It helps control the movement of the spring.

Nonindependent Suspensions

Nonindependent suspension uses a solid or rigid connection between the two wheels on an axle. This is usually the rear-drive axle assembly. Many four-wheel-drive vehicles also have a rigid axle housing at the front. In a nonindependent system, any movement of one wheel is transmitted to the opposite wheel. In most cases, **leaf springs** are used in this suspension system to support the sprung weight of the vehicle. Some nonindependent systems use coil springs instead of leaf springs. Leaf springs are attached at one end to the frame and run over or under the axle (Figure 10-7). Shock absorbers are mounted between the axle and the frame.

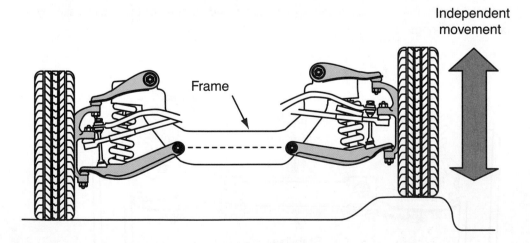

Figure 10-4 The control arms attach the steering knuckle to the vehicle and allow for the vertical movement of the wheel. One of the arms will support the spring and one end of the shock absorber.

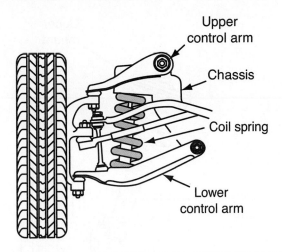

Figure 10-5 The coil springs support the weight of the vehicle and dampen road shock. In this illustration, the spring is retained between the lower control arm and frame (not shown).

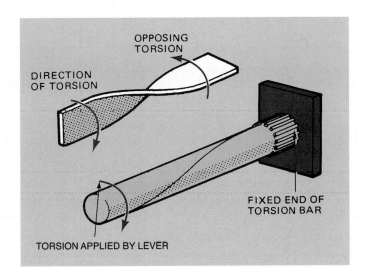

Figure 10-6 Twisting of the torsion bar acts like a coil spring.

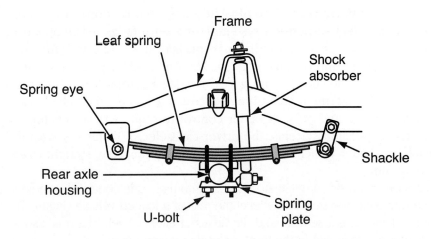

Figure 10-7 Leaf springs are usually found on trucks and may be used in the front and rear. Only the top leaf is connected to the body or frame.

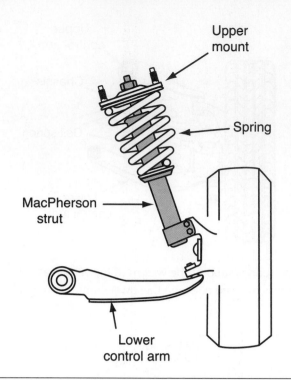

Upper
mount

Spring

MacPherson
strut

Lower
control arm

Figure 10-8 The strut performs the same function as a shock absorber, but also provides mounting for the spring and eliminates the upper control arm.

MacPherson Struts

A third, more recent suspension system uses **MacPherson struts.** The MacPherson strut is basically a shock absorber with mounting brackets for a coil spring (Figure 10-8). The strut assembly also eliminates the upper control arm by using a rotating mount at the top of the strut. The MacPherson strut assembly provides better support of the vehicle, controls the wheel assembly better, and reduces the weight of the vehicle. MacPherson struts can be found on all wheels, but are most common on the front wheels.

Electronic Suspensions

In recent years, **automatic ride control (ARC)** has become popular on more expensive cars. It is an independent suspension system. Using sensors, an air compressor, air valves, and a computer, the strength of the individual springs can be adjusted. Air can be added to or removed from the air spring (Figure 10-9). The onboard suspension computer typically is programmed to raise the vehicle during low speeds and stops. The vehicle is raised at the halt so the door can clear any road curbing. At cruising speeds, the vehicle can be lowered to reduce air resistance and provide better control through a lower center of gravity. On some automatic suspension systems, the driver can select a stiffer suspension for narrow, winding roads (for better vehicle feel) or a softer setting for wide, straight roads like the interstate road system.

The number of sensors and actuators in the automatic ride control depends on the type of system. Some ride control systems only level the rear of a loaded vehicle (Figure 10-10). Full automatic systems have height sensors and air valves at each wheel. There is also an electrically driven air compressor mounted in the trunk or engine compartment.

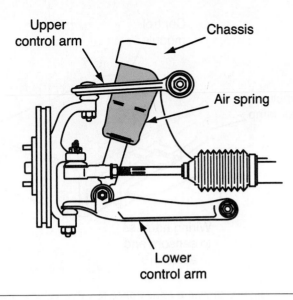

Figure 10-9 Air springs are heavy-duty balloon-type bags that can be inflated or deflated to raise or lower the vehicle.

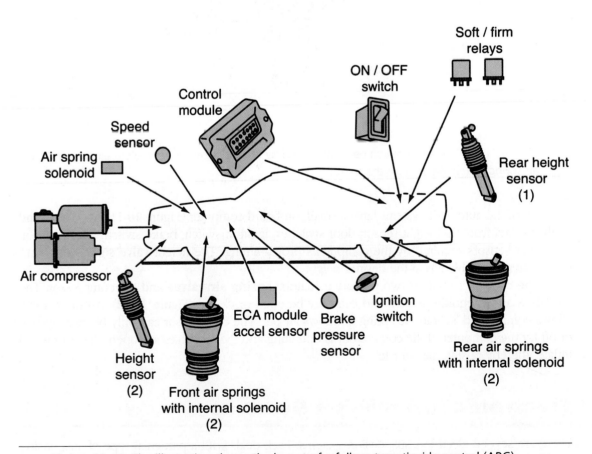

Figure 10-10 The illustration shows the layout of a fully automatic ride control (ARC).

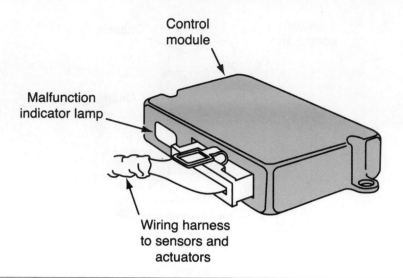

Control module

Malfunction indicator lamp

Wiring harness to sensors and actuators

Figure 10-11 This computer is similar in appearance to other vehicle computers but only has the programming for ride control.

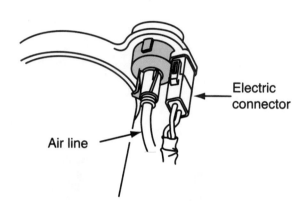

Electric connector

Air line

Figure 10-12 The air valve can be commanded by the ARC computer to allow air to enter the air spring or release air from the spring.

All of the automatic systems have a small, on-board computer (Figure 10-11). It collects and analyzes data from the height sensor, door switches, ignition switch, brake switch, speed sensor, and other sensors required by the computer's programming. The height valve signals the computer when the vehicle is at the desired height.

The computer controls two primary actuators: **spring air valves** and the compressor. The air valves are solenoid-operated and can only be open or closed (Figure 10-12). An open valve allows compressed air into the spring or allows air out. The compressor can only be switched on or off by the computer. If the compressor is operating and the air valves are open, the spring will inflate, thereby raising the vehicle.

Electronic Suspension Systems

There are several electronic suspension systems. Each deals with a particular type of suspension and its purpose. The earlier one came from an aftermarket idea for light trucks and so-called performance cars. This is the rear load leveler system. The idea came from adding air shock absorbers that were capable of raising the rear of the truck when it was loaded. This assisted in steering control and ride. It was also used on some cars in an attempt to gain more acceleration from one stop light to the next. In that case, it didn't work very well, plus the passengers took a beating on any road that wasn't smooth.

Automatic load leveling grew from this attempt to level the loaded vehicle. In place of the "air shocks," air springs were installed at the rear axle. When the vehicle was loaded, height sensors transmitted this information to a basic computer module. The module switched on the compressor and opened the air valves at the springs. When the rear of the vehicle reached its level height, the valves were closed and the compressor shut down. When unloaded, the computer would open the air valves to deflate the air springs until the vehicle was level again. This system is available on many vehicles today, but is not that common. It has largely been replaced by more versatile and better systems.

One of those newer systems is termed the automatic ride control (ARC) system, but this also is the accepted term for many variations. The ARC provides leveling capability at each wheel, lowers the vehicle during long, flat cruising, raises the vehicle when the door is opened (to miss the curb), and makes other adjustments. It can adjust for the tilt of the vehicle during a turn, provides information to the antilock and traction control systems to control wheel sliding or spinning, and can sometimes help control the vehicle on high-speed turns by measuring the body tilt at one wheel or one side. Features such as ride selection, where the driver can opt for a smooth easy ride or a sport's ride with its better cornering ability, are also available. This system is commonly referred to as programmable ride control (PRC). It is ARC with added features made possible with the newest family of electronics and their application in a vehicle. Each of these systems is used in many vehicle models with different features, but all deal with keeping the vehicle's body level and keeping the center of gravity close to the designer's intention.

Suspension System Components

Shop Manual
pages 297–300

The components of the suspension undergo a great deal of stress during operation. They must keep the tires on the road and maintain vehicle stability during all road and driving conditions. The suspension components are designed for durability with a minimum amount of maintenance.

Coil Springs

Coil springs are commonly used on the front of passenger cars and light trucks. They have also been used on the rear of many vehicles, particularly on the smaller types. They come in various strengths according to the type of vehicle and its maximum payload. The spring is made of a high-tempered, steel rod that is twisted into a coil. The diameter, overall length, and manufacturing treatment of the rod determine the strength or rate of the spring.

Variable rate coil springs are made so the spring can support more weight per inch of compression (Figure 10-13). Some variable springs use the same-diameter wire and unequally spaced coils. Other types have a wire machined in different diameters. The smaller-diameter portion of the wire supports lighter loads. The vehicle tends to stay more level, even under moderate loads with variable-rate coil springs. This type of spring action provides a better ride at light payloads and better support as the vehicle is loaded.

Air springs have been used for many years on over-the-road trucks. The springs are actually large air bags that can be inflated or deflated to support different payloads (Figure 10-14). This ability to control the spring's capacity makes it possible to employ electronics to control a vehicle's ride. The bags are made of very tough, thick, flexible material similar to rubber. There is actually little wear on the bags unless they are underinflated. The main points of wear occur at the top and bottom where the bags are mated to the vehicle.

The wire in a **variable rate** spring may be tapered, truncated cone, double cone, or barrel-shaped.

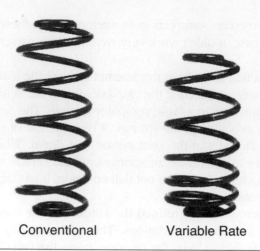

Conventional   Variable Rate

Figure 10-13 The variable-rate spring provides more vehicle control as the load becomes heavier.

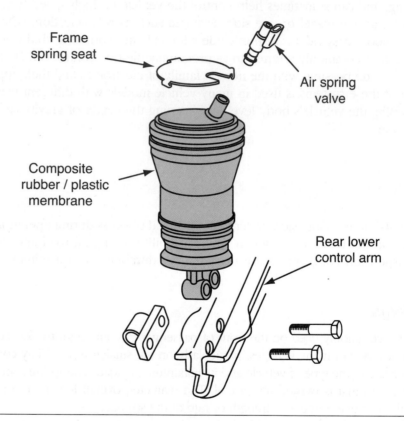

Frame spring seat

Air spring valve

Composite rubber / plastic membrane

Rear lower control arm

Figure 10-14 The only real difference between front and rear air springs is the method of mounting the spring to the control arm and body/frame.

Leaf Springs

Leaf springs are basically a set of long, curved, metal strips or leaves that are manufactured to bend at a certain rate. A leaf spring typically has three to five leaves. The leaves are placed on top of each other with the longest on top. Clamps are used to hold the stack of leaves in alignment with each other. Sometimes an extra leaf is added to better support heavier loads (Figure 10-15). One end of the top leaf is anchored to the lower side of a frame rail. The other end, usually the rear, is mounted in a mechanism that allows the top leaf to extend or shorten in length (Figure 10-16). This is necessary because as the leaf flattens under load, its ends move away from each other,

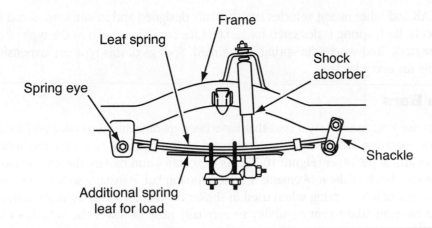

Figure 10-15 Leaf spring lift capability can be increased by adding an additional leaf under the standard set. This will cause a rough ride when the vehicle is not loaded.

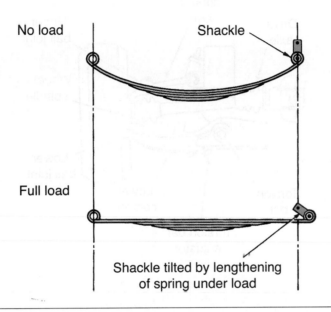

Figure 10-16 The rear shackle is pushed to the rear as the spring is flattened under load.

thereby increasing the linear length of the leaf. The opposite action occurs as the leaf is bowed and becomes shorter. The flattening and bowing of the leaf spring absorbs the shock of road conditions much as coil springs do. Although leaf springs are usually stiffer than coil springs, they will still flex until all energy is expended.

A **monoleaf spring** has one steel leaf. It is thicker in the middle and tapers down to each end. The tapering provides a variable rate similar to that of a variable coil spring. The spring may mount to the vehicle in a style similar to a multileaf spring or it may be mounted transversely. Some vehicles use a monoleaf spring made of fiberglass. Shock absorbers are used with leaf springs the same as coil spring systems.

Leaf and coil springs are designed and manufactured to support the expected weight of the vehicle and its payload. However, standard coil springs tend to be slightly less supportive of the vehicle during turns. Coil springs tend to offer a more comfortable ride. Standard leaf springs are harsher in their movements, resulting in a rougher ride for the passengers and load but affording a little better support during a turn. Both types can be designed to provide a smoother or firmer ride. However, the typical spring is usually a compromise between the two requirements.

NASCAR and other racing vehicles use specially designed and manufactured coil springs on all four wheels. Each spring is designed for certain race conditions such as the type of track, condition of the track, and where the spring is mounted. Springs of this type are expensive and are impracticable for everyday use.

Torsion Bars

The **ride or curb height** is the distance an "at rest" vehicle sits above the ground. It is measured at specific points on the vehicle.

Torsion bars are long, round, steel bars that have been manufactured to twist and then return to their original condition. Torsion bars are anchored to the frame at one end and splined to the lower control arm at the other (Figure 10-17). As the control arm moves, the torsion bar is twisted and absorbs the shock of the movement. A typical torsion bar is usually stiffer than a coil spring, but less than that of a leaf spring when used in similar conditions. Many vehicles with torsion bar suspensions have an adjustment capability to manually raise or lower the vehicle's **ride height** (Figure 10-18).

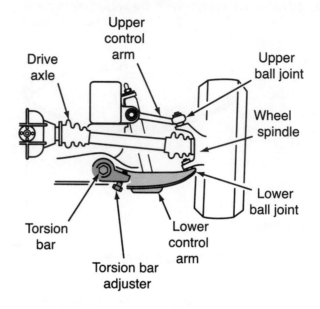

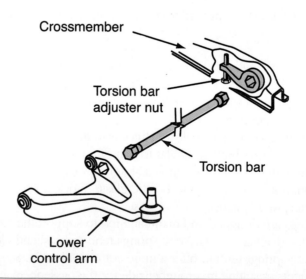

Figure 10-17 The torsion bar acts as a spring as it is twisted rotationally, but it uses less space and the vehicle's ride height can be adjusted easily.

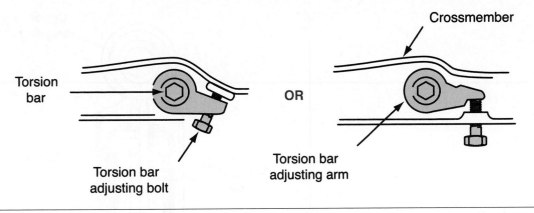

Figure 10-18 This is a torsion bar adjustment mechanism commonly found in light trucks and cars.

Shock Absorbers

Shock absorbers do not support any weight of the vehicle. They are designed to control spring movement and are used in some manner on all suspension systems. The absorber is a long, extendable cylinder that houses several valves, a piston, and a piston rod (Figure 10-19). Hydraulic fluid in the cylinder is used to slow or stop the piston movement.

The lower end of the shock is normally attached to the lower control arm or axle. The upper end is attached to the vehicle's frame. As the control arm or axle moves with the wheel

Special **shock absorbers** with air bags can be used to support a heavy load. The air bag inflates or deflates to keep the vehicle level.

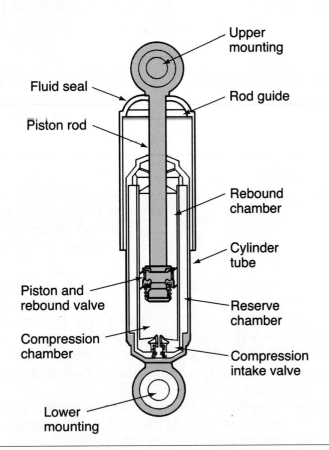

Figure 10-19 Fluid moves between the rebound and compression chamber to control spring movement.

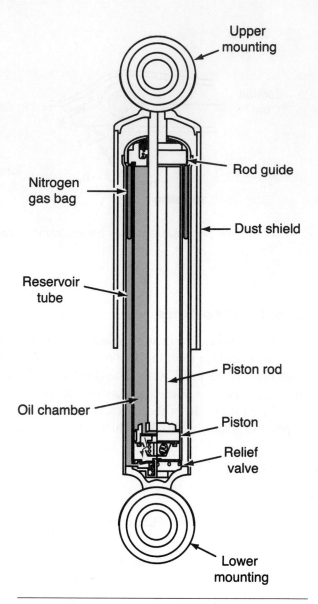

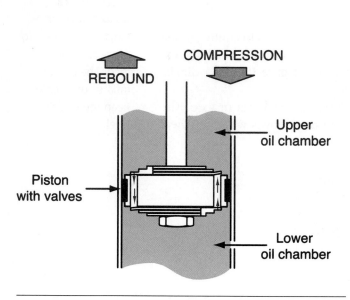

Figure 10-20 The metering valves in a shock absorber or strut control the flow of hydraulic fluid between the upper and lower chambers.

Figure 10-21 Older shocks with unpressurized oil had a lag time before they would damper shocks. Adding pressurized nitrogen gas to the top chamber pressurizes the hydraulic fluid, causing instant reaction to road shocks.

assembly, the upper or rod end of the absorber moves in and out of the cylinder. Inside the housing at the lower end of the rod is a valve assembly mounted within a piston. There is hydraulic fluid on both sides of the piston. The fluid must move from one side of the piston to the opposite side as the piston/rod moves up and down. The piston valve assembly meters the fluid's rate of movement (Figure 10-20). The resistance of the fluid against the piston assembly absorbs the energy of the spring. This restricts the oscillations of the spring and controls the up-and-down motion of the vehicle.

A problem with earlier shock absorbers was the delayed initial action of the absorber. The cylinder could not be completely filled with fluid. The fluid did not restrict piston travel until it was pressurized and forced to move through the valves. Newer shocks have pressurized nitrogen gas pumped into the cylinder after the fluid has been installed (Figure 10-21). This eliminates the initial lag, and the restricting action of the fluid is instantaneous upon any spring movement.

Control Arms

Control arms are the connection between the vehicle's frame and the steering knuckle. Control arms make it possible to use independent suspension systems. They may be located at each wheel or just on the two front wheels. On the rear of some vehicles, *trailing* or *radius arms* connect the wheel assembly to the frame. This type of arm will not keep the tire's tread flat on the ground. Other suspension components have to be used with trailing or radius arms to properly support and guide the wheels.

The control arms are mounted with rubber bushings at the frame and extend outward from the frame to the steering knuckle (Figure 10-22). The inner bushings allow the control arms to pivot up and down with the movement of the wheel. The SLA system has the shorter arm at the top. As the wheel moves upward over a hump, the shorter arm draws the top of the wheel inward. The same action happens when the wheel drops into a hole. This arc movement keeps the tire tread flat on the road. The two arms are designed to complement each other as the wheel moves over various road conditions.

If a coil spring is used, it is mounted between the frame and one of the control arms. This arm is the load-bearing arm and may be either upper or lower. At the ends of the control arms are ball joints that attach to the steering knuckle. The ball joints provide a pivot point for the steering knuckle to rotate during steering and while moving the wheel assembly over the road. Control arms used at the rear of the vehicle perform basically the same as front arms, although a steering operation is not usually present on the rear wheels.

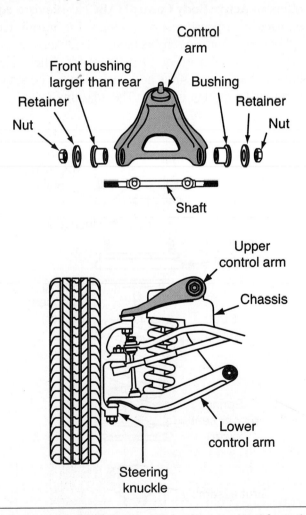

Figure 10-22 Control arm bushings lessen the shock transmitted from the control arm to the body/frame.

Struts

The MacPherson strut assembly houses a shock absorber internally and a coil spring mounted on the exterior (Figure 10-23). The coil spring mounted on the strut is not considered as part of the strut assembly. The shock absorber action of the strut is very similar to that of a regular shock absorber. The primary advantage of the strut is its use as the upper control arm and weight reduction.

AUTHOR'S NOTE: When a strut is used on the rear wheels of an independent suspension, the strut may be referred to as a **Chapman strut.** It is so named for Colvin Chapman, a well-known engineer for Lotus. The first application of the Chapman strut was on the Lotus 12.

Since it is mounted to the frame or body with a rotatable bearing at the top, the entire strut turns with the steering knuckle (Figure 10-24). The strut and spring are kept in alignment with the knuckle in this manner. The road shocks and steering stresses are more easily absorbed when all components are kept in line. A smaller and lighter-weight lower control arm can also be used. In addition to the suspension advantage, better steering control can be maintained for the same reasons. Pressurized nitrogen gas is also added to the strut to pressurize the strut's fluid.

The assembly and operation of the valves and piston in the strut is very similar to that of a shock absorber. The strut's rod is attached to the piston. The fluid within the cylinder must move from one side of the piston to the other. Metering valves control the rate of flow and restrict the oscillations of the coil spring.

Mercedes-Benz offers an **Active Body Control (ABC)** as standard equipment on its SL and CL classes. ABC is a semiactive suspension system designed to greatly reduce body roll in cornering, acceleration, and braking. Thirteen sensors monitor body movements and body level, providing the ABC computer information every 10 milliseconds. The computer commands four hydraulic servos mounted on four spring struts near each wheel. The strut is connected in parallel to a hydraulically controlled adjusting cylinder. The adjusting cylinder can be lengthened to-

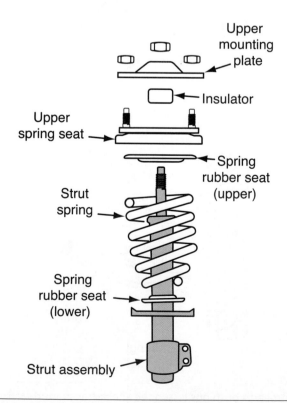

Figure 10-23 The strut acts like a shock absorber, provides mountings for the coil spring, and eliminates the upper control arm.

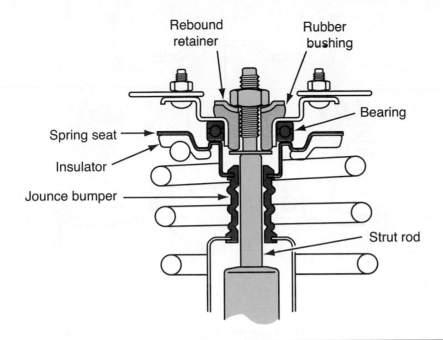

Figure 10-24 The strut's upper mounting mechanism allows the strut to rotate with rubber cushions to seat the coil spring and reduce shock.

ward the strut, thereby changing the suspension length. This change in suspension affects the suspension and damping effect of road surface on the vehicle. The damping effect can operate at a frequency of up to 5 **hertz.** In other operational areas, the ABC system functions much like other automatic ride systems.

Active suspension systems are showing up on a lot of newer-model vehicles. This system has been wedded into the ABS and TRC systems and some steering systems. This "total" active system can apply certain wheel brakes to overcome cornering problems, prevent or reduce wheel spin or sliding, reduce driver and passenger fatigue, and increase driver control, all while providing a comfortable ride. In some vehicles, the systems can command a reduction in engine output by changing the fuel or spark control. *Today's Technician Automotive Suspension and Steering* and *Automotive Brake Systems* texts provide in-depth studies of the active systems.

Sway Bars

Sway bars are designed to restrict or reduce vehicle lean during turns. The bars extend across the undercarriage of the vehicle from control arm to control arm. The sway bar is anchored to the frame or body and attached to each of the lower control arms with links (Figure 10-25).

As the control arm moves, it bends or deflects the sway bar. The bar resists and reduces the amount of lean. As the vehicle leans, the outer side of the vehicle drops, pushing the lower control arm and the end of the sway bar upward. The inner side of the vehicle lifts and the springs push the lower control arm and the end of the sway bar downward. The opposing actions at each end of the sway bar, along with the manufacturing process, help prevent excessive leaning of the vehicle. This gives better control to the driver. The springs, regardless of type, must also support the vehicle as it leans. The sway bar alone cannot effectively reduce vehicle lean.

Ball Joints

The ball joint is the connection between a control arm and the steering knuckle. It has a stud that is ball-shaped on one end and is treaded from near the ball to the end of the stud. The ball fits into a molded socket (Figure 10-26). The treaded end fits through a hole in the steering knuckle

Hertz is a measurement of how many times a function cycles in 1 second. In the 5-hertz instance, the suspension can change up to five times per second.

Sway bars can be purchased in different strengths to provide better support during turns.

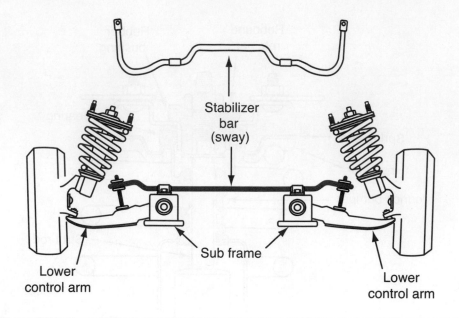

Figure 10-25 The isolator bushings prevent metal-to-metal contact between the strut and body/frame.

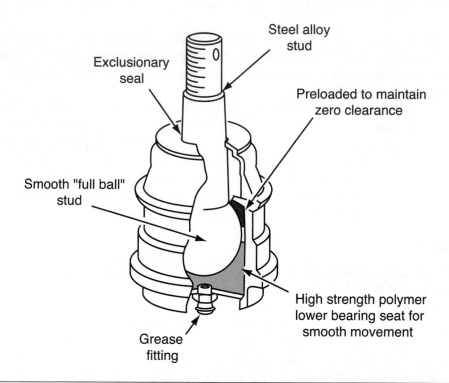

Figure 10-26 The ball joints allow the wheel assembly to vertically move up and down due to the road surface and allow the wheels to be steered horizontally.

and is secured with a nut or through bolt. Some ball joints are fastened to the steering knuckle with the threaded end through the control arm. The ball can rotate within limits to allow vertical and rotational movement of the steering knuckle assembly. Each control arm has a ball joint. The ball joint is one of the few suspension items needing routine maintenance.

Types and Components of Steering Systems

The steering system provides the driver with the means to steer the vehicle. There are two types of steering systems: **parallelogram** and **rack and pinion.** Power boost systems are available on both.

Parallelogram System and Components

The parallelogram steering system has been used for years on almost every vehicle made. The linkage in this system is designed to remain parallel to the control arms during all steering movements (Figure 10-27). There are several ball-joint-type devices connecting different parts of the linkage. The driver is connected to the steering linkage and gear through the steering wheel, steering shaft, and steering gearbox. The steering wheel and most of the shaft is mounted within the passenger compartment while the linkage and gearbox is located in the engine compartment or under the vehicle.

The steering wheel allows the driver to move the linkage to steer the vehicle. Steering wheels on manual steering systems are larger in diameter than the ones used on a power-steering system. This gives the driver some mechanical advantage during low-speed steering. The horn control button and air bag are located within the steering wheel. Some newer vehicles have switches for cruise control and other accessories on the steering wheel. The steering column supports and houses the steering wheel and shaft along with any switches and wiring mounted on or in the column (Figure 10-28). The steering shaft extends through the column to the steering gearbox.

The most common type of steering gearbox in the parallelogram system is the **recirculating ball and nut.** Large, steel balls are placed in grooves on the nut and **worm gear** (Figure 10-29). The steering shaft rotates the worm gear. As the worm gear turns, it uses the balls as a treading device to move the nut up and down the length of the worm gear. Teeth on the nut mesh with teeth on a **sector shaft.** The lower end of the sector shaft is splined to a **pitman arm.** The gear arrangement multiplies the force delivered by the steering shaft, thereby reducing driver effort.

Shop Manual
pages 300–305

The term **recirculating ball and nut** refers to the steel balls that move around in a continuous passage to move the nut that is meshed with the sector shaft gear.

A **sector shaft** will only rotate within a limited arc. On steering gear boxes this arc is approximately 60 degrees.

The **pitman arm** changes the rotational movement of the sector shaft to a linear motion for the steering linkage.

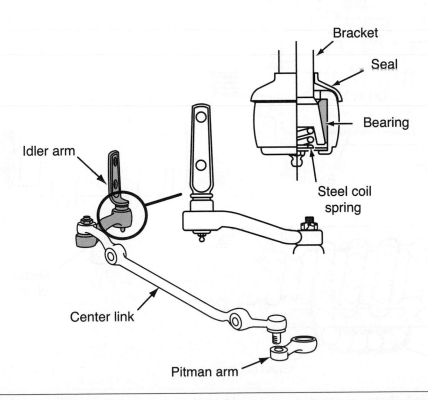

Figure 10-27 The steering linkage is moved by the pitman arm and sector shaft. The linkage will remain parallel to the control arms during all turning movements.

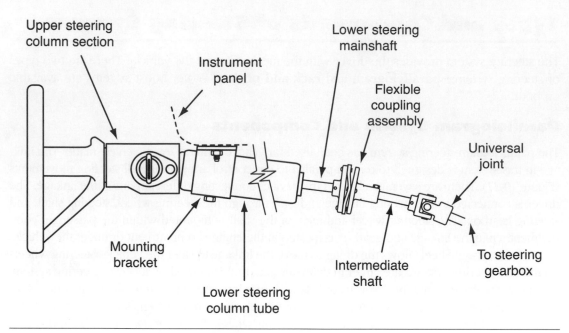

Figure 10-28 The steering shaft is inside the steering column and connects the steering wheel to the steering gear.

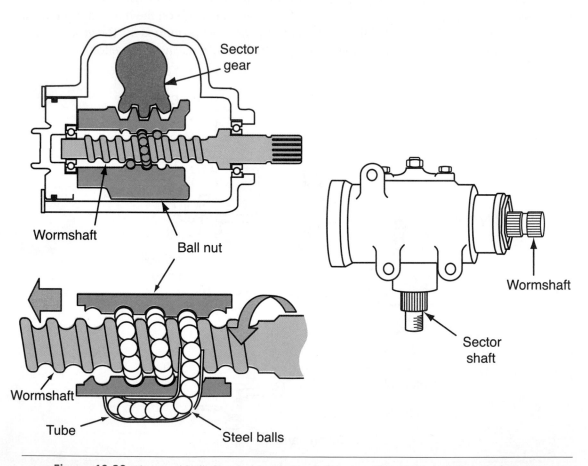

Figure 10-29 The steel balls located inside the ball nut act like the internal threads on a nut. They follow matching grooves in the worm gear and move the sector shaft in and out of the ball nut.

The pitman arm is attached to a **center link.** An **idler arm** at the opposite end of the center link from the pitman acts as a rotating anchor for the link (Figure 10-30). Sockets similar to the ball joint attach the center link to the idler arm and two tie-rods. A threaded, split, pipe-like adjusting sleeve is used to connect a tie-rod end to each of the tie-rods. The **tie-rod** ends are attached to the steering knuckle with sockets. Any rotary movement of the worm gear results in a linear movement of the pitman arm, center link, tie-rods, and ends to turn the steering knuckle and the wheel assembly from side to side. The adjusting sleeve is used in adjusting the wheel during alignment. It is clamped to the tie-rod and tie-rod end (Figure 10-31).

In some vehicles, the **center link** is known as the "drag link."

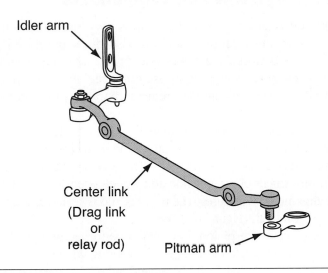

Figure 10-30 The idler arm only supports one end of the center link. Note that the center link may be referred to as a drag link or relay rod.

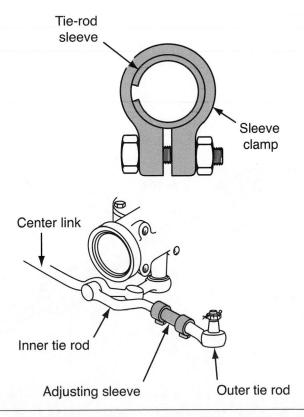

Figure 10-31 The tie-rods and tie-rod ends connect the center link to the steering knuckle. The adjusting sleeve connecting the end to the rod is threaded for alignment adjustments.

The steering knuckle is attached to the steering linkage and the suspension system. Attachment points at the top and bottom are used to attach the knuckle to the suspension. Another attachment point is used to connect the steering linkage. The steering knuckle may also have mounting holes or brackets for brake hardware. A spindle is a forged portion of a RWD steering knuckle. The spindle provides a machined surface (journal) for the wheel bearing, which in turn supports the wheel and tire assembly. On FWD or four-wheel-drive (4WD) vehicles the steering knuckle has a machined bore to accept a bearing for the outer end of a drive axle.

Rack-and-Pinion Systems and Components

The steering knuckle on a rack-and-pinion system is very similar to the parallelogram system. The steering column and steering wheel are also the same for all practical purposes. The main differences are between the lower end of the steering column and the steering knuckle. The rack-and-pinion system is comprised of three major components: a rack, a pinion gear, and the linkage. The linkage also remains parallel to the control arms during all phases of steering.

The rack is a long, straight bar of steel with teeth along a portion of one side (Figure 10-32). At each end of the rack are the inner tie-rod sockets. The tie-rod sockets perform the same function as the tie-rods and are attached to tie-rod ends. The sockets can pivot, allowing the outer end to follow the vertical movement of the knuckle and wheel assembly. The tie-rod end can be adjusted during wheel alignment procedures. The rack fits and is sealed in a housing where the rack teeth mesh with the pinion teeth (Figure 10-33).

The steering shaft rotates the pinion. The pinion performs the same function as the worm gear. The pinion moves the rack left and right as the steering wheel is turned. The rack and pinion occupies less space, is lighter, has fewer parts, and generally gives the driver better road feel. The rack-and-pinion system has very little, if any, torque multiplication. Steering even a small vehicle in traffic with manual rack and pinion can be tiring.

Power Steering

Both types of steering systems can be equipped with power boost. The boost systems for each are very similar. A pump supplies pressurized fluid to increase the driver's steering force. The pump is equipped with a pressure relief valve to prevent overpressurizing the system. Automatic transmission fluid is used many times as a power-steering fluid on older systems. Many newer

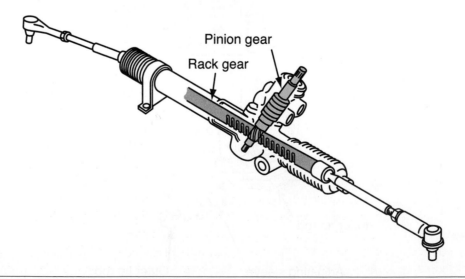

Pinion gear

Rack gear

Figure 10-32 The rack-and-pinion gear are mounted and sealed inside the housing. Normally two rubber-padded clamps fasten the housing to the body or frame.

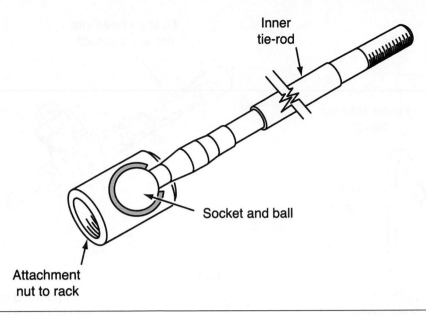

Inner
tie-rod

Socket and ball

Attachment
nut to rack

Figure 10-33 The inner tie-rod is fitted with a ball-socket section that is threaded onto one end of the rack. The outer end of the tie-rod is threaded to accept the tie-rod end.

models use specially blended power-steering fluid. The service manual should be consulted before topping off power-steering fluid.

A crankshaft belt-driven pump provides the pressurized fluid to the system. The fluid is delivered to the power-steering valves through steel lines and flexible hoses (Figure 10-34). The power-steering valve in a recirculating ball and nut gear is located at the top of the steering gear. The steering shaft rotates the valve to open a passageway to the correct side of the steering gear piston. Pressure applied to the piston assists the driver in turning the worm gear.

Pressurized fluid on a rack and pinion is delivered to the valves located at the top of the pinion gear. As the shaft and pinion are rotated, a valve opens and directs fluid to the correct side of a piston mounted to or as part of the rack (Figure 10-35). In this system, the power boost is applied directly to the steering linkage. In the ball and nut system, the boost is delivered to the steering gears.

Most power-steering systems have a power-steering switch in the pressure line. The switch signals the PCM to raise the engine idle speed during low-speed conditions such as parking.

Electric Power Steering

There are two versions of electric power steering: electro-hydraulic and full electric systems. The electro-hydraulic system merely replaces the belt-driven pump with an electric pump. The pump can be switched on/off based on steering load. During parking or other slow-speed maneuvering the electric motor will be on and full hydraulic pressure will be available. During cruise conditions the motor may be switched off until pressure is needed. The determination of when to switch on or off is signaled by the power-steering pressure sensor and a rotational sensor measuring steering shaft rotation speed and amount of rotation. The system can be tied into the PCM to allow for speed and body angle changes.

The full electric systems like the one found on the Acura NSX use electronic sensors, a controlling module, and electric motors. Sensors mounted on the steering shaft measure the rotation motion and speed of the shaft plus the torque being applied to the shaft. Based on this data, the computer commands electric motors to provide steering assistance. The current to the motors is controlled, determining how much torque and speed is applied to the steering linkage. At the time of this text printing, there is not a full-blown mass-production electric steering system. The steering shaft is still connected to the gearbox and the electric components are

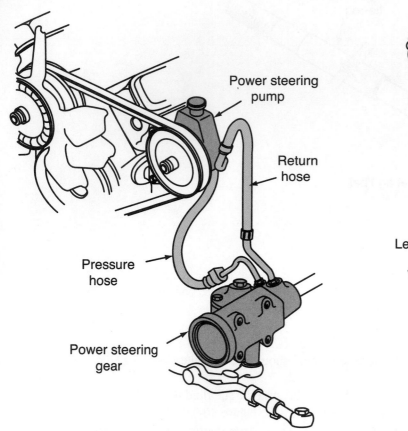

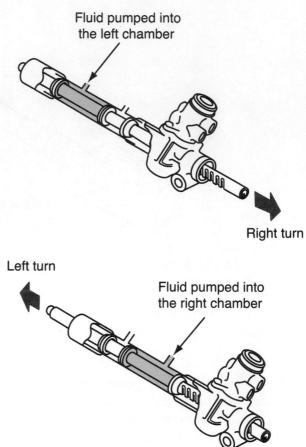

Fluid pumped into
the left chamber

Right turn

Left turn

Fluid pumped into
the right chamber

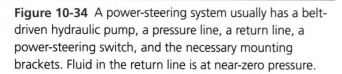

Figure 10-34 A power-steering system usually has a belt-driven hydraulic pump, a pressure line, a return line, a power-steering switch, and the necessary mounting brackets. Fluid in the return line is at near-zero pressure.

Figure 10-35 A rack-and-pinion power-steering operation is simple: Pressurized fluid is applied behind a piston forged to the rack. Fluid on the opposite side of the piston is allowed to return to the reservoir.

more for power assist than actual steering. This will probably change in the next few years after 2007.

Four-Wheel Steering (4WS)

There are several versions of 4WS steering systems. The system allows for the normally non-steering wheels, usually the rear, to be turned so the rear of the vehicle swings in an arc similar to that of the front of the vehicle. Typical of all systems is the reduced turning arc of the rear wheels. They cannot be turned as far as the front wheels.

Earlier Honda systems used a long worm shaft extending from the front steering gear to the rear steering unit mounted online with the rear wheels. This worm shaft had a different gear ratio to reduce the degree of rear wheel turn with regard to the front wheel turning arc. At slow speeds the rear wheels are turned in opposite direction to the front wheel. The front wheels are turned left and the rear wheels turn to the right. This reduces the amount of road space needed to complete a turn or the space needed to ease the vehicle into a parking space. Some sports cars use 4WS at higher road speeds. The turning of the rear wheels slightly into a turn in the same directional arc as the front wheels reduces the transition from straight-ahead driving to a curve. Some larger vehicles with four-wheel-drive, such as tow trucks, use 4WS steering to enable them to better retrieve their load.

Fully electrical 4WS is similar to the electric steering on the front wheels. Sensors measure the steering shaft's rotation and applied torque. Information from the PCM on vehicle speed determines if 4WS is activated or not. If activated, usually at low speeds, the 4WS computer processes the sensor data and commands the electric motor(s) attached to the rear wheel steering linkage to move the linkage in the correct direction, the correct amount, and the degree of turn. Sensors in the steering motor(s) report back to the controlling module on the degree of turn achieved and where the steering linkage is in relation to the centerline of the vehicle. In this manner, the computer compares the accomplished actions with the issued commands and the signals from the front steering to determine any further actions.

4WS-equipped vehicles that are offered today are not selling well. This is due to customers' complaints of a decrease in driver sensitivity or the feel of the road. Part of this loss is caused by the increased weight and complexity of the systems. In fact, General Motors will not offer their Quadrasteer[R] after the 2006 model year. The reason: only 16,500 Silverados/Sierras and Suburbans/Yukons with this system were sold between model years 2002 and 2005. Other manufacturers, primarily Honda and Mazda, are having the same marketing problems.

AUTHOR'S NOTE: One of my students purchased a Honda with 4WS and I test drove it on the commercial truck driving range. The most unsettling sensation I encountered was the feeling that the rear end was sliding out from under me. Admittedly, I was driving near the maximum 4WS speed allowed. It felt strange to me and I kept wanting to turn the steering wheel back to correct the slide. However, it is one of the greatest systems I've seen lately for parallel parking. We quickly parked in an area that was only about a foot longer than the car, and it came out of that space even easier.

Wheels and Tires

Wheels

The tire is the contact between the vehicle and the road. Since the tire is **pneumatic** (air-filled), a wheel or rim is needed to support it. Most wheels are made of steel or an alloy of magnesium and aluminum. Steel wheels are made by welding the rim to the centerpiece. Magnesium and aluminum wheels are cast molded. Passenger car wheels are listed as **dropped center safety rims.** The rim flange is set above the center (Figure 10-36). Holes around the center fit over studs fitted into the hub, and lug nuts are used to mount the wheel to the hub. Wheels are measured in diameter (across the wheel) and width (between edges) (Figure 10-37). The tire is held to the wheel by air pressure that forces the **tire's beads** over the wheel's center humps and seals them against the rim flanges. The two tire beads are tightly wrapped steel wires formed into loops slightly smaller in diameter than those of the rim flanges. Tires are matched to the wheel by diameter and width.

Many of the newest vehicle models offer electronic tire pressure monitors. The newest version of this monitor is a sensor installed into the wheel and is similar in appearance and mounting to a typical valve stem. The sensor radiates a low-power radio signal to a receiver on the vehicle. The signal is then transmitted to the body control module (BCM). The BCM switches on the instrument-panel-mounted warning light to alert the driver. This system is usually fitted with run-flat tires, which will be discussed in the next section. The sensor can be damaged by improperly removing or installing a tire onto the wheel if care is not exercised. There are very specific procedures to follow when using a tire machine or tire balancer.

The present-day tire is a result of research and development conducted at the racetrack and at tire manufacturing facilities. There are all types of tires, from the large racing slicks used for drag racing to the small ones for lawn mowers and go-karts. Passenger car tires range from 13 inches to 18 inches in diameter. There are different types of **tread** and **sidewall** designs to meet almost all road and driver situations. Tires can be selected for different types of weather, payload, and driving conditions. Some vehicle manufacturers have normal, sport, or touring tires that can

Shop Manual
pages 309–316

Tread is the molded portion of the tire that touches the road. It is supported by the **sidewall,** which extends from the bead to the tread area.

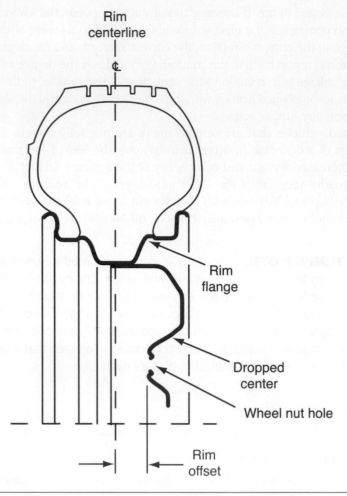

Figure 10-36 Shown is a dropped center rim. This places the mounting portion to the right of the vertical center line.

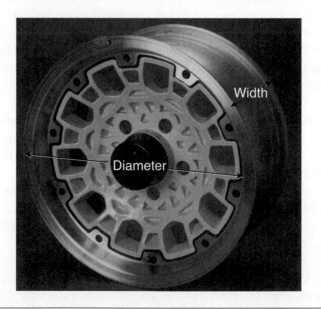

Figure 10-37 The diameter and width measurements of a wheel.

be placed on the same model car for different driving styles. Tires come in two basic configurations: **bias ply** and **radial ply.**

Bias Ply

Bias ply tires are no longer used on most vehicles. They can still be found on larger trucks and trailers. The fleet service technician may deal with bias ply tires regularly. The word "bias" refers to the direction the cords are laid under the tread (Figure 10-38). Cords run from one bead to the other, forming the sidewall and supporting the tread. Bias ply tires have the cords installed at an angle to the radii of the tire and tread. The bias ply tire is stiff and can carry heavy loads. As the wheel tilts during a turn, the bias sidewall picks up the inner portion of the tread from the road, thereby reducing traction and increasing wear. Bias tires are not fuel efficient and can be rough riding because of the lack of flexibility in the sidewall.

Radial Ply

The radial tire is currently the most popular tire in use today. The cords are laid 90 degrees to the tread and parallel to the tire's radii (Figure 10-39). This provides a softer ride, better fuel mileage, and better traction. The radial sidewall stiffens the tread while maintaining a flat tire/road contact. As the wheel assembly tilts during a turn, the tire's sidewall flexes instead of pulling the tread from the road. The soft sidewall flexes more readily over various road conditions. The tire's flexing sidewall results in a smoother ride and less rolling resistance. The lower resistance increases the vehicle's fuel mileage.

The greatest drawback of the radial tire design is the lack of driver education on its inflation characteristics. With a bias ply tire, it is fairly obvious that the tire is low on air pressure. However, with the soft sidewall of a radial tire, the tire tends to look a little underinflated even when pressurized to 35 psi. Many drivers do not recognize when the tire is low. As a result, many radial tires are worn out prematurely.

A tire that became available from several manufacturers in 1997 is commonly known as the run-flat or **extended mobility tire (EMT).** It has extra strong sidewalls that will support the

Extended mobility tires (EMTs) have strengthened sidewalls to support a vehicle for up to 125 miles in the event of tire damage. Some have temporary self-sealing capabilities.

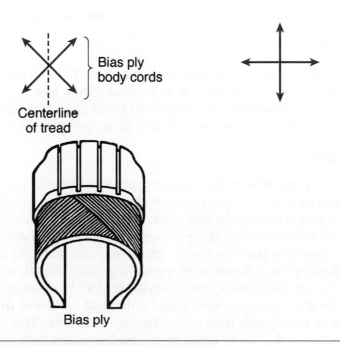

Figure 10-38 Bias cords make a stiff sidewall, but tend to lift the tread during cornering.

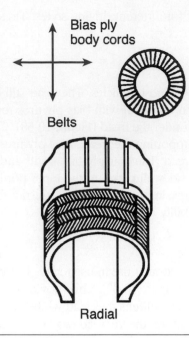

Bias ply
body cords

Belts

Radial

Figure 10-39 Radial cords cause a softer sidewall, allowing the tread to remain flat on the road.

vehicle for up to 125 miles. These tires are usually mounted on vehicles that are equipped with electronic tire pressure monitors to alert the driver of low tire pressure. Another type of EMT has self-sealing capabilities, but it is not designed to run forever with a puncture. The tire was designed to seal the hole and hold air until a repair facility could be reached. With this tire, there is no loss of control that often happens during a sudden deflation of the tire.

Tire Construction and Labeling

The tire is made up of several layers of different materials, each designed for a specific task. The cord layout was discussed in the last section. Other major components of the tire are the belts and tread.

Belts

The **carcass** is the completed tire minus the tread.

Belts of reinforcing materials are placed under the tread for better support. The belts are laid under and run in the same direction as the tread (Figure 10-40). The belts help hold the tread in position on the **carcass** and give better tread/road contact. The belts commonly used today are made from layers of steel wires. Belts are used in bias and radial tires.

Treads

There are many different tread designs, each for a particular purpose. They can be molded to bias or radial carcasses, although some work better with one type or the other. **Straight treads** are most common on over-the-road trucks, which require traction for starting and stopping the vehicle with little concern for heavy cornering. **All-weather treads** are made for the typical light vehicle. They have good traction on dry or wet roads and have some mud and snow capabilities. The shortcoming of the all-weather tread is its ability to work well in most situations, but not perfectly in any. Like most equipment on the vehicle, it is a design compromise.

Another tread commonly found is the **mud and snow tire.** The tread consists of knobby lumps of rubber with large spaces between the knobs. This allows the mud and snow to be pushed out of the way so the knobs can reach a traction surface. The mud and snow tread is generally used only on four-wheel-drive vehicles and in those areas of the world where mud or snow

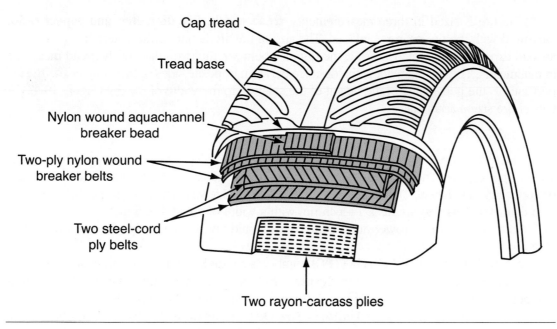

Figure 10-40 Note how the belts support the tread. This particular tire has a water (aqua) shedding channel supported by a special belt.

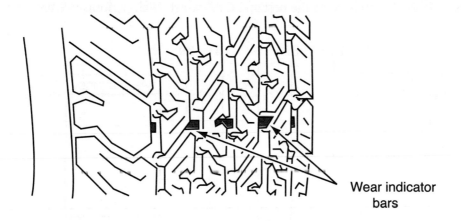

Wear indicator bars

Figure 10-41 Wear indicators are a quick, visible method of determining tire wear.

are common conditions. The northern parts of the United States have tire treads designed for snow only. While this tire works well in snow, it, like the mud/snow tread, is loud and wears quickly on dry pavement. A common solution in areas with regular snowfall is to use an all-weather tread and install tire chains when necessary.

A special tire is also available for ice. This tire has a mud/snow tread with small steel spikes protruding from the tread area. The spikes dig into the ice for traction. However, the spikes will not last long when driven on pavement without ice.

Each tire is equipped with tread wear indicators (Figure 10-41). Several solid strips of rubber run across the width of the tread area near the bottom of the grooves. When the tread wears to a depth where the indicators become readily visible, the tire should be replaced.

Tire Labeling

Each tire is identified by markings etched or molded into the sidewall. The markings include the manufacturer, tire brand, tread size, and test certifications. The most recognized marking is the tire size.

The tire is sized in three measurements: **tread width, tire diameter,** and **aspect ratio,** measured with the tire properly inflated. The size of the tire is one of the determining factors in its load capability. The tread width is the distance between the two edges of the tread measured in millimeters. The tire diameter is the distance between opposite sides of the same bead. The aspect ratio is the percentage of the height of the sidewall to the width of the tire (Figure 10-42). A typical tire size marking is:

P185/70R14

The first group, 185, means the tread is 185 millimeters wide. The letter "P" prefix designates a passenger car tire. The number 70 indicate the sidewall height is 70 percent of the tire's width or about 129 millimeters. The letter "R" indicates a radial sidewall. The last group, 14, is for a 14-inch diameter. This size tire fits a 14-inch rim with a width of about 4 to 5 inches. The tire could be fitted to a wider or narrower rim but the beads would not seal correctly and might result in a blowout.

A **Tire Performance Criteria (TPC)** number is molded into the sidewall near the tire size (Figure 10-43). This number indicates the tire's traction, noise, handling, rolling ability, and endurance characteristics. Replacement tires should have the same TPC number.

Other markings include the **Uniform Tire Quality Grading (UTQG),** which outlines the tire's traction, tread wear, and temperature resistance. The grades are shown by letters and numbers stamped on the sidewall. Tread wear is based on a percentage of 100. Percentages range from 50 (poor) to 270 (long) relative wear. Resistance to skidding (traction) and temperature are listed by letters A, B, or C, with A being the best and C the worst. High-performance tires can be rated

The **Tire Performance Criteria (TPC)** is a number, molded in the sidewall, that lists important information about the tire.

Uniform Tire Quality Grading (UTQG) is a system developed to provide the customer a way to compare, select, and trace tires.

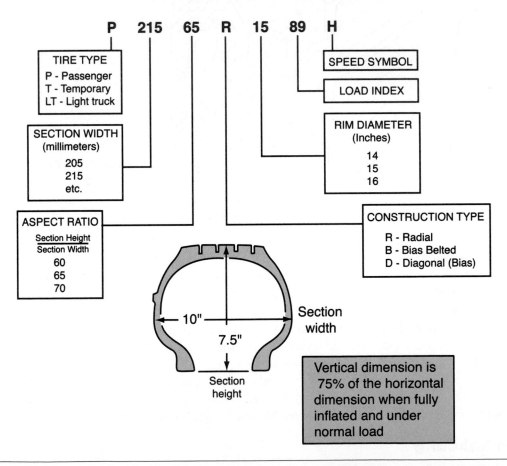

Figure 10-42 The aspect ratio is the height of the tire as a percentage of tire width. The measurements apply when the tire is at recommended pressure, cold, and loaded by the vehicle.

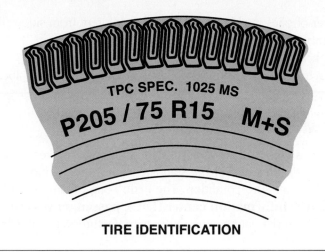

TIRE IDENTIFICATION

Figure 10-43 The type of tire and the tire's performance are indicated by a TPC marking molded into the tire's sidewall.

TIRE SPEED RATINGS

Letter Symbol	Maximum Speed
F	50 mph
G	56 mph
J	62 mph
K	68 mph
L	75 mph
M	81 mph
N	87 mph
P	93 mph
Q	100 mph
R	106 mph
S	112 mph
T	118 mph
U	124 mph
H	130 mph
V	149 mph
Z	+149 mph

Figure 10-44 Tire speed ratings chart.

for speed using the codes F (50 mph) to Z (149 mph). The codes would show up as part of the tire size markings such as 225/70R 16 Q for a tire rated at 100 mph (Figure 10-44).

The Federal Department of Transportation requires certain information to be placed on the tire. The required information is:

- Tire size
- Maximum permissible inflation pressure
- Maximum load rating
- Generic name of each cord material in sidewalls and tread

- Actual number of plies in sidewall and tread if different from sidewall
- Tubeless or tube type as applicable
- Radial as applicable
- DOT manufacturing code listing the tire manufacturer, factory location, and date of manufacture

Tire Pressure Monitors (TPMs)

Tire pressure monitors (TPMs) have been installed on some newer vehicles. The TPM sensor is mounted into the rim in place of the standard valve stem. It monitors tire pressure and alerts the driver to a low-pressure condition. The main components of a General Motors TPM system consist of a **driver information center (DIC), passenger door module (PDM),** four pressure sensors, and the serial data circuit. Other manufacturers use similar system operations and similar components.

The pressure sensor mounted on each wheel transmits a radio signal to a module in the vehicle (Figure 10-45). The sensor has an internal roll switch that is opened/closed by centrifugal force or the lack of that force. When stationary, the switch is open and in a stationary mode. In this mode, the sensor monitors tire pressure every 30 seconds and transmits that data to the module every 60 minutes. When the vehicle is moving (tire is rolling), the switch is forced closed, putting the sensor into drive mode. In this mode, the pressure is sampled every 20 seconds and the data transmitted every 60 seconds.

The PDM is the receiver for the transmitted data. The data is translated into sensor presence, sensor mode, and the tire pressure. This interpreted information is forwarded to the DIC where it is processed and displayed.

The DIC provides tire pressure reading displays plus two other messages: *Check Tire Pressure* and *Service Tire Monitor.* The meaning of the first message is to alert the driver to a low-pressure condition in one or more of the tires. This condition will not set a diagnostic trouble code (DTC). The second message, Service Tire Monitor, should set a DTC and is indicative of a problem within the TPM system. The code(s) can be retrieved with a scan tool. The procedures for servicing, replacing, or performing a relearn are covered in the service manual. The relearn procedures are

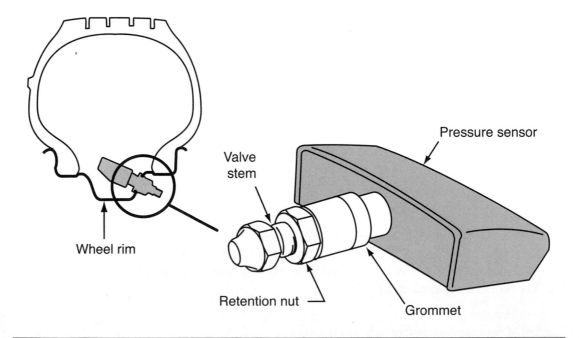

Figure 10-45 Note how the TPM sensor extends into the tire's cavity. It can be damaged by not following repair procedures. The cap and valve are designed to withstand corrosion and must not be replaced with regular caps or valves.

required after tire rotation, before replacing a suspected sensor, and after replacing a sensor. In the case of the General Motors TPM system, there are two means to conduct a relearn. One is tire pressure changes when the tire pressure is lowered and increased to force the sensor to remeasure. The changes in air pressure must be a specified amount in each direction. The specific instructions are contained in the service manual. The second method requires a special TPM diagnostic tool to command the system into the relearn procedure. This is probably not a good task for an unsupervised, entry-level technician.

Wheel Alignment

Shop Manual
pages 316–320

A vehicle with perfect alignment of the four wheels will steer straight ahead and return to center after a turn. Wheels that are not in alignment with the vehicle and other tires affect the driver's control. Tires that are not in balance could also affect steering and ride. Alignment of the wheels places all wheels parallel to each other and the center or thrust line of the vehicle.

There are usually three angles that can be adjusted by the technician, but other measurements may affect tire wear and steering. **Steering axis inclination (SAI), setback angle,** and **toe-out-on-turns** can only be repaired or adjusted by replacing worn or damaged parts. SAI is the angle created by the steering knuckle and control arms or struts as they are installed in the vehicle (Figure 10-46). If the SAI is wrong, the damaged parts, including the frame, must be replaced or repaired.

Setback is the measurement of one wheel to the other wheel on the same axle (Figure 10-47). In other words, the two wheels on one axle should be straight across from each with relation to the frame/body. Extreme setback could cause steering and possibly suspension problems. Some vehicles may have enough adjustment capability to offset some error in SAI or setback. While this may prevent tire wear and retain good steering, the vehicle will look as though it moves sideways. This "dog tracking" is a common condition on older, half-ton pickup trucks that carried loads across rough terrain and had a bent or twisted frame.

A bent or twisted frame can change the **setback, SAI,** and **toe-out-on-turn** settings without damaging suspension or steering components.

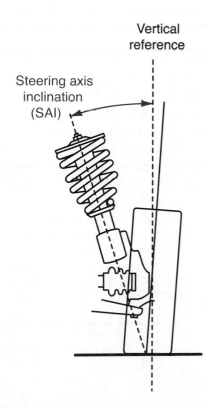

Vertical reference

Steering axis inclination (SAI)

Figure 10-46 The steering axis inclination is manufactured into the suspension and steering assembly. It is not adjustable without a body shop.

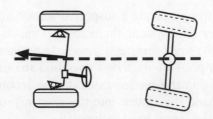

Figure 10-47 Setback occurs when the wheels or axles are not aligned to each other.

Toe-out-on-turn is required because the turning angles between the inside wheel and the outside wheel are different. When a vehicle travels around a curve, the inside wheel travels a shorter distance at a sharper angle than the outside wheel (Figure 10-48). The shaping and positioning of the steering linkage and knuckle sets the outside wheel's turning angle. If the toe-out-on-turns is wrong, some or all of the components must be replaced.

Three alignment angles that may be adjusted are **caster, camber,** and **toe.** Many new FWD vehicles are not equipped for adjustment of caster or camber or both. The rear wheels on many FWD vehicles can be adjusted for caster, camber, toe, or all three. All vehicles should be checked for four-wheel alignment even though an adjustment of the wheels may not be possible. A four-wheel alignment check gives the technician an idea of the frame or unibody condition by comparing the caster, camber, and toe on all wheels.

Caster is the forward or rearward tilt of the steering axis from vertical (Figure 10-49). Caster is measured as positive or negative. Most light vehicles require a positive caster or a rearward tilt of the top of the steering axis centerline. A positive caster tends to make the wheels steer straight and assists in the return of the steering to center after making a turn. Positive caster does require more steering effort when negotiating a turn.

Negative caster is used on large, heavy trucks to assist in turning the front wheels, but the wheels have to be steered back to straight. Poor caster will not normally wear the tire but may affect steering quality. Positive caster is usually expressed as a positive (+) number on the alignment machine.

Camber is the inward or outward tilt of the steering axis centerline (Figure 10-50). A positive camber tilts the top of the wheel away from the center of the car. Camber is used to place the load of the vehicle at the center of the tire/road contact area. This prevents tire wear and assists in keeping the tires straight ahead. Poor camber will wear the tires and affect steering quality. Positive camber is usually expressed as a positive (+) reading on an alignment machine.

Toe is the difference distance between the front edges and the rear edges of tires on the same axle (Figure 10-51). Older vehicles usually had some toe-in while most FWD cars have zero toe. All vehicles have some type of toe adjustment. **Toe-in** means the leading (front) edges of the tires are closer than the rear edges. **Toe-out** is the opposite. Toe sets the wheels straight ahead when the vehicle is loaded and moving. Note that the measurement is total toe, but toe adjustments are made at individual wheels so the wheels need to be set at equal toe. For instance, if the total toe is (+) 0.12 inch, each wheel would be set at (+) 0.06 inches. Poor toe wears the tires

When two wheels are not set equally in **toe,** the customer may complain that the steering wheel is off-centered.

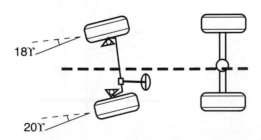

18Υ

20Υ

Figure 10-48 The outside wheel does not turn at the same angle as the inside wheel.

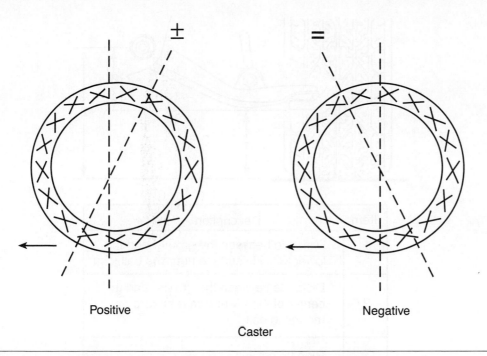

Positive   Negative

Caster

Figure 10-49 Positive caster is the rearward tilt of the steering knuckle or suspension when viewed from the side.

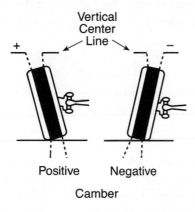

Vertical
Center
Line

+   −

Positive   Negative

Camber

Figure 10-50 Camber is the inward or outward tilt of the wheel as viewed from the front.

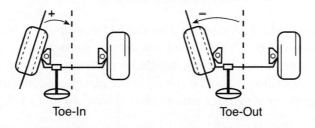

+   −

Toe-In   Toe-Out

Figure 10-51 Shown is the toe on a wheel that is half of the total toe.

and affects the steering. Toe-in is usually expressed as a positive (+) number on an alignment machine.

A fourth measurement that may have to be adjusted before setting the caster, camber, and toe is the vehicle's **ride or curb height** (Figure 10-52). Many vehicles require the ride height to be measured and adjusted before beginning alignment procedures. The height of the vehicle

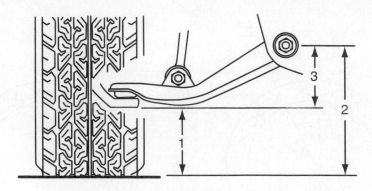

Item	Description
1	Distance between the ground and the lower knuckle surface near the ball joint
2	Distance between the ground and the center of the lower arm rearward mounting bolt
3	Ride height = 2 - 1

Figure 10-52 Ride height is critical to many vehicle wheel alignment procedures. It is usually calculated by measuring between two different points and the ground (floor) and finding the difference between the two.

affects the angles of the steering and suspension components. Earlier, in the discussion on torsion bars, it was noted that some torsion bars have an adjustment to change ride height. Other systems may require suspension components to be replaced to correct ride height. The amount of load a vehicle carries may change the ride height and suspension angles.

A vehicle that normally carries a heavy load, such as a large tool unit with repair equipment in the bed, should have a wheel alignment performed with the load onboard. The load tends, depending on its place in the bed, to add or remove weight from the steering wheels. Another example is a sales representative's vehicle, which usually carries numerous product samples and sales material in the trunk or rear seat. This removes weight from the front steering wheels. In each case, the load can force the alignment out of adjustment and result in excessive tire wear and poor steering.

When performing a wheel alignment, caster should be completed first, followed by camber. While caster and camber can be adjusted simultaneously on some vehicles, toe is always adjusted last. Toe may affect camber to some degree, but camber adjustments always affect the toe. Many vehicles require that a specified weight be placed in the passenger compartment to simulate the driver's weight. In theory, most vehicles are assumed to have a 150- to 175-pound driver without any other load.

Tire Balance and Rotation

Shop Manual
pages 313–315

A balanced tire provides a smooth ride, better steering, and smoother braking. An unbalanced tire may cause the vehicle to shake or vibrate at certain speeds and it may cause tire wear. Tires that are not rotated regularly will wear unevenly and shorten the tire's life. They will also become unbalanced quickly.

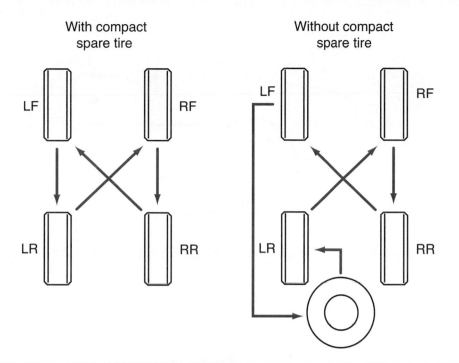

Figure 10-53 The movement of the tires during a rotation service is set by the vehicle manufacturer. Shown is a typical rotation pattern with and without a compact spare.

Most vehicle and tire manufacturers recommend tire rotation about every 5,000 miles. All tires should be the same size and type. The direction of rotation depends on vehicle and tire manufacturer instructions, but generally, the rear tires go to the front axle and the fronts to the rear. Some manufacturers recommend that the tires be crossed on the vehicle as they are moved from axle to axle (Figure 10-53). Moving the tires to the opposite sides of the vehicle reverses the rolling movement of each tire. This helps even the wear across the tire. The tires should be balanced during the rotation.

Regular tire wear will cause some unbalancing of the tire. Rough road surface, striking curbs, or other road hazards can damage a tire or wheel, thereby causing an unbalanced condition. Balancing basically means adding weights to offset a heavy spot in the wheel assembly. Most wheels are balanced by the wheel manufacturer and will remain balanced unless damaged. If a wheel is bent or damaged in some way, it should be replaced. Even if a bent wheel assembly can be balanced, it will still wobble and cause a vibration. The wheel assembly is balanced using either static or dynamic procedures.

Static Balancing

A wheel assembly is considered to be **static balanced** if its weight is distributed evenly around the center of rotation as viewed from the side (Figure 10-54). The wheel will tend *not* to rotate if balanced. A heavy spot will cause the tire to rotate until the spot is resting at the bottom. The heavy spot would also cause a tramping or thumping noise as the vehicle was driven. Static balancing is performed by adding a weight equal to the heavy spot on the opposite rim of the wheel. This would stop the hopping or tramping of the wheel but it would not eliminate a side-to-side wobble if the wheel was not dynamically balanced.

Static balancing is usually done on heavy truck or trailer tires using a bubble balancer.

Dynamic Balancing

A wheel assembly that is evenly balanced in four quadrants (viewed from the front) is considered to be **dynamic balanced.** Not only is the tire evenly balanced around its center, it is also balanced from side to side on the tire's vertical centerline. To accomplish this task, the front of the

Dynamic balancing is almost always used on cars and light trucks.

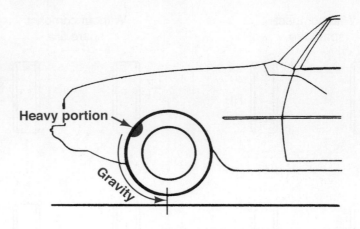

Figure 10-54 This wheel is statically unbalanced. It will rotate by itself until the heavy spot is at the lowest point.

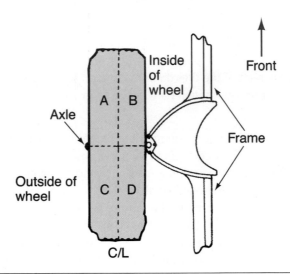

Figure 10-55 The tire is theoretically divided into four quarters and each quarter is balanced to the others in dynamic balancing. A computerized balancer is required.

tire is divided into four equal sections and each section is balanced to the other three (Figure 10-55). The entire wheel assembly is now balanced.

Alignment and Tire Balance Machines

Shop Manual
pages 316–320

There are two machines used by repair shops to align the suspension and steering and balance the wheel assemblies. Both are fairly expensive pieces of equipment and should be used only by trained technicians.

Alignment Machines

Most shops use a special lift for wheel alignments. It is usually a drive-on lift with drop-down legs to hold the vehicle level during alignment (Figure 10-56). There are turntables for the front wheels so they can be moved for adjustment. Most lifts of this type have long rear portions on each side

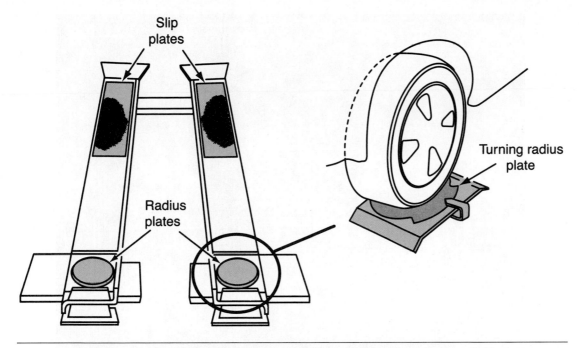

Figure 10-56 The rear skid plate can move for adjustment of the rear wheels. The turntables are for the front wheels. (Courtesy of Snap-on Tools Company)

that can be unlocked to allow adjustment of the rear wheels. The lift may have jacks at the front and rear so the vehicle can be raised from the lift for compensating the **wheel sensors** and to make repairs or adjustments.

The **alignment machine** consists of the computer, receiver, and four wheel sensors (Figure 10-57). Older versions used electrical cables to carry the sensor data to the computer. The newest machines use sensors with radio transmitters to send the data to a receiver. The receiver translates the signal so it can be used by a personal computer. Most systems use Windows 95/98 or newer versions to display the information in a readable form for the technician. The computer uses software that is designed and written for a particular brand of alignment machine. Hunter, Bear, and other machine manufacturers use the same basic data, but interpret and display it according to their software. Each manufacturer usually includes some type of technician training package on its computer. Hunter Machine Company has video/audio segments that a technician can select to show exactly how an adjustment is done on a particular vehicle. Other manufacturers use similar video, audio, or graphics to show methods of adjustment. Most of the newer units also offer a vehicle safety checklist and a repair order that is completed on screen and then printed. The computer can store customer and vehicle data, before and after measurements, and other data the repair shop may desire. In this manner, the shop can recall data about a vehicle on the customer's next visit to the shop. Software can be added to connect the alignment machine to the cashier, service writer, other departments, and the Internet.

The **wheel sensors** must be compensated to allow for the angle of the wheel assemblies.

Tire Balancers

The tire balancer is another computerized machine used in most repair shops (Figure 10-58). The tire is mounted onto the machine's spindle and clamped down. Data on the wheel diameter, wheel width, and distance from the machine is entered into the computer using buttons or dials (Figure 10-59). The balancer is set up to spin the tire for a set time at a set speed. In this manner, it can locate the heavy spots and select the positions for the counter weights. The location of a weight is usually displayed using a stationary marker or arrow and a light that moves with the

Wheel weights are made of lead and should be treated and disposed of as hazardous waste when no longer usable.

Figure 10-57 Shown is a Hunter P-410 alignment machine. Other brands and models are similar.

rotation of the wheel assembly. The light is aligned with the mark and the weight is installed at the top dead center of the wheel's rim. Weights may be added on one side or both sides of the wheel assembly.

The balancer can be used for static and dynamic balancing. However, most machines will only be able to balance tires up to a certain size and type. Some machines can balance a car tire up to 18 inches in diameter, but may not be able to balance a medium truck tire because of the tire weight and type.

Figure 10-58 Shown is a Hunter balancer. Other brands and models are similar.

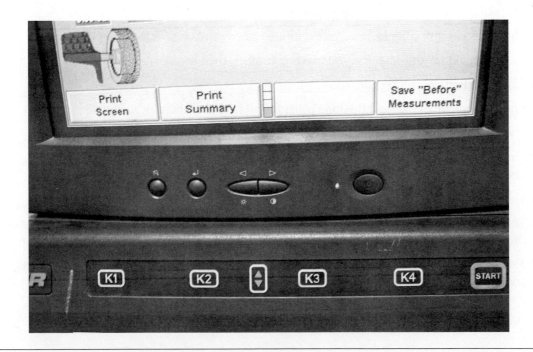

Figure 10-59 The display on a typical tire balancer.

Summary

❑ The Laws of Motion set forth the principles of the movement of objects.

❑ The Laws of Motion directly affect the suspension and steering systems because of their normal operations.

❑ Independent suspension means one wheel can react to road conditions without affecting the opposite wheel.

❑ Nonindependent suspension usually means the two opposing wheels are connected solidly to the same axle.

❑ Control arms are used on independent suspension systems.

❑ Ball joints connect the control arms to the steering knuckle.

❑ Springs support the vehicle's weight while the shock absorber controls spring movement.

❑ Jounce is the upward movement of a wheel assembly; rebound is the downward movement.

❑ Parallelogram steering systems require a system of linkages that remain parallel to the control arms.

❑ Rack-and-pinion steering uses a direct linkage between the input steering gear (pinion) and steering knuckle to turn the wheels.

❑ Power assist units are used to reduce driver steering efforts.

❑ Electric steering uses electronics to turn the steering gear.

❑ Four-wheel steering provides better vehicle control.

❑ Wheels fasten to the hub and support the tire.

❑ Many wheels have tire pressure sensor installed in them.

❑ The tire size includes tread width, diameter, and aspect ratio and determines the tire's load capacity.

❑ Tires are graded for traction, tread wear, speed, and temperature.

❑ SAI, setback, and toe-out-on-turns cannot be adjusted.

❑ The three possible alignment adjustments are caster, camber, and toe.

❑ Caster should be adjusted first, followed by camber, then toe.

❑ Camber adjustments affect the toe.

❑ Checking alignment should be done using four-wheel alignment procedures.

❑ The vehicle should be aligned while carrying its normal load.

❑ Tire balance can wear tires, cause the vehicle to shake or vibrate, and cause an uncomfortable ride.

❑ The wheel assembly can be balanced statically or dynamically.

❑ The alignment machine uses four wheel sensors and a computer to check the alignment of each tire individually and each tire to the center or thrust line of the vehicle.

❑ The tire balancer spins a tire to a set speed, uses a computer to calculate the heavy spots, and pinpoints the spot to add a weight.

Review Questions

Short Answer Essay

1. Describe the parallelogram steering system.

2. Explain how power assist is used on a rack-and-pinion steering system.

3. Describe the use of the short and long arm configuration in a suspension system.

4. Explain the purpose of caster and camber.

5. List the major components in a rack-and-pinion steering suspension.

6. Define the elements of the following data:

   P235/70R15

7. Define static and dynamic balance.

8. Describe the common purpose of coil springs, leaf springs, and torsion bars.

9. Explain the use of adjusting sleeves on a parallelogram steering suspension.

10. List the wheel data that is entered into the tire balancer.

Fill-in-the-Blanks

1. A(n) _____ is used to connect the pinion gear to the wheels.

2. Shock absorbers _____ the movement of _____.

3. Tie-rod ends are attached to the steering _____ by a(n) _____.

4. Control arms are mounted to the _____ and attached to the steering knuckle with _____ _____.

5. Tire wear is not caused by misadjusted _____.

6. A tire can be balanced using _____ or _____.

7. The parallelogram steering system normally uses a(n) _____ gear system.

8. The control arm resting on the coil spring is the _____ _____ arm.

9. The _____ , _____ , and _____ can be adjusted using an alignment machine.

10. Pressurized _____ gas is used to cause immediate action in the shock absorber.

ASE-Style Review Questions

1. The steering system is being discussed.
   Technician A says a parallelogram system uses a rack for connection to the steering knuckle.
   Technician B says the steering knuckle is used to mount the hub.
   Who is correct?
   A. A only
   B. B only
   C. Both A and B
   D. Neither A nor B

2. Each of the following may be part of a fully independent suspension EXCEPT:
   A. coil springs.
   B. shock absorbers.
   C. struts.
   D. multileaf springs.

Terms to Know

Shock absorber
Sidewall
Spring air valves
Sprung weight
Static balance
Steering Axis Inclination (SAI)
Straight treads
Strut
Sway bar
Tie-rod
Tire bead
Tire diameter
Tire Performance Criteria (TPC)
Tire Pressure Monitor (TPM)
Toe
Toe-in
Toe-out
Toe-out-on-turns
Torsion bar
Tread
Tread width
Uniform Tire Quality Grading (UTQG)
Unsprung weight
Variable rate springs
Wheel sensors
Worm gear

3. Wheel alignment is being discussed.
   Technician A says ride height is a factor in alignment procedures.
   Technician B says the total toe must be divided by two during the adjustment procedures.
   Who is correct?
   A. A only
   B. B only
   C. Both A and B
   D. Neither A nor B

4. Wheel alignment is being discussed.
   Technician A says caster adjustments may affect toe.
   Technician B says the toe adjustments should be made after the caster and before the camber.
   Who is correct?
   A. A only
   B. B only
   C. Both A and B
   D. Neither A nor B

5. Steering angles are being discussed.
   Technician A says setback results from the manufacturing and assembly of the vehicle.
   Technician B says SAI angles can be adjusted on some vehicles.
   Who is correct?
   A. A only
   B. B only
   C. Both A and B
   D. Neither A nor B

6. *Technician A* says SAI angles include the three adjustable alignment angles.
   Technician B says toe-out-on-turn angles are only effective during sharp turns.
   Who is correct?
   A. A only
   B. B only
   C. Both A and B
   D. Neither A nor B

7. The suspension system is being discussed.
   Technician A says a torsion bar suspension requires a shock absorber.
   Technician B says a leaf spring system does not use a shock absorber.
   Who is correct?
   A. A only
   B. B only
   C. Both A and B
   D. Neither A nor B

8. A nonindependent suspension system is being discussed.
   Technician A says this type suspension is commonly used on rear-wheel-drive trucks.
   Technician B says coil springs can be used on this type of suspension.
   Who is correct?
   A. A only
   B. B only
   C. Both A and B
   D. Neither A nor B

9. Tire data is being discussed.
   Technician A says actual speed test data is molded directly onto the sidewall.
   Technician B says the required tire information is determined by the Department of Transportation.
   Who is correct?
   A. A only
   B. B only
   C. Both A and B
   D. Neither A nor B

10. Wheel alignment is being discussed.
    Technician A says a good alignment will help the driver control the vehicle with little effort.
    Technician B says adjusting the toe before adjusting the camber could cause the final toe setting to be incorrect.
    Who is correct?
    A. A only
    B. B only
    C. Both A and B
    D. Neither A nor B

Powertrains

Upon completion and review of this chapter, you should be able to:

❏ Discuss the purpose of the driveline.

❏ Discuss using gears to change torque ratio.

❏ Describe the operation of a manual transmission/transaxle.

❏ Explain the purpose of the clutch.

❏ Explain the purpose of the torque converter.

❏ Describe the general operation of an automatic transmission.

❏ List and explain the components of drive shafts and axles.

❏ Describe the purpose and operation of the differential and final drive assemblies.

CHAPTER 11

Introduction

The engine produces the power that drives the wheels. However, a direct connection between the engine and drive wheels is not a satisfactory arrangement. There has to be a method of delivering high torque to get the vehicle moving and then lowering the engine speed once under way. The transmission and other components of the driveline accomplish that task. For the purpose of this chapter, the term transmission refers to a RWD transmission and a FWD transaxle. The term differential refers to the RWD differential and the FWD final drive. Specific operational and design differences will be noted for each.

Purpose and Types of Drivelines

Shop Manual
pages 337–342

The driveline delivers the power of the engine to the drive wheels. A transmission is used on most rear-wheel drives while a transaxle is used with front-wheel-drive vehicles. Some vehicles, like the older VW Beetles and Pontiac Fiero, used a transaxle to drive the rear wheels. The 1999 Chevrolet Corvette has a transmission mounted near the rear drive wheels.

The primary difference between the transmission and the transaxle is the method of delivering transmission/transaxle output power to the drive wheels. The transmission uses a drive shaft, which drives the input gear of a differential. The differential turns the torque line 90 degrees and drives the axles. The axles are attached to the drive wheels.

A transaxle has an integral final-drive assembly. The final drive performs the same function as the differential except the transaxle output is delivered directly to the input gear of the final drive. There is no drive shaft in this arrangement.

The transmission and transaxle use gears to change the engine's output torque. A manual transmission requires the driver to select, through mechanical linkage, a gear that is best suited for the engine load. A typical gear setup is reverse, first, second, third, fourth or direct drive, and fifth gear. Some vehicles have a sixth gear.

An automatic transmission uses hydraulics, valves, sensors, and actuators to achieve gear changes. The driver selects a drive range and drives away. A typical drive range is reverse (R), which has one gear ration, and drive (D). Drive usually has three forward gear ratios and often has a fourth or overdrive gear. Each system has a neutral and the automatic has a position for park.

Gears and Gear Ratios

Different gears and gear ratios are used to harness the torque of the engine as it is needed for different loads. The transmission and differential provide the means to do this.

Gears

Gears in a transmission are normally circular-toothed components. The teeth of one gear meshes or interacts with another gear. The diameters of the gears determine if the input torque is increased or decreased. If torque is increased, the speed is decreased and vice versa. A simple gearset consists of two gears working together (Figure 11-1). The gear delivering the torque into the gearset is the **drive gear.** The gear delivering the torque out of the gearset is the **driven gear.** Gear sets of three or more gears are called **compound gear** sets and may have more than one gear ratio within the set. Gear teeth may be on the outside of the gear, external, or on the inside of the gear, internal. Internal gears are usually found in planetary gearsets. Planetary gearsets are most commonly found in automatic transmissions but may be used for fifth or sixth in a manual system.

Ratios—Torque and Speed

Ratios are computed by dividing the driven gear by the drive gear. A gear set with a 1-inch drive and a 2-inch driven has a gear ratio of two to one or 2:1. If the drive was delivering 10 foot-pounds of torque, the driven would be delivering 20 foot-pounds of torque to the output. The example here shows a doubling to the available torque (Figure 11-2).

Using the same example, we can show the speed ratio for this gearset. Speed ratios are computed by dividing the drive by the driven, in this case, one is divided by two or .5 to 1. The driven gear will turn half a revolution to each complete revolution of the drive. If the drive turned at 10 revolutions per minute, the driven would only turn 5 revolutions per minute. In some vehicles, this would be low gear: high torque with a low speed.

Direct drive gear in most vehicles has the transmission output shaft turning at the same speed as the input. The gear ratios between low and high are determined by the gears being used. An overdrive gear would have a larger drive gear than the output gear (Figure 11-3). Assuming that the overdrive drive gear is 2 inches in diameter and the output is 1 inch, divide the drive by the driven for a gear ratio of 2:1. In this case, each time the drive gear completed one revolution, the output gear would rotate two times.

Note that seems to be the same gear ratio mentioned in the first paragraph of this section. Both are 2:1. However, in that instance the ratio caused an increase torque, but a decrease in speed. The situation noted in this paragraph causes a decrease in torque and an increase in speed. When calculating gear ratios, always remember that a large gear driving a smaller gear will always result in a speed increase with a reduction in torque.

Note that the high torque reduces the speed by the same proportion. This is true of any gearset. If the torque is high, the speed will be low. As the vehicle nears cruising speeds, different gearsets are selected to provide a high speed, which in turn lowers the amount of torque being delivered to the drive wheels and lowers the engine speed.

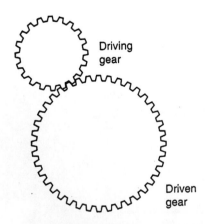

Figure 11-1 A small drive gear and a large driven gear increase the torque and decrease the speed.

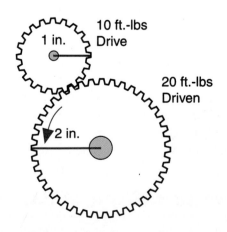

Figure 11-2 The 2:1 gear ratio in this gearset doubles the amount of torque available at the output.

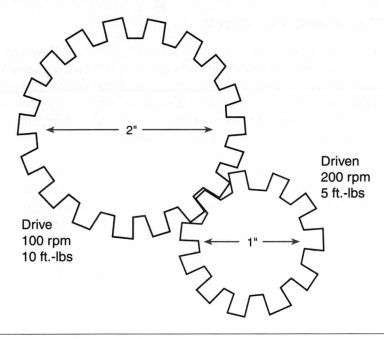

Figure 11-3 Using the small gear as the driven gear will result in higher output speed but less torque.

Clutches

In order to apply the engine power smoothly to the transmission or transaxle, there has to be a device to couple and uncouple the two components. The most common clutch used in a typical vehicle is a manual clutch assembly for a manual transmission and hydraulic clutches within an automatic transmission. Hydraulic clutches will be discussed with automatic transmissions. The manual clutch assembly is made up of three subassemblies and the linkages (Figure 11-4). The subassemblies are the **pressure plate, friction disc,** and **release bearing.**

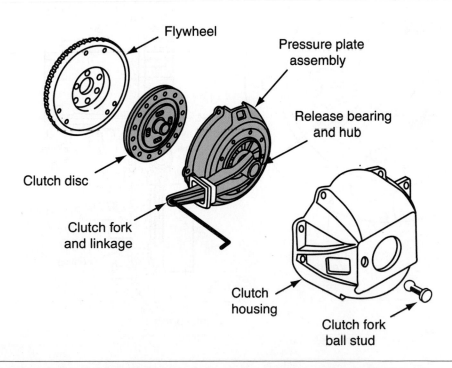

Figure 11-4 The major components of the clutch assembly are the clutch disc, pressure plate, and release bearing.

Shop Manual
pages 342–344

The **pressure plate** assembly is used to clamp the clutch disc to the flywheel.

The **friction disc,** once clamped, provides the connection between the engine and the transmission.

The **release bearing** provides a point of contact between a stationary component (linkage) and a moving component (pressure plate assembly).

Pressure Plates and Flywheels

The **flywheel** is driven by the starter motor during cranking to spin the crankshaft.

Bolted to the rear or output end of the crankshaft is a heavy, balanced **flywheel,** which is considered to be part of the engine assembly. According to one of Newton's Laws of Motion, anything in motion tends to stay in motion. The flywheel, once rotating, tries to keep rotating. This helps rotate the crankshaft and tends to reduce the vibrations generated by the start/stop movements of the pistons. The flywheel also provides a solid crankshaft-driven platform to mount the clutch's pressure plate (Figure 11-5). Some manufacturers refer to the pressure plate as the clutch cover. The flywheel is machined smoothly on its outward or rear face.

A steel frame supports the pressure plate, springs, and release levers or fingers (Figure 11-6). The frame is bolted to the flywheel. The pressure plate is finely machined on the flywheel side. High-tension springs apply force to hold the machined side firmly against the flywheel. This force clamps the clutch disc between the flywheel and pressure plate. The clutch is now engaged and engine power is transferred to the clutch disc.

Release levers extend from pivot points along the outer edge of the frame up and toward the center (Figure 11-7). If force is applied against the inner tips of the levers, the levers' other ends pull the pressure plate away from the flywheel and clutch disc. The clutch disc spins free and no power is transferred. The clutch is disengaged.

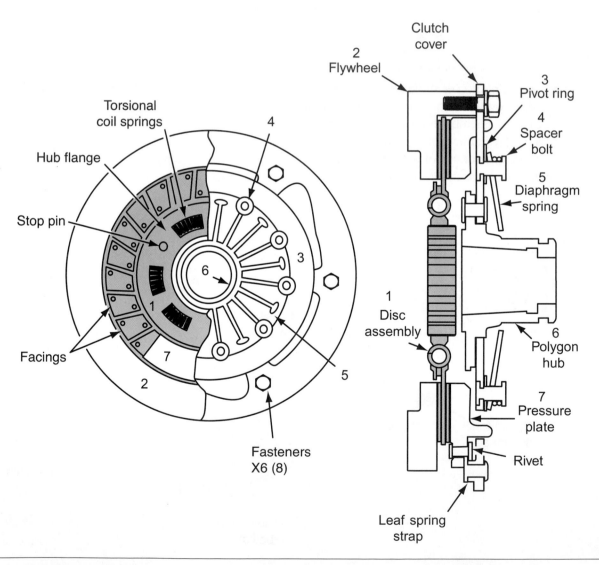

Figure 11-5 The pressure provides the clamping force to hold the clutch disc firmly to the flywheel.

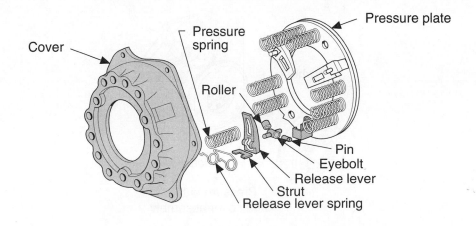

Figure 11-6 The pressure plate is moved rearward (disengaged) by the levers and clamped forward (engaged) against the flywheel by springs.

Figure 11-7 The release levers pull the pressure plate from the flywheel, thereby releasing the clutch disc. The machined flywheel side of the pressure plate is visible in this photo.

Friction Discs

The friction disc is an assembly of springs and friction materials (Figure 11-8). There are two friction areas, one on each side of the plate. Between the two opposing friction areas are cushion springs that help absorb the shock of engagement. Mounted around the center of the plate are torsion springs. The torsion springs are designed to reduce the sudden rotation impact of engagement. In the center of the plate is a splined hole that fits the splines of the transmission's input shaft. When the clutch is engaged and the clutch disc is rotating with the flywheel, power is transferred to the transmission at this point.

Release and Pilot Bearings/Bushings

The release bearing, like all bearings, provides a point of contact between stationary and moving components (Figure 11-9). The outer shell of the bearing snaps into a clutch fork and moves forward and backward on a guide extending from the transmission. The bearing shell does not rotate. The rotating portion is mounted within the shell and rotates when it contacts with the release levers. When the clutch is engaged, most release bearings totally break contact with the release levers and spin to a stop. Other systems use a continuously running bearing that maintains contact

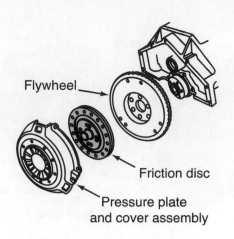

Flywheel

Friction disc

Pressure plate
and cover assembly

Figure 11-8 The friction material is laid at an angle to the clutch radius and is cushioned by springs.

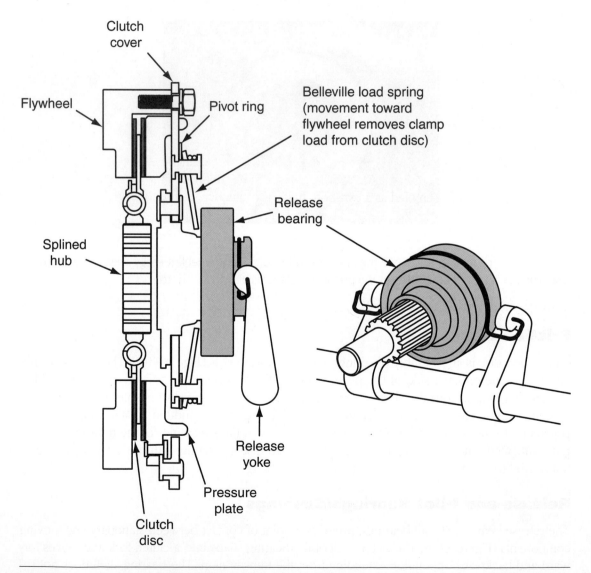

Clutch
cover

Flywheel

Pivot ring

Belleville load spring
(movement toward
flywheel removes clamp
load from clutch disc)

Release
bearing

Splined
hub

Release
yoke

Pressure
plate

Clutch
disc

Figure 11-9 The release bearing provides the contact point between a rotating member, pressure plate, and a stationary member, the release fork.

with the release levers at all times. This eliminates any free movement in the linkages. The bearing is sealed and permanently lubricated.

A **pilot bearing** or **pilot bushing** is installed in the rear of the crankshaft. The bearing or bushing is used to support the forward end of the transmission's input shaft. The pilot bearing is sealed and permanently lubricated. A pilot bushing is made of bronze and needs no lubrication.

Mechanical Clutch Linkages

Older cars and many present-day trucks use a manual linkage. The linkage starts at the clutch pedal and extends to the clutch fork at the **bell housing.** The linkage is a system of rods and ball joints that transfer the linear force of the driver's foot through the floorboard and around obstacles to apply a linear force to the fork (Figure 11-10). The fork extends through the side of the bell housing and moves the release bearing. There are usually several springs to ensure that the clutch linkage returns to the disengaged position when the pedal is released. When the driver releases the clutch pedal, the clutch is engaged. Mechanical linkage usually requires periodic adjustment to compensate for clutch wear.

Hydraulically Operated Clutches

New cars and many light trucks use a hydraulic system to apply the clutch (Figure 11-11). This requires less force from the driver, removes any free travel, and normally uses a continuously running release bearing. The hydraulically controlled clutch uses a small master cylinder to apply hydraulic pressure against a piston in the slave cylinder. The **slave cylinder** is mounted to the outside of the bell housing. The piston pushes a small rod that moves the clutch fork. A release spring pulls the slave piston to the release position when the clutch pedal is released. Hydraulic controls do not normally require adjustment for clutch wear. The hydraulic system keeps the release bearing in contact with the pressure plate fingers as the clutch disc wears.

Some manufacturers use an internal slave cylinder mounted inside the bell housing. The release bearing guide is assembled as a two-bore hydraulic cylinder. The inner bore fits around the

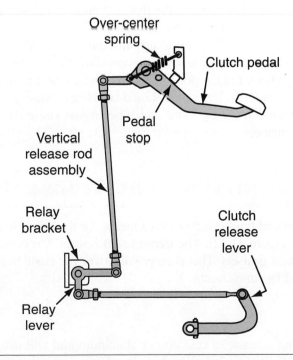

Figure 11-10 Mechanical linkage are still in use, particularly on heavier vehicles.

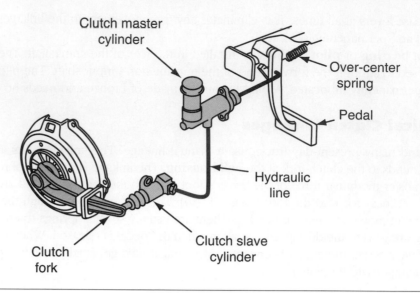

Clutch master cylinder

Over-center spring

Pedal

Hydraulic line

Clutch slave cylinder

Clutch fork

Figure 11-11 The hydraulic clutch control functions much like a hydraulic brake system.

transmission's input shafts bearing retainer. Inside the outer bore is the slave cylinder's piston shaped roughly like a doughnut. Fluid applied behind the piston directly applies the driver's force to the release bearing.

Clutch Operations

The clutch is released when the driver applies force to and pushes down on the clutch pedal. Through linkage or hydraulics, the release bearing is forced forward against the inner ends of the pressure plate release levers. The levers pull the pressure plate rearward, releasing clamping force on the clutch disc. The connection between the flywheel and the transmission input shaft is broken and no engine power is transferred to the driveline.

When the clutch pedal is slowly released upward, the linkage allows the release bearing to move rearward. As the bearing moves, the release levers allow the pressure plate springs to clamp the clutch disc to the flywheel. Engine power is transferred to the transmission through the flywheel, clutch disc, and input shaft. The clutch should be released slowly to allow the connection to be made smoothly. A sudden release of the clutch pedal may cause damage to any part of the driveline, including the engine.

Manual Transmissions and Transaxles

Shop Manual
pages 344–346

The manual transmission and transaxle provides a means for the driver to select a gear ratio that best suits the situation (Figure 11-12). The transmission houses a reverse gearset, neutral, and three, four, or five forward gearsets. This chapter will cover the basic five-speed, synchromesh, manual transmission and its components.

Housings

The transmission housing is made of cast iron or aluminum and will usually consist of at least two pieces (Figure 11-13). It is sealed at each end and may have a gasket between individual

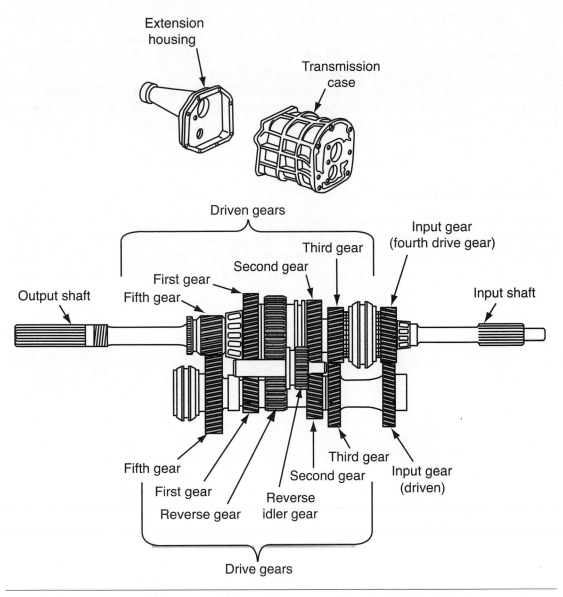

Figure 11-12 The different sizes of drive/driven gears allows the driver to select a gear ratio suitable for either torque or speech.

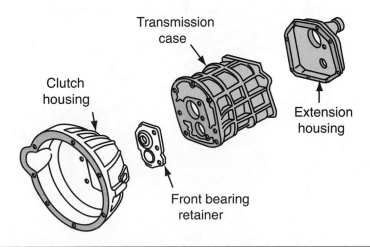

Figure 11-13 Both RWD and FWD transmissions usually have two main housings or cases.

pieces. **Shifter rails** extend though the housing and are connected to the driver's shift lever (Figure 11-14). All of the gears are located within the housing. The major differences between the transmission and transaxle are the location and shape within the vehicle. Both use a similar gear layout and linkage and operate in about the same manner.

The two pieces of the housing are the **transmission housing** and the **extension** or tail shaft housing. All of the gears and some of the linkage are located in the transmission housing (Figure 11-15). The extension housing extends to the rear and covers the tail of the output shaft. Also enclosed in the extension housing is the speedometer drive mechanism. Some four- or five-speed transmissions may have the fourth or fifth gear in the extension housing.

Gear Arrangements

The nose of the **input shaft** extends through the clutch disc and into the pilot bearing.

The gear layout in a RWD transmission is basically three shafts with gears placed along them (Figure 11-16). The **input shaft** and **input gear** form a one-piece unit. It extends from inside the transmission housing and is supported by the input shaft bearing. Behind the input shaft is the output shaft. The forward end of the output shaft is inserted into a pilot bearing mounted in a cavity in the rear of the input shaft.

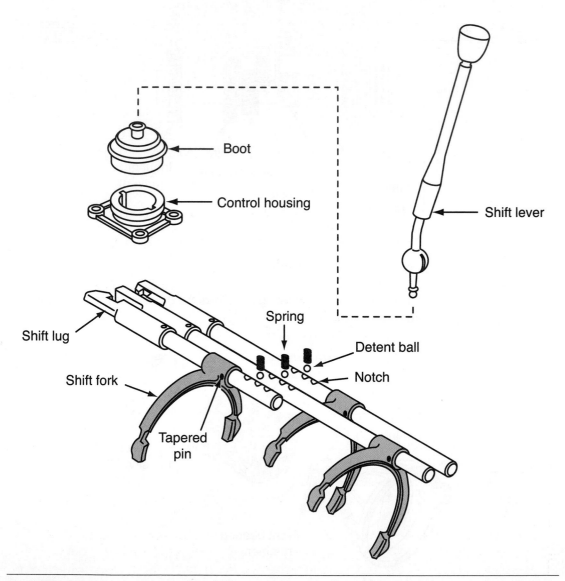

Figure 11-14 The driver changes gears by moving a shifter that moves the selected shift rail and fork.

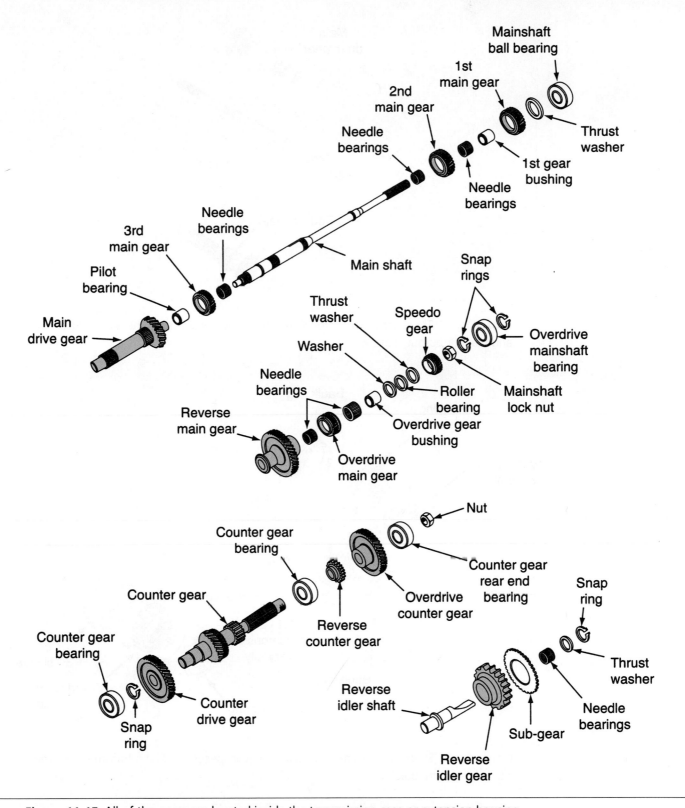

Figure 11-15 All of the gears are located inside the transmission case or extension housing.

First and second driven gears are bearing-mounted on the output shaft. Between the first and second and between the second and third or direct are the **synchronizers.** Normally, the first driven gear is at the rear end of the shaft but within the transmission housing (Figure 11-17). The reverse driven gear is directly in front of first or just behind first with second normally just forward of reverse. Some systems have a portion of the first/second synchronizer machined into a reverse driven gear. There is no actual third gear, but direct or third is obtained by locking the

Synchronizers are used to adjust the speeds of the mating gears so they can be engaged without damage.

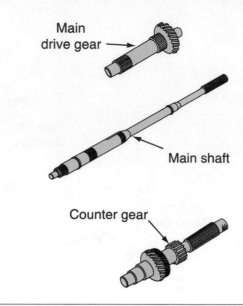

Main drive gear

Main shaft

Counter gear

Figure 11-16 There are usually three shafts in a RWD transmission: the input, cluster (counter), and main or output shafts.

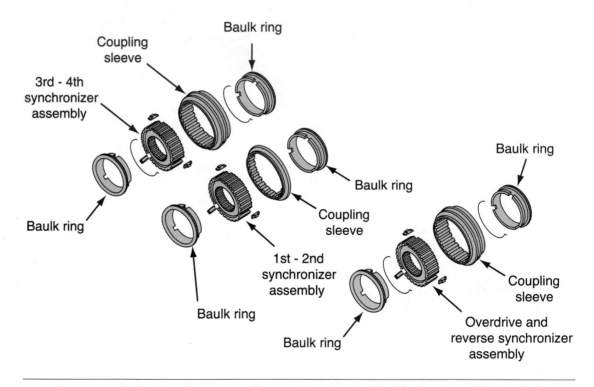

Coupling sleeve

Baulk ring

3rd - 4th synchronizer assembly

Baulk ring

Baulk ring

Baulk ring

Coupling sleeve

Baulk ring

1st - 2nd synchronizer assembly

Coupling sleeve

Overdrive and reverse synchronizer assembly

Baulk ring

Baulk ring

Figure 11-17 Synchronizers are used to match gear speeds inside the transmission, which allows a smooth, quiet shift.

Constant mesh means the mating gears' teeth are running together constantly.

output shaft directly to the input shaft. Each forward gear is in **constant mesh** with a gear within the cluster gear. The first, second, and third gears turn freely on the shaft unless locked by the synchronizer. Fifth gear is usually outside the main transmission housing with fifth drive on the counter shaft and fifth driven on the output or main shaft.

Mounted to one side and aligned with the reverse drive and reverse driven is the **reverse idler gear.** The idler gear is a sliding gear in that it must be moved along a shaft to mesh with the two reverse gears. It is not a synchronized gear. The vehicle should be stopped and the clutch disengaged before selecting reverse. Once in place, the idler is driven by the reverse drive gear and drives the reverse driven gear (Figure 11-18). The idler gear rotates in the opposite direction of the drive gear. This causes the input rotation to change direction and the output shaft rotation is reversed.

Between first and second gears and between third and the fourth gears are synchronizer assemblies (Figure 11-19). They are splined to the output shaft and are used to make gear changing easier and quieter. They lock the driven gear to the output shaft. The synchronizer assembly consists of a hub, sleeve, two **blocker rings,** three centering locks, and lock springs. The hub is the portion splined to the output shaft. The sleeve moves along external splines on the hub. The sleeve also has teeth to engage the driven gear. The blocker rings engage the gear first and is used to bring the gear speed to synchronizer speed. The center locks and springs lock the sleeve in neutral and drive the blocker rings. A shifting fork fits over the synchronizer sleeve.

Cluster Gears

The **cluster gear** is a one-piece unit. There are usually five molded gears in the cluster, which is mounted on bearings or a counter shaft (Figure 11-20). The forward (from front of transmission) gear is meshed with the transmission's input gear. The third and fifth gears in the cluster mesh

> **Blocker rings** make the first contact with a gear and initially allow the gear to slip a little. As the blocker ring is forced against the gear, the connection becomes tighter and the gear speeds up to blocker-ring speed, allowing a smooth, quiet gear change.

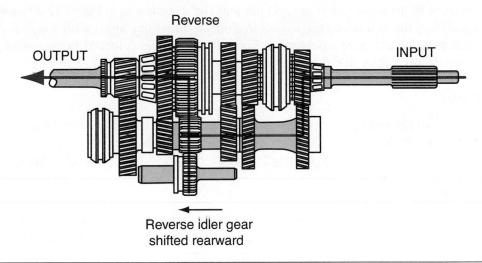

Figure 11-18 Reverse power flow is from the input gear through the cluster gears, the reverse drive, the reverse driven, and out the main shaft.

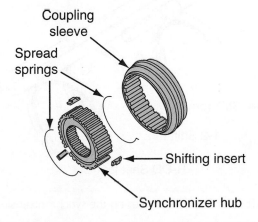

Figure 11-19 The synchronizer hub is splined to the output shaft (normally) while the sleeve can move over and lock onto the driven gear teeth.

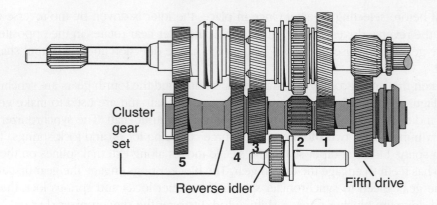

Cluster gear set

5   4   3   2   1

Reverse idler

Fifth drive

Figure 11-20 The cluster or counter gear is a one-piece gear except for the fifth gear, if equipped, which can be removed from the shaft.

and drive the second and first driven gears on the output shaft. The fourth cluster gear is the reverse drive gear. Notice that the reverse drive and reverse driven are not meshed. Also notice that the entire cluster gear rotates anytime the transmission's input shaft is rotating. This means that all gears meshed with the cluster gears are also turning. The transmission in Figure 11-20 shows the gear marked "6" is the fifth drive gear. It spins free on the cluster or counter shaft until it is locked to that shaft with the fifth gear synchronizer. The fifth driven gear is locked to the output shaft, but can only deliver torque after the fifth synchronizer is engaged.

Linkages

The shift lever is operated by the operator and may be mounted on the floor or the steering column. Most new vehicles have the lever on the floor. The movement of the lever is transmitted through mechanical linkages to the shifting fork in the transmission or transaxle (Figure 11-21).

Operation

With the engine running, clutch engaged, and the transmission in neutral, the input shaft is turning and driving the cluster gear. The cluster gear is turning the first, second, and third driven

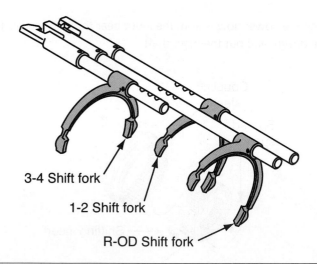

3-4 Shift fork

1-2 Shift fork

R-OD Shift fork

Figure 11-21 The shift forks fits into grooves on the synchronizer sleeves and are pinned to the shift rails.

gears (Figure 11-22). Since the synchronizers are not locked to the gears, the output shaft is not being driven. To select a gear, the clutch is disengaged and the shift lever is moved by the operator to the first gear position. The first gear synchronizer sleeve is pushed by the shifting fork into position with the first driven gear (Figure 11-23). As the clutch is slowly engaged, power from the engine is delivered to the transmission's input shaft and gear. The power flows through the input gear and cluster gear to the driven first gear. With the synchronizer in place, the power flow continues through the synchronizer assembly and rotates the output shaft. First gear has a gear ratio that provides high torque and low speed. The other forward gears are selected in the same manner. In direct drive or third gear, the 2-3 synchronizer locks the input and output shaft together giving a gear ratio of 1:1. In reverse, the two synchronizers are in neutral and the reverse shifting fork slides the reverse idler gear into position between the drive and driven reverse gears (Figure 11-24).

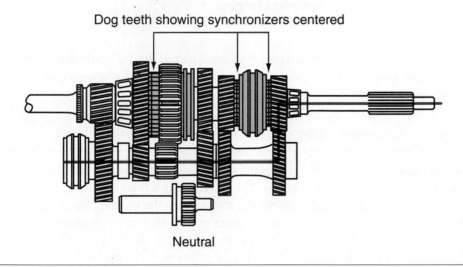

Dog teeth showing synchronizers centered

Neutral

Figure 11-22 In neutral gear with the clutch engaged, all of the gears inside a transmission or transaxle are spinning except for the reverse idler gear.

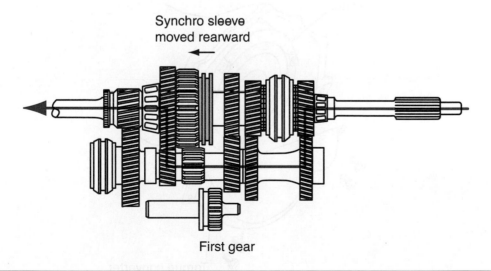

Synchro sleeve moved rearward

First gear

Figure 11-23 First gear power flow is from input through the cluster, first drive and driven, through the synchronizer and the main shaft.

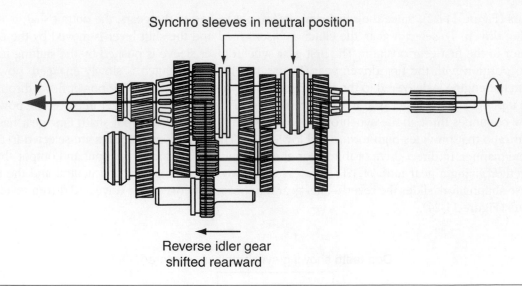

Synchro sleeves in neutral position

Reverse idler gear
shifted rearward

Figure 11-24 This reverse driven gear is a forged part of the 1-2 synchronizer. Note how the reverse idler gear has been moved forward to mesh with both the reverse drive and the reverse driven gears.

Torque Converters

Shop Manual
pages 346–351

The **torque converter** assembly replaces the clutch assembly when the vehicle is equipped with an automatic transmission (Figure 11-25). The converter is a fluid coupling. Automatic transmission fluid is used to transfer and multiply the torque from the engine. There are four subassemblies in the converter, three required for operations and another one to improve fuel mileage. Also, the flywheel may be replaced with a flexplate (Figure 11-26). The flexplate allows for the swelling of the torque converter during operation. Torque converters are almost the same for automatic transmissions and transaxles.

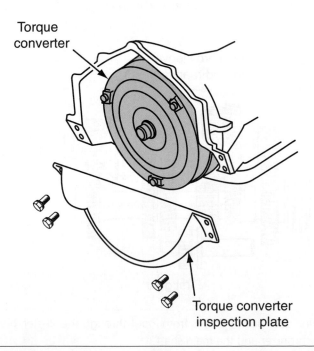

Torque
converter

Torque converter
inspection plate

Figure 11-25 The torque converter is a hydraulic coupler. It is bolted to the flexplate and splined to the transmission pump.

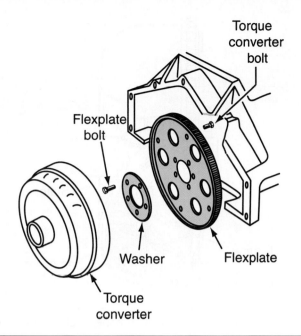

Figure 11-26 The torque converter expands and contracts during normal operation. The flexplate is flexible to absorb this ballooning action.

The torque converter is shaped like an inflated balloon that has been squeezed inward on two sides. It has enough weight to perform the same inertia action as the flywheel. The front of the converter is bolted to the flexplate so the outer shell rotates with the crankshaft (Figure 11-27). The rear of the shell has an extended lug that slides into the transmission and locks to the transmission pump. The converter subassembly components are sealed within the shell.

Inside the shell are the **impeller, turbine,** and **stator** (Figure 11-28). The impeller is made of vanes that are welded to the inside of the shell. They are angled in the direction of rotation. When fluid is in the shell and the engine is running, the impeller directs the fluid against the turbine blades. Engine torque is delivered to the turbine, but since a fluid instead of a mechanical connection is used, some or all of the torque is lost and slipping occurs. If the vehicle is stopped with the engine idling, slipping is the desired condition. The fluid flow is classified as a **rotary flow** (Figure 11-29).

The **turbine, stator,** and lock-up clutch are not visible without cutting the two halves of the converter apart. The **impeller** is welded to the inside of the shell.

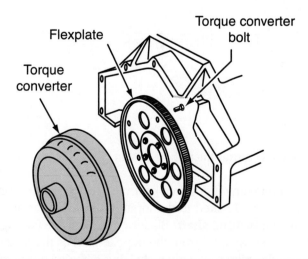

Figure 11-27 The flexplate is bolted to the crankshaft and the torque converter is bolted to the flexplate.

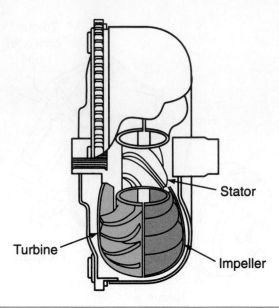

Stator

Turbine

Impeller

Figure 11-28 Internally, there are four major components of a modern torque converter: impeller fins welded to the converter housing, turbine mated to the transmission turbine (input) shaft, a one-way clutch known as the stator, and a torque converter lockup clutch for a mechanical connection between the converter housing and the turbine shaft.

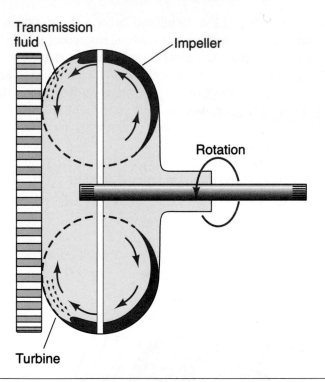

Transmission fluid

Impeller

Rotation

Turbine

Figure 11-29 Rotary flow causes the turbine to spin.

The turbine is bearing-mounted and free of a mechanical connection with the converter's outer shell. The turbine is splined-mated to the input shaft of the transmission. In some makes, this shaft is referred to as the **turbine shaft.** Blades on the turbine face against the rotation and opposite to the impeller vanes. As fluid from the impeller strikes the face of the blades, the turbine is rotated in the same direction as the outer shell but with much less speed and force. At this point, the engine torque is delivered to the input shaft of the transmission. However, the turbine could turn backward if a holding device was not in place.

The stator is the holding device. The stator is a **one-way clutch** that allows the turbine to rotate in one direction only (Figure 11-30). If the turbine attempts to rotate backward, the stator locks and the turbine stops moving or reverses its rotation. To accomplish this task, the stator is bearing-mounted and splined to a transmission extension that is similar to the release-bearing guide. A one-way clutch is a mechanical lock that is released or locked depending on the rotational direction of the turbine and stator. The outer edge of the stator locks into a cavity on the turbine. Under normal conditions, the stator rotates with the turbine. If the turbine moves backward, the stator locks drop into place and the turbine and stator are prevented from moving by the stationary stator mount.

In addition to its holding task, the stator has blades that redirect the fluid from the turbine. The stator blades are angled in the opposite direction of the turbine blades. As the fluid strikes the stator blades, it is forced back against the turbine blades. This creates higher pressures and increases the torque between the impeller and turbine. The fluid flow at this point is a **vortex flow** (Figure 11-31).

The impeller, turbine, and stator are all required for torque converter operations. The next component may or may not be included in a particular torque converter. It is the **lockup clutch** (Figure 11-32). The lockup clutch may be mechanical, hydraulic/mechanical, or electrical/hydraulic. The clutch, when operated, locks the turbine to the outer shell of the torque converter. When this occurs, the turbine and the input shaft are moving at the same speed as the crankshaft similar to direct in a three-speed manual. Usually, the lockup clutch is not applied until the vehicle is in direct (high) gear and the engine load has decreased. Obviously, if the lockup clutch was applied at a low speed, there would be no slipping within the torque converter and the engine

The **turbine** can be forced in reverse if the impeller fluid strikes the backside of the turbine blades.

Without the stator and **vortex flow,** there would not be enough torque to power the gear train.

A **lockup clutch** locks the turbine to the outside of the torque converter and is not applied until the vehicle is in high gear and engine load has decreased.

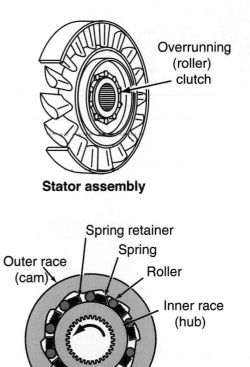

Figure 11-30 The stator will allow the turbine to spin in one direction only. It also reverses fluid flow within the converter.

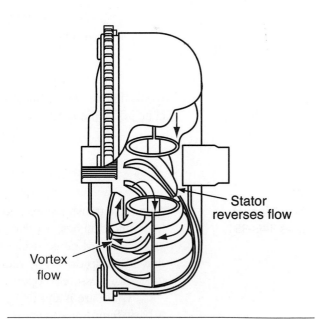

Figure 11-31 The fluid flow caused by the stator creates a vortex flow. This multiplies the torque from the impeller.

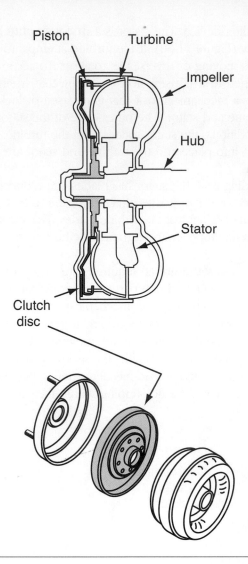

Figure 11-32 The TCC mechanically links the turbine to the converter housing. This provides direct drive from the crankshaft to the transmission turbine shaft.

would lug or stall. Almost every torque converter manufactured since 1988–1989 has a lockup clutch of some type.

Automatic Transmissions and Transaxles

Shop Manual
pages 346–351

Automatic transmissions and transaxles shift up and down through the forward gear ranges using hydraulic clutches, a valve body, and various controls. The driver selects park, reverse, neutral, and a forward range using a shift lever similar to the one used for manual transmissions. A hydraulic pump driven by the outer shell of the torque converter supplies the necessary fluid volume and pressure to operate the transmission.

There are many different transmission designs that are based on the **Simpson gear train.** A basic Simpson system uses two hydraulic clutches, two bands, two planetary gearsets, an overrunning clutch, and a valve body (Figure 11-33). This chapter will discuss the Simpson gear train design since it is basic and is the simplest to understand at this point.

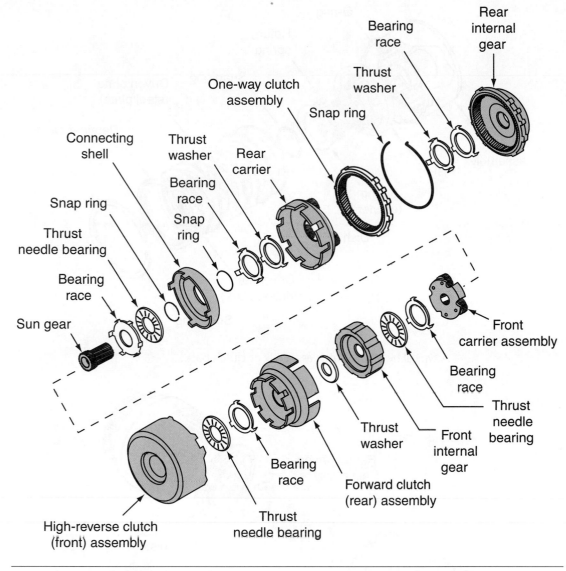

Figure 11-33 The Simpson gear train in its simplest form uses two hydraulic clutches, two planetary gearsets, two holding bands, and a one-way mechanical clutch.

Hydraulic Clutches

A hydraulic clutch or **clutch pack** is composed of a piston assembly, a drum, and a number of steel and friction clutch plates (Figure 11-34). The center hub of the drum fits around a gear or shaft and provides the mounting area for the piston. The piston is a thin, aluminum plate with a seal around the outer and inner edges. The piston is held to the bottom of the drum's bore with return springs and a retainer. Above the piston and within the bore are alternating steel and friction clutch plates. The steel plates have lugs at the outer edge to lock into grooves on the drum. The inner edge of the clutch discs is splined to fit either a gear or shaft. The plates are held in place loosely with a retaining snap ring.

As pressurized fluid is directed behind the piston, it moves forward and clamps the friction and steel plates together (Figure 11-35). This operation is similar to the manual pressure plate clamping the clutch's clutch disc to the flywheel. In hydraulic clutches, input power may be driving the clutch discs or the clutch drum. As the plates are clamped together, they begin to turn together and the power flow is directed through the clutch pack from input to output.

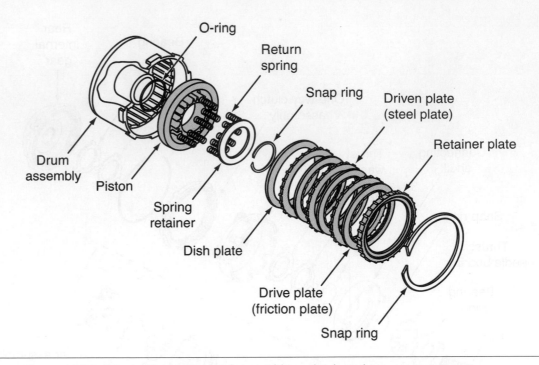

Figure 11-34 A typical hydraulic clutch assembly or clutch pack.

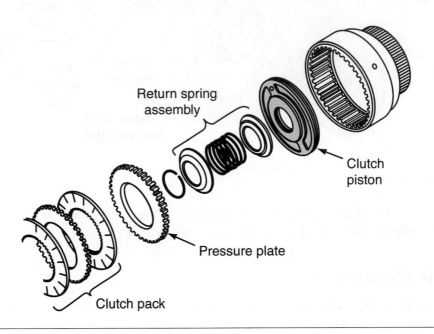

Figure 11-35 The hydraulically activated piston clamps the clutch disc together similar to the pressure plate clamping the manual clutch disc to the flywheel.

Bands

Bands are wrapped around the outside of the clutch drum and act as a holding device. The band is anchored at one end while the other end is attached to a hydraulic **servo** (Figure 11-36). When fluid is applied to the servo, the band is applied and prevents the clutch drum from moving. If the clutch is not applied, the power flow passes through the clutch pack without any action. If the clutch is applied when the band is also applied, the gear or shaft splined to the clutch's clutch disc is held and prevented from rotating.

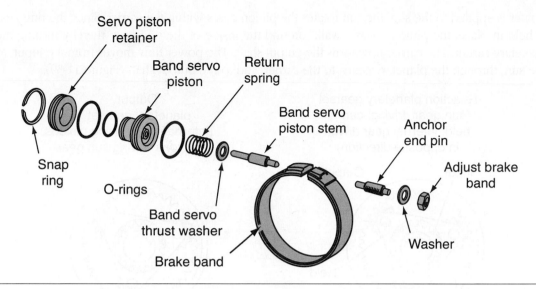

Figure 11-36 The servo piston is activated by hydraulic pressure to clamp or release a band around the exterior of a clutch drum or gearset.

Planetary Gear Sets

A **planetary gear** has three gears. The outer gear, called a **ring gear,** is an internal gear. The gear teeth are on the inside of the ring. Fitted and meshed inside the ring gear is the **planetary carrier.** The planetary carrier is actually a set of three or four pinion gears mounted to a platform or carrier. The carrier is designed to allow the pinion gears' teeth to extend past both edges of the carrier. This allows the gears to mesh with the ring gear around the carrier and the **sun gear** mounted through the center of the carrier (Figure 11-37). The external sun gear extends completely or partially through the planetary carrier and meshes with the pinion gears (Figure 11-38).

Any of the three major gears in this set can be used as the drive or driven. However, in order for the set to work, one of the three must be held in some manner so the other two have a point to rotate around. For a quick example, use the following scenario. The ring gear is mechanically connected to the clutch pack drum. The planetary carrier is splined to an output shaft and an input gear drives the sun. The sun gear is also splined to the clutch discs of the clutch. The band is applied and holds the drum and ring gear in place. The clutch is released freeing the clutch discs. As

The two **planetary gearsets** In a Simpson gear train are used together to obtain the correct gear ratio for a given situation.

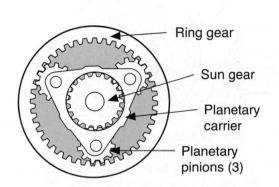

Figure 11-37 A simple planetary gearset consists of a sun, planetary carrier and pinions, and ring gears.

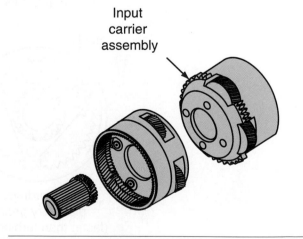

Figure 11-38 Looking closely at the ring gear, a second set of planetary gears are visible. This set would mesh with another sun gear or with a shaft.

power is applied to the sun, the sun rotates the pinion gears within the carrier. Since the ring gear is held in place, the pinion's gears "walk" around the inside of the ring gear, thereby turning the planetary carrier. The carrier now turns the output shaft. The power flow moves from the input, to the sun, through the planetary gears, to the carrier, and to the output shaft (Figure 11-39).

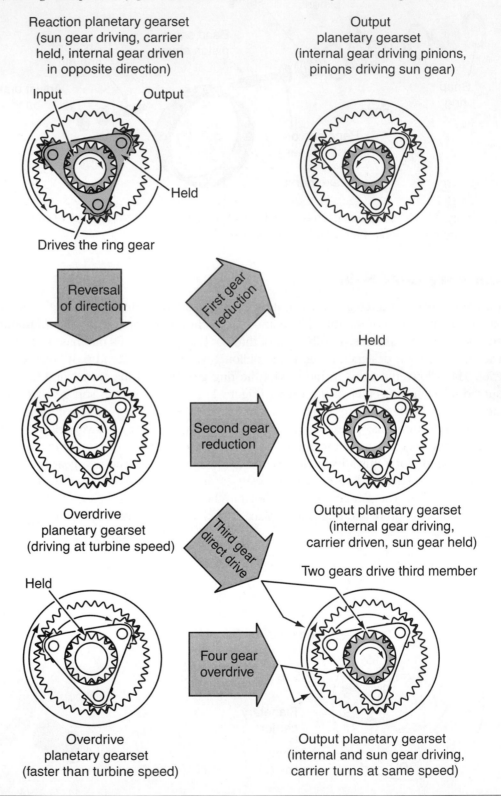

Figure 11-39 The use of planetary gearsets provides a variety of gear ratios in a small, lightweight package.

Since the sun gear is smaller than the carrier, there is an increase in torque and a reduction in speed. The pinion gears are not counted in computing the ratio because they are used as idlers. If the power flow is reversed using the carrier as the input and the sun as output, the torque would decrease and speed would increase.

Overrunning Clutches

The overrunning clutch in most Simpson gear trains is a mechanical clutch. The inner ring or race is splined to a gear or shaft and the outer race is locked to the transmission housing or some other anchor. There are two basic types of overrunning clutches: **sprag** and **roller** (Figure 11-40). Between the two races there are diagonals that can tilt outward (long) to lock the races or tilt inward (short) to release the races. In a roller clutch, roller bearings are fitted between the races. One race has ramps. Springs hold the bearings in the release position. If the inner race tries to turn backward, the bearings roll up the ramps and lock the two races together, thereby locking the shaft or gear. The overrunning clutch is used in an automatic transmission to redirect the power flow.

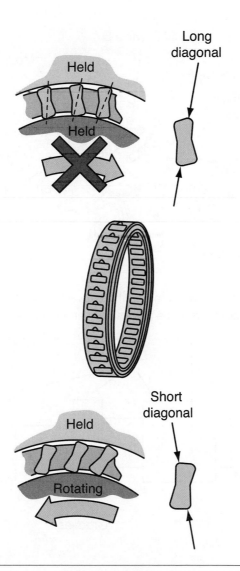

Figure 11-40 A sprag clutch is a mechanical one-way locking clutch commonly used to hold the ring gear of a planetary set.

Valve Bodies

The valve body consists of many valves and fluid passageways (Figure 11-41). This chapter will not attempt to discuss the hundreds of designs and operations of the various valves and valve bodies. We will discuss the spool valve design, the major valves, and the three major controls.

Spool Valves

The **spool valve** is a highly machined steel rod that fits tightly and smoothly into a highly machined bore in the valve body. The rod is grooved at different points along its length. The width of the grooves may be different. The remaining high points on the rod are called **lands** (Figure 11-42). There are fluid passageways entering the valve body bore at different points. With the spool valve inserted into the bore and a land moved directly over a fluid passageway, the fluid is blocked. Moving the spool until a groove lines up with the passageway, fluid flows through the groove and enters another passageway. In this manner, the fluid can be directed to or cut off from clutches, servos, other valves, or a control device. Springs or fluid move the spool valves. The major valves are the 1-2 and 2-3 shift, reverse, pressure regulator, pressure booster, and the manual valve. The operator controls the manual valve.

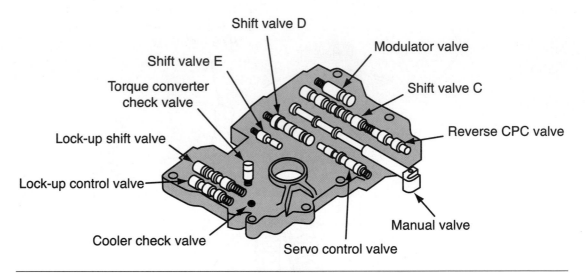

Figure 11-41 The valve body holds many different valves and their springs. They control fluid to and from hydraulic elements in the transmission and regulate fluid pressure.

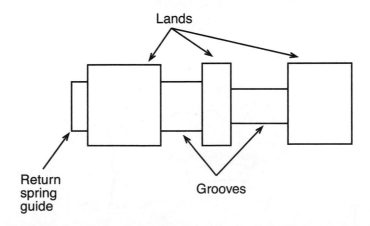

Figure 11-42 The spool valve is used to block or open fluid passageways in the valve body passages.

There are two primary control systems: throttle and governor. They work together to control up- and downshifting through the forward gear ranges. The throttle system controls the position of the **throttle valve** (Figure 11-43). The **governor** controls the **governor valve** (Figure 11-44). Both valves are in the valve body. Shifting is the balancing of the two systems against each other. The throttle valve senses engine load while the governor senses the vehicle's forward speed. The vehicle starts in low gear with high throttle valve pressure and low governor valve pressure. This prevents an **upshift** to second gear. As the vehicle speed increases and engine load drops, the governor pressure increases and the throttle pressure decreases. As the governor pressure exceeds throttle pressure, the two valves move and fluid is directed to the 1-2 shift valve. The transmission shifts to second gear, engine load is increased, and throttle pressure rises to prevent upshifting to third until the next balance is reached.

Note that this is a simplistic manner of explaining the shifting actions of an automatic transmission. There may be several different valves actually involved in the shift process, but all are based on the relation between the vehicle speed and engine load.

There are many mechanical and operational designs for automatic transmissions. The newer transmissions/transaxles use electrical solenoids instead of spool valves to direct fluid and shift the gears. The solenoids can be used as shift solenoids or they can control the lockup clutch in the torque converter. Most new vehicles use engine and transmission sensors to supply data to the PCM, which in turn operates the shift valves, torque converter lockup clutch, and the many other devices needed to provide a smooth and comfortable ride. Almost all of the new systems have four forward gears and may drive all four wheels. Like the manual system, there is not much difference between an automatic transmission and transaxle. Both operate on the same theory and use very similar components. The space available for the transaxle did require some engineering to fit the components into a smaller housing.

The **throttle valve** senses engine load by mechanical linkage or by engine vacuum.

The governor pressure is generated by the **governor** mounted on the output shaft of the transaxle or by the electronic vehicle speed sensor.

As the **upshift** occurs, the pressure regulator and boost valve adjust the hydraulic pressure to ensure a firm clutch or band application.

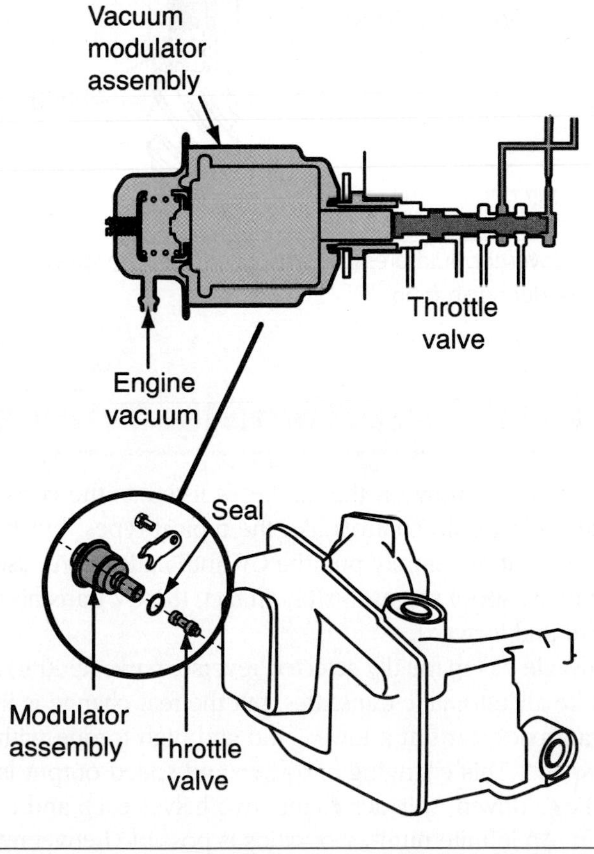

Figure 11-43 The throttle valve and pressure on this older THM Turbine 350 is operated by engine vacuum through a vacuum motor.

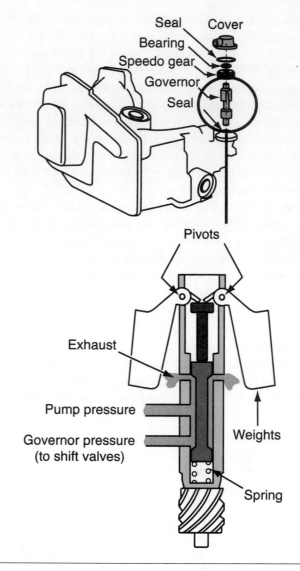

Seal Cover
Bearing
Speedo gear
Governor
Seal

Pivots

Exhaust

Pump pressure

Governor pressure
(to shift valves)

Weights

Spring

Figure 11-44 The governor fluid pressure responds to vehicle speed and is a counterbalance to the throttle pressure during shifting.

Constant Variable Automatic Transaxles

The newest automatic transaxle now on the market is listed as the constant variable (CV) type. They use hydraulics and electronic controls like the typical types, but have very distinct differences in other areas of operation. Simply put, the CV automatics don't use gearsets to change ratios and do not use as many spool valves to shift. In fact, the CV transmissions don't actually shift gears except for reverse and forward.

A typical CV transaxle has manually selected reverse, park, neutral, and forward gears. The first three are almost like all automatic transaxles, but the real change is in the forward selection. In forward gear the transaxles starts at a lower ratio and high torque with the ratio increasing as the vehicle comes to speed. This changing of torque and speed output is basically done by two pulleys, one drive and one driven, split across into two halves each and connected by a steel belt or a multiple-link chain. An infinite number of ratios is possible between the lowest gear and the highest gear. If you have ever driven a go-kart or an all-terrain-vehicle, you get the same sensation of ratios changing without the gear change.

The electronic control unit causes the two pulleys to change their internal diameters based on engine load and vehicle speed (Figure 11-45). This is calculated very similar to all other types of automatic transaxle controls, but the hydraulics in a CV unit is used to split the pulleys instead of engaging a gearset through hydraulic clutches. The driven pulley is splined to the transaxle input shaft. One side or half of the pulley can be moved back and forth along the shaft. The driven

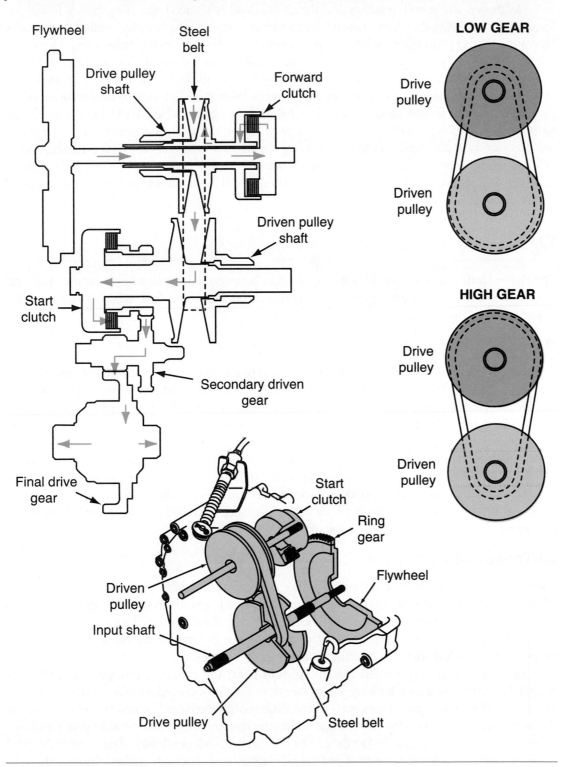

Figure 11-45 In low on this CV transaxle, the drive pulley is narrow, creating a larger pulley while the driven is spread apart, a small pulley. The pulleys change sizes based on engine load and vehicle speed.

pulley is set up the same way except it is splined to the output shaft. At low speed the two halves of the drive pulley are pressed together so the area where the chain runs is narrow and the chain rides near the outside edges of the pulley. In this instance the two halves of the driven pulley are pulled apart so the chain rides low. In effect there is a big pulley driving a small pulley with a torque increase at the output shaft.

As the vehicle speed increases and the engine load is reduced, the drive pulley halves are forced apart, allowing the chain to drop lower and now running on a "smaller" pulley. At the same time and in direct proportion to the outward movement of the drive pulley halves, the sides of the driven pulley are the forced in, creating a "larger" pulley. This reduces the torque ratio and has a corresponding increase in speed ratio. This system, when controlled by a well-programmed computer with accompanying sensors and actuators, means there is no "physical" feel of an up or down gear shift, provides innumerable ratios within the maximum high and low fixed limits, provides almost instant response to load and speed, and provides a quiet and comfortable ride to the driver and passengers. It also reduces the total number of parts that can wear out. With the exception of belt and pulley wear on the first-generation CV transaxle introduced in the late 1990s, the maintenance of second- and third-generation units is favorable and the models appear to be well received by the automotive public.

Drive Shafts and Differentials

Shop Manual
page 351

The **drive shaft** assembly provides the connection between the transmission's output shaft and the differential in the rear axle housing. The **differential** allows the drive wheels to rotate at different speeds when the vehicle changes direction.

Drive Shafts

A typical drive shaft assembly has two **U-joints,** a **slip joint,** and the drive shaft tube (Figure 11-46). The drive shaft is a hollow, balanced, steel pipe designed to withstand high rotational forces. Each end has a **yoke** to mount the U-joints. The rear U-joint connects the drive shaft to the companion flange on the differential. The front U-joint mates the drive shaft to the slip joint. The slip joint is internally splined to fit over the output shaft of the transmission. The slip joint can slide in and out on the output shaft to allow for linear distance changes between transmission and differential. When the rear wheels go over a high spot, the differential moves upward, in effect, getting closer to the transmission. The opposite happens when the wheel drops into a hole. As the distance changes, so does the angle between the transmission and differential. The two U-joints rotate within their mounts to adjust for the different angles.

Differentials

The differential is mounted in the rear housing and uses a system of gears to let the two drive wheels turn at different speeds (Figure 11-47). As the vehicle rounds a curve, the outer wheel has to travel a longer distance than the inner wheel. The inner wheel has to reduce speed as the outer wheel increases speed. Both wheels use the center of the housing as the pivot point. A second purpose of the differential is to turn the power flow 90 degrees.

The input **pinion gear** is a long gear with spiral teeth. The pinion gears in the carrier are small, round gears with ten to fourteen teeth.

From the engine to the differential's **pinion gear,** the power is traveling basically in a straight line. At the rear axle housing, the power flow must be changed 90 degrees (Figure 11-48). The pinion drives a ring gear mounted to the **differential carrier.** The entire carrier rotates in the same direction as the pinion. Mounted within the carrier are two axle or **side gears** and two pinions. The pinions are attached to the carrier but can spin independently of each other. Meshed with and at right angles to the pinion are the side gears. A solid axle is splined through the center of each side gear. As the vehicle moves in a straight line, the differential carrier rotates and carries the pinion with it. Since the pinions are meshed with the side gears they rotate with the carrier and turn the axles. The pinions are not rotating or spinning at this time. During a vehicle turn, the

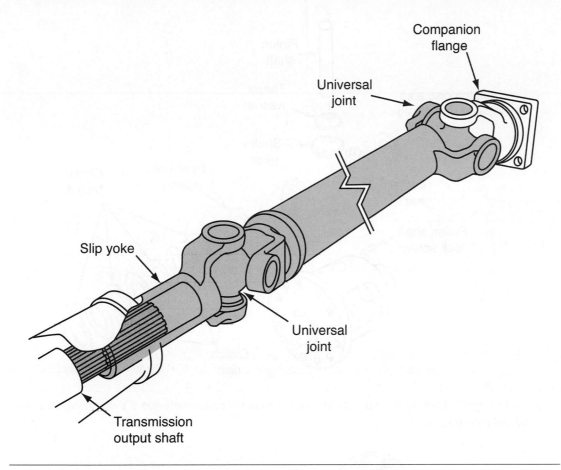

Figure 11-46 The drive shaft assembly consists of a hollow tube, two yokes, two U-joints, and a slip yoke that fits over the transmission output shaft.

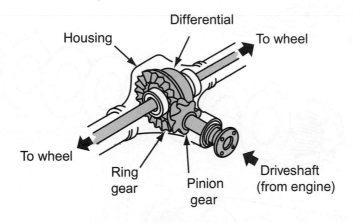

Figure 11-47 The final drive gears turn the power flow 110 degrees and the differential gears allow the drive wheels to rotate at different speeds.

inner side gear slows down and the two pinions begin to spin. The outer side gear speeds up for two reasons: the inner wheel's action and the actions of the pinions. The two side gears change speed at the same rate. For example, if the inner one slows five RPMs, then the outer one must speed up five RPMs.

The differential carrier is mounted on bearings and must be adjusted for proper pinion gear and ring gear backlash. Some differentials may have a clutch that can lock or slow an axle (Figure 11-49).

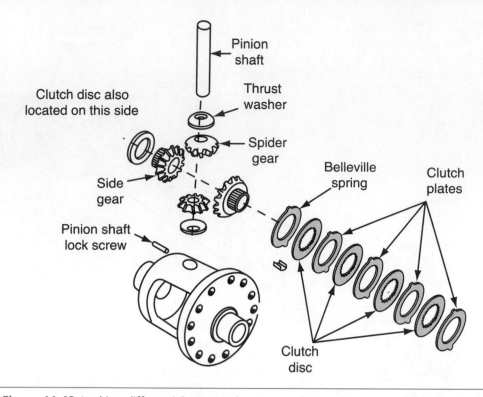

Figure 11-48 Locking differentials are used to manage power between a slipping wheel and a wheel with traction.

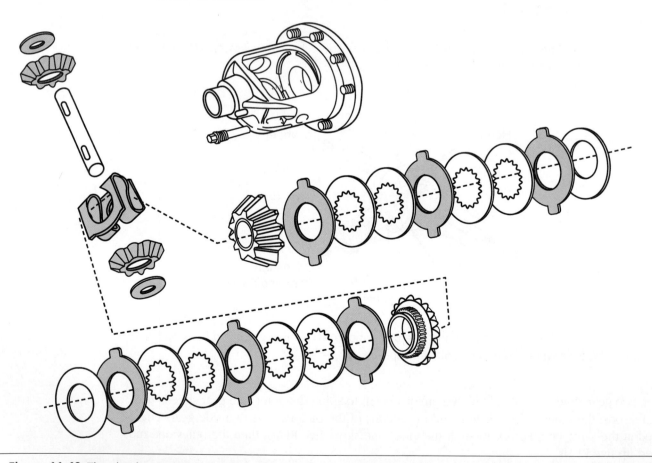

Figure 11-49 The clutches connected to the side gears have two different plates: metal plates fitted into slots in the axle housing and friction plates splined to the side gears. When sufficient rotation force is applied the two types of plates lock together through friction and effectively lock the side gear to the axle housing.

This is known by various names, but is actually a locking differential. Under normal circumstances the differential allows the wheels to speed and slow as needed. However, if a wheel is spinning, more power is sent to that wheel. The result is no vehicle movement. The locking differential uses a clutch to lock the spinning axle so the power is sent to the wheel that has the traction. Assuming that wheel has enough traction, the vehicle will move. As the traction improves on the spinning wheel, the power is more evenly divided between the two wheels.

Final Drives and Drive Axles

Shop Manual
pages 356–361

The **final drive** is the most important difference between the transmission and transaxle (Figure 11-50). Its purpose is the same as the differential and it is mounted inside the transaxle housing. The ring gear is driven directly by the transaxle's output gear. There is no drive shaft used with transaxles.

The FWD drive axles are also different from the rear wheel drive axles (Figure 11-51). RWD axles deliver power directly to the wheels that do not turn when negotiating a curve. With FWD, the front wheels are not only the drive wheels, but they are the steering wheels also. The drive axles must transfer the power to the wheels on sharp turns while moving over road irregularities. To do this, each axle has two constant velocity (CV) joints. The outer or outboard CV joint allows the wheel to be steered and lets the wheel and suspension travel over the road (Figure 11-52). The inner or inboard CV joint also allows for the changing vertical angles and adjusts for the linear distance between the final drive and the wheel (Figure 11-53). This joint performs the same function as the U-joint and slip joint on the drive shaft.

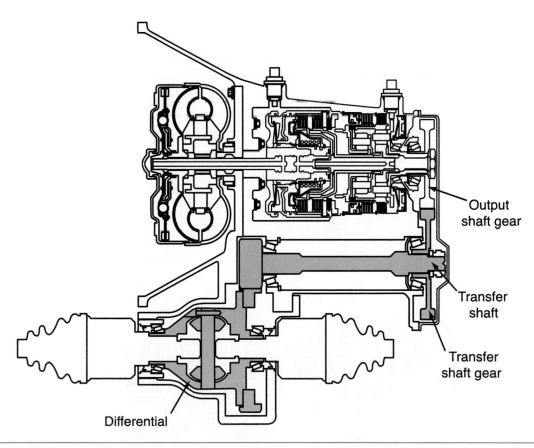

Figure 11-50 The final drive in a transaxle allows the drive wheels to turn at different speeds, but may not change power flow direction on certain transaxle designs.

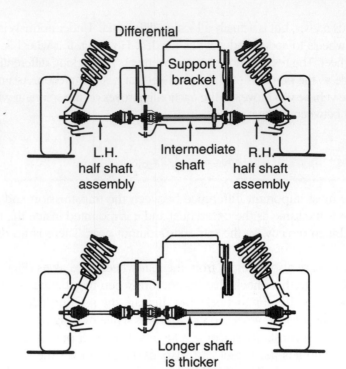

Figure 11-51 Generally, the right- and left-side FWD drive axles are different lengths and may have additional mounts on the longer one.

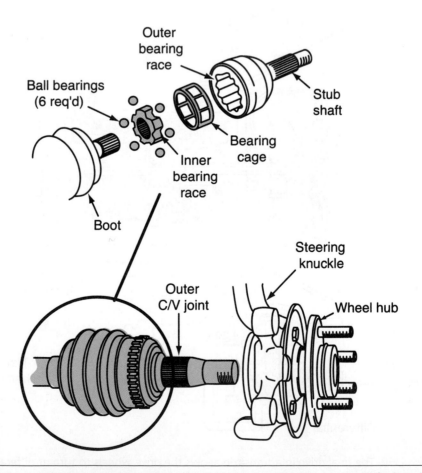

Figure 11-52 The outer CV joint allows for horizontal and vertical angle shifts but does not allow for changes in drive axle length.

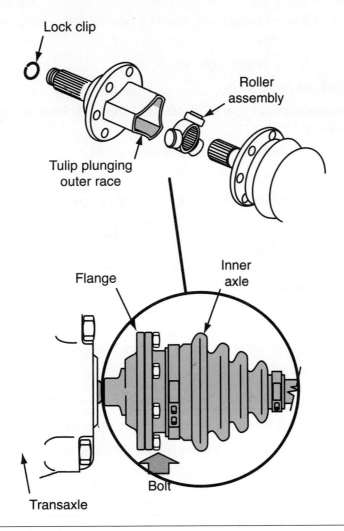

Figure 11-53 The inner CV joint allows for vertical angle shifts and changes in axle length, but does not allow horizontal angle shifts.

Summary

❏ The driveline is used to transfer engine torque to the drive wheels.

❏ Gear ratios are used to match the engine load with the vehicle speed.

❏ A clutch provides the driver a method to separate the engine from the driveline.

❏ The pressure plate clamps the clutch disc to the flywheel for transferring engine torque to the transmission.

❏ Manual transmissions/transaxles provide the necessary gear ratios to move the vehicle.

❏ The driver makes the gear selection in a manual transmission/transaxle.

❏ Torque converters are used to connect the engine to automatic transmission/transaxles.

❏ The impeller drives the turbine.

❏ The stator acts as a one-way clutch.

❏ A driver with an automatic transmission/transaxle selects the gear range.

❏ The automatic transmission/transaxle selects the forward drive gear automatically.

❏ The Simpson gear train uses two hydraulic clutches, two gearsets, two bands, and an overrunning or one-way clutch.

❏ The bands are hydraulic/mechanical-holding devices.

❏ A typical drive shaft has two U-joints.

❏ The differential turns the power flow 90 degrees and allows the drive wheels to turn at different speeds during turns.

❏ Final drives are used with transaxles in place of differentials.

❏ FWD drive axles use an outer and an inner CV joint.

❏ CV transaxles have an infinite number of ratios within their upper and lower limits.

Review Questions

Short Answer Essay

1. Explain how a hydraulic clutch operates.

2. Describe the purpose and use of the torque converter stator.

3. Describe the two fluid flows in the torque converter during operation.

4. Explain how shifting is controlled in an automatic transmission or transaxle.

5. List and explain the purpose of each component of a drive shaft assembly.

6. Describe the purpose of the two CV joints commonly found in a drive axle.

7. Explain the purpose of transmission bands.

8. Explain how power is transferred from the flexplate to the transmission's input shaft.

9. Describe the actions of the 1-2 synchronizer during shifting from first to second gear.

10. Explain how ratios are changed in a CV transaxle.

Fill-in-the-Blanks

1. The _____ _____ is located in the center of a planetary gearset.

2. The steel plates of a hydraulic clutch are fitted into grooves in the clutch _____.

3. The _____ of a spool valve closes the fluid passageway.

4. The transaxle powers _____ drive axles.

5. Movement of the _____ _____ against the _____ _____ of a clutch pulls the pressure plate away from the flywheel.

6. The rotary motion of the flywheel reduces the _____ of the engine.

7. The differential ring gear is directly driven by the _____ _____ .

8. If the drive gear is larger than the driven gear, then the speed of the driven is _____ than the speed of the drive.

9. The torque is increased if the _____ gear is smaller than the _____ gear.

10. If the drive pulley is spread apart and the driven pulley is completely closed, then the torque output is at its _____ ratio.

ASE-Style Review Questions

1. Automatic transmissions are being discussed.
   Technician A says the band drives the clutch drum.
   Technician B says a basic Simpson gear train uses three clutches.
   Who is correct?
   A. A only
   B. B only
   C. Both A and B
   D. Neither A nor B

2. A drive gear with 9 teeth and a driven gear with 27 teeth results in a gear ratio of:
   A. 33:1
   B. 3:1
   C. 3.3:1
   D. 1:.33

3. FWD drive axles are being discussed.
   Technician A says the outboard CV joint provides vertical and horizontal movement.
   Technician B says the inboard CV joint provides vertical and steering movement.
   Who is correct?
   A. A only
   B. B only
   C. Both A and B
   D. Neither A nor B

4. Manual transaxles are being discussed.
   Technician A says the reverse idler and reverse gears are connected with a synchronizer.
   Technician B says the reverse idler is moved with a shifting fork.
   Who is correct?
   A. A only
   B. B only
   C. Both A and B
   D. Neither A nor B

5. Differentials are being discussed.
   Technician A says the carrier lets the drive wheels travel at different speeds.
   Technician B says the pinions do not spin when the vehicle is traveling in a straight line.
   Who is correct?
   A. A only
   B. B only
   C. Both A and B
   D. Neither A nor B

6. Hydraulic clutches are being discussed.
   Technician A says the piston is applied by spring force.
   Technician B says the clutch discs are splined to a gear or shaft.
   Who is correct?
   A. A only
   B. B only
   C. Both A and B
   D. Neither A nor B

7. Final drives are being discussed.
   Technician A says the ring gear is driven by the transmission output shaft.
   Technician B says power is delivered to the final drive by a drive shaft.
   Who is correct?
   A. A only
   B. B only
   C. Both A and B
   D. Neither A nor B

8. Automatic transaxles are being discussed.
   Technician A says the transaxle pump is driven by the torque converter.
   Technician B says the shift valves control shifting.
   Who is correct?
   A. A only
   B. B only
   C. Both A and B
   D. Neither A nor B

9. Drive shafts are being discussed.
   Technician A says the slip joint corrects for distance changes between the transaxle and differential.
   Technician B says U-joints allows angular changes in the drive shaft's movement.
   Who is correct?
   A. A only
   B. B only
   C. Both A and B
   D. Neither A nor B

10. Planetary gear sets are being discussed.
    Technician A says that if the ring is driven and the carrier is the drive, the torque will decrease.
    Technician B says if the sun is driven and the carrier is held, the ring is the drive.
    Who is correct?
    A. A only
    B. B only
    C. Both A and B
    D. Neither A nor B

Auxiliary Systems and Climate Control

Upon completion and review of this chapter, you should be able to:

- ❑ Discuss the different and overlapping features of convenience and comfort systems.
- ❑ Discuss the warning systems found on most vehicles.
- ❑ Discuss some of the information systems available.
- ❑ Discuss the electrical circuits used with instrumentation.

- ❑ Discuss the interior and exterior lighting system.
- ❑ Explain the general operation of typical accessory systems.
- ❑ Describe the general operation of heating and air-conditioning systems.
- ❑ Explain the purpose of the Clean Air Act and the control of refrigerant.
- ❑ Discuss restraint systems.

Introduction

The increasing use of electricity and electronics in the automobile has provided an almost unlimited range of accessories and informational devices. The car or truck can be equipped with systems that memorize seat positions for a certain driver, adjust individual temperature controls, provide antitheft protection, and remote start the engine and heater on a cold morning. More and better protection devices for the passengers are possible with electronics. With the advances comes an increasing need for electronically proficient technicians who can diagnose and repair sophisticated electronic circuits quickly and accurately. To the customer, the convenience of electronics brings the possibility of hugely increased repair costs.

This chapter, and its corresponding chapter in the Shop Manual, are not designed to make the entry-level technician an electronic or climate control repair expert. Instead, the chapters will give the reader a working knowledge of the various auxiliary systems' operations and some basic diagnostic testing. Since some of the systems are married together, a failure of other systems may affect engine and transmission operation.

Passenger Comfort and Convenience

Back in the good old days of 1957 when this author received his driver's license, vehicle comfort consisted of a decent pad between the passenger and the seat spring! The only convenience item was blue oil smoke from the exhaust that killed half the mosquitoes in south central Virginia! Today, there are devices installed on the typical vehicle that were not dreamed possible in 1957.

Passenger comfort devices are items such as heat, air, conditioning, padded and heated seats, and others that make a passenger's ride in the vehicle enjoyable. Warning and information systems provide the driver with engine and vehicle operation and indications of problems before they get too serious. Systems like lighting, rearview mirrors, heated back glass, and automatic ride control give the passenger a comfortable ride while the driver has an almost unrestricted view of the actions around the vehicle. Of course, many of the so-called comfort items are required just to operate the vehicle. Comfort comes from the improved processes and equipment that accomplish the job.

Convenience features are those that make the comfort items easier or simpler to use. Remote controls and heated outside mirrors eliminate the need to lower the windows on cold, rainy days. The driver can change seat positions during a long drive to relieve or at least postpone fatigue. There are driver seat systems offered today that will massage the driver's lower back. In the old days, if a driver got lost, the solution was to use a map or ask directions. Today, the punch of a button displays the vehicle's position on a small screen or digital display. In the old days, if a car broke down on a lone, dark road, the driver and passengers either slept in the vehicle or walked. Now, just the punch of a button summons help.

All of the features taken for granted today place a tremendous load on the battery and alternator, not to mention the training and operational costs to maintain nationwide repair facilities and technicians. The increased cost is passed to the customer in the vehicle's purchase price and maintenance costs. The only real drawback to the increased number of electronic devices offered is the cost of a mistake by the vehicle or part manufacturer, repair technician, or the owner.

AUTHOR'S NOTE: Many customers complain of the high cost of routine preventive maintenance and repairs. Today it is not unusual for a scheduled engine service to cost around $400.00. However, let's take a look at that cost. In 1960 this same service cost about $35.00 to $40.00, but it had to be done about every 6,000 miles. Assuming a cost of $40.00 each tune up and extending that over 60,000 miles (typical mileage of this service type today), the total cost would be about $400.00. So overall the long-term cost is about the same. The main problem is that now we, as technicians, want that $400.00 at one time. Considering the average salary in 1960 was about $1.35 per hour versus today's salary, the impact on the customer's budget is roughly the same. The other point to consider is that if you had a 1960 automobile that even lasted much more than 60,000 miles, it could be considered a miracle in itself.

Instrumentation

Shop Manual
pages 374–380

A **light-emitting diode (LED)** is a diode that gives off light when current passes through it.

Gauges and lights on the dash are the means used to convey information to the operator. They are usually in an area of the dash referred to as the **instrument panel.** The gauges may be **analog** or **digital.** Analog gauges use a needle moving over a fixed scale to display the data much like the hands on a clock (Figure 12-1). Other systems use numbers or digits that change based on sensor input (Figure 12-2). The numbers are displayed by **light-emitting diodes (LEDs).**

One system uses an illuminated band or bands against a fixed scale. The bands act like an analog gauge but are actually LEDs that are turned on and off by a computer module based on incoming data. The bands can be programmed as different colors to attract the driver's attention. For instance, the band representing vehicle speed could be programmed to turn red for any speeds over 70 miles per hour (mph). It should be noted that certain red and amber warning lights are present in every dash regardless of the type of instruments used. There are more colored warning lights in today's instrument panel than just 10 years ago. They will be discussed in the section describing warning systems.

ANALOG GAUGES

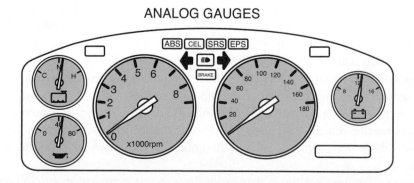

Figure 12-1 Three typical analog gauges found on most modern vehicles.

Figure 12-2 The small dots of light seen in digital displays are the individual LEDs in the display.

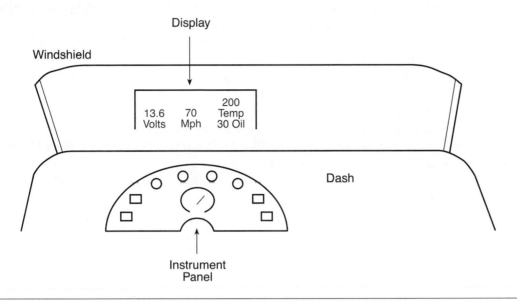

Display

Windshield

13.6 Volts 70 Mph 200 Temp 30 Oil

Dash

Instrument Panel

Figure 12-3 Some of the same data shown on the instrument panel can be projected onto the windshield with HUD.

The **heads-up-display (HUD)** was adopted from military fighter aircraft. The information most commonly required is projected onto the windshield in the driver's line of sight (Figure 12-3). The information usually consists of vehicle and engine speed, oil and temperature conditions, and warning lights. In the future, the driver could possibly scroll through other data using steering-wheel-mounted controls. The object of HUD is to keep the driver's eyes and attention on the environment around the vehicle while providing necessary vehicle data. Some instrumentation systems can display data on an overhead console.

The term **heads-up-display (HUD)** was adopted because the driver does not have to look down to read operational data.

Warning Systems

Warning systems and information systems (next section) would appear to be basically the same; however, they have distinct purposes. Warning systems are found on every vehicle and alert the driver to a fault or failure that could damage the vehicle or cause a loss of control. Information systems generally provide data that can assist the driver in some way.

Red and amber lights are used because of their visibility, relationship to danger, and eye-catching properties. Many times, gauges supplement the lights. Red means the danger is serious and failure to heed the **warning light** may cause serious damage or an accident. Amber lights indicate some repair is needed but the vehicle can be operated. Lights are used in place of or with

Shop Manual pages 374–378

gauges. Drivers tend not to check gauges like oil pressure or engine temperature until it is too late. The sudden glare of a red dash light will attract the driver's attention immediately.

Typical red warning lights are low oil pressure, high engine temperature, the alternator, door ajar, seat belts, brake, or low brake fluid (Figure 12-4). On some vehicles, the oil pressure and engine temperature sensors or switches are wired in parallel to the same lamp. The parking brake, low fluid level, and brake failure sensors are also normally wired to the same lamp. A sensor or switch is mounted into the component/assembly being checked. Sensors will normally send data to the vehicle's PCM or another computer. The computer will turn on the warning light as needed. Older vehicles and some newer models use a simple switch that is opened or closed based on temperature, pressure, or other **trigger** conditions. The switch completes the circuit and the warning light is turned on.

The switch in some warning circuits is a one-wire **grounding switch** (Figure 12-5). In an oil pressure warning circuit, the switch may be held open (NO) or closed (NC) by oil pressure. When pressure drops below a specific setting and allows the switch to function, the circuit activates. The alternator is switched on by low voltage in the sensing circuit. Mechanical switches may be used to control other warning lights.

A grounding switch at the parking brake pedal or lever usually switches on the red brake warning light (Figure 12-6). A fluid sensor in the brake's reservoir may also turn it on as well as a switch in the brake's hydraulic system. A switch built into the doorpost or trunk latch controls door and trunk ajar lights. The door switch will activate interior lights.

There are several amber-colored lights that alert the driver to a condition that is not life threatening or a problem that will not result in serious damage to the vehicle if corrective action is taken soon (Figure 12-7). One of the most prominent is the **malfunction indicator light (MIL).** It is

A **trigger** is a condition that will cause an action to be taken (e.g., low oil pressure or fire a spark plug).

A **grounding switch** is a switch that, when closed, completes the circuit to ground.

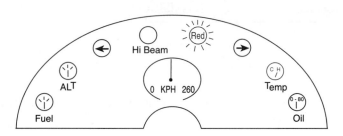

Figure 12-4 This lit warning light alerts the driver to a high engine temperature. Note the temperature gauge.

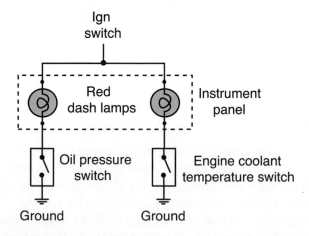

Figure 12-5 This red warning light may be switched on by either the oil pressure or the engine coolant temperature.

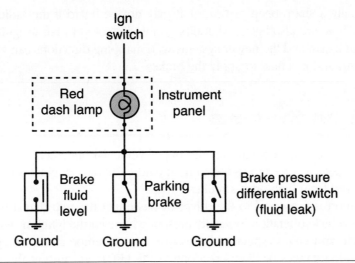

Ign
switch

Red dash lamp — Instrument panel

Brake fluid level — Ground

Parking brake — Ground

Brake pressure differential switch (fluid leak) — Ground

Figure 12-6 A typical red brake warning light may be switched on by a switch on the parking brake lever/pedal, low brake fluid in the reservoir, or by a leak in one of the brake hydraulic circuits.

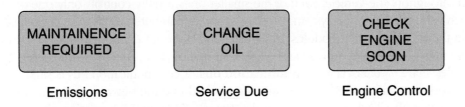

MAINTAINENCE REQUIRED

CHANGE OIL

CHECK ENGINE SOON

Emissions   Service Due   Engine Control

Figure 12-7 Three possible maintenance reminder lights. The one on the far right is required by the EPA and deals with engine and transmission controls that affect exhaust emissions.

switched on by the PCM based on data received from the engine and transmission sensors and is commonly known as the check engine light. The light is a warning that some condition is present that interferes with the proper operation of the vehicle and should be corrected at the earliest possible time. This allows the driver to safely drive the vehicle until a repair shop can be reached. The lens of this light may reflect the words "Service Engine Soon" or "Check Engine." Many vehicles have another amber light that indicates a service is due or the light may indicate which service is required. Labeling on these lights may be "Change Oil" or "Check Emissions." Another prominent amber light is the warning light for the ABS. A lit ABS warning light indicates two things: the ABS has a fault or the ABS is deactivated. The regular (service) brakes will work fine and there is no safety concerns as long as the driver understands that the ABS will not function in an emergency stop. Most vehicle electronic systems that have some type of self-diagnosis capability have the ability to alert the driver to the fault. In many cases, the technician will need a scan tool and other electronic testers to isolate and diagnose the problem.

☑ **SERVICE TIP:** Though the amber ABS light or the red brake warning light may not indicate an immediate danger to the driver, if both lights are lit there is a serious problem involving the service brakes and the ABS. The fault(s) must be corrected before the vehicle is allowed on the road even if the brakes seem to be working.

Other amber lights include low fuel, traction control, overdrive off, electronic suspension, low tire pressure, supplemental restrain systems (SRS) or air bags, shift now, and others either present today or that will exist in the near future. Most warnings lights are backed up by an audible alert, usually a sharp beep or buzz that may not be heard if the radio or road noise is above "normal." There are also light and audible reminders of keys left in ignition, seat belt not fastened, door/trunk ajar, and the newer systems with mapping directions can even alert you that a stop sign is ahead and it is time to apply the brakes.

Information Systems

Shop Manual
pages 374–380

Information is conveyed to the operator on almost any vehicular operation desired. The data available ranges from a vehicle speed to its location on earth. The data originates from sensors on the vehicle or is received from high-orbit satellites.

The information is presented to the driver or passenger through digital or text display. Information such as vehicle operation is usually present whenever the ignition switch is on. The basics include vehicle and engine speed, oil pressure, engine temperature, A/C generator output, and fuel level. Analog gauges or digital readouts using LEDs can present the information in an easy-to-read format, normally a number.

Newer vehicles may use a section of the PCM or a separate computer module to compute information. The 2005 model cars may have up to fifty **microprocessors** or computer modules. Generally the microprocessor can be physically as large as the PCM but most are small black boxes scattered throughout the vehicle. Each of the smaller ones usually controls only one or two functions, such as the information system. When a separate computer or module is used, it collects data from the PCM and other modules to calculate and present the information. A PCM may have a separate internal circuit to process and display the information, however, it is typical that separate modules share all data via a **communication bus.** A communication bus is similar to a computer network in a large office. Each computer is wired into the network and can communicate and share data with all other computers in the system. Additional information about the vehicle and its surroundings may include:

A **communication bus** has all sensors, computers, and modules connected to shared circuits. In this manner, each computer and module can monitor data.

- Outside and inside temperatures
- Continuous or accumulated fuel mileage
- Time of departure and expected time of arrival
- Miles left on the remaining fuel
- Mileage completed or left on this trip
- The location of the vehicle at any single time
- How to get from here to there
- Road conditions and suggested detours

Some of this information is displayed continuously or at the touch of a button (Figure 12-8). The operator can select the desired information and even select USC or metric measurements. Like operating data, this information can be displayed by analog or digital method, but digital is the most common. Many options offered on the 1999 high line models had become standard features on most vehicles by 2006.

One option that is becoming more common came from military and space exploration. The **Global Positioning Satellite (GPS)** is actually a network of satellites deployed in orbit around the earth. A vehicle-mounted receiver translates data from several satellites and calculates the exact location of the vehicle within a few yards. The resulting display provides the vehicle's position on the earth in an easy-to-read and easy-to-understand format. Using this information and an on-board mapping system, the operator can plan a route to an unfamiliar destination, around an unexpected detour, or to find out how to get back on route.

The satellite network known as the **Global Positioning Satellite (GPS)** consists of hundreds of satellites that communicate with each other and transmitters/receivers on the earth.

The future implications of GPS are not easy to define, but a vehicle with an automatic electronic "take me home, James" chauffeur feature is not unrealistic. The GPS is used by manufacturer-sponsored roadside assistance options. The OnStar® system, which is standard on many General Motors 2006 vehicles, allows the driver to request the vehicle's doors be unlocked (if proper ID is presented), provides breakdown assistance via cell phone, and even monitors air bag deployment.

Figure 12-8 An overhead console module usually displays "nice to know" information that is not essential to vehicle operation.

In case of an air bag incident, an OnStar® operator will attempt to contact the vehicle driver and offer assistance in notifying emergency personnel. This system is available because of signals sent from the vehicle via satellite and GPS locators to alert the system's human operators. According to the General Motors OnStar® Web site, during November 2004 through January 2005, it responded to 900 airbag notifications, 400 stolen vehicle location requests, and 20,000 roadside assistance calls. This seems to prove that this system and similar operations were worth the time and cost of development. More sophisticated systems are in the research and developmental stages that will provide even more assistance or information with a lower consumer cost. At some point in the near future, repairs to this type of system will be routine provided the shop and the technicians have spent the time and money to continuously upgrade their equipment and training. This upgrade applies to all electronic systems now available or that will be available on future vehicles.

One manufacturer of large trucks has a vehicle/dispatcher satellite transmitter/receiver. If the truck breaks down, the driver calls the dispatcher and gives the code for that truck. The maintenance department communicates via satellite with the truck's PCM. The PCM transmits operating data and **snapshot** data back via the satellite. The technician selects the most probable repair parts based on that data, proceeds to the downed vehicle, and makes the necessary repairs. Vehicle down time is greatly reduced.

A **snapshot** is stored data that can be retrieved and studied at a later time. It is usually data of a particular happening at a particular time, such as a misfiring cylinder under heavy load. It may be referred to as a "movie."

Shop Manual
pages 374–380

Instrument Cluster Operation

The instrument panel houses most of the gauges and warning lights. Most informational data can be displayed in an overhead console or near the midpoint of the dash (Figure 12-9). This keeps the small space in the panel clear for essential operating data. A printed circuit board, or **integrated circuit,** is located behind the instrument panel. This board supplies the necessary circuits between the gauges and warning lights to their individual sensors or computers. A printed circuit board is used to reduce or eliminate the number of wires needed to connect all the different components. A printed circuit board is basically a very early, crude model of an integral circuit. It has been replaced by much better versions that are smaller and capable of many more circuits.

The panel's power and all other electrical connections to the remainder of the vehicle are made using one or two large, multiple-pin connectors. Large, molded, or shell terminals are plugged into the connector on the vehicle side. The panel side of the connector is an integral part of the board. The conductors are downsized and attached to the board's circuits using small

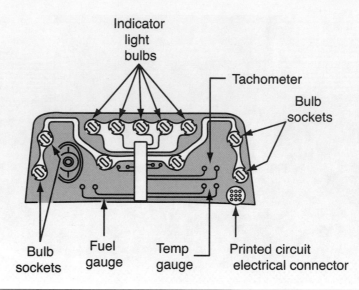

Figure 12-9 Small, flat strips of conductors connect the various gauges and lights in the instrument panel. Note the suggested circuit indicated by the highlight. (Courtesy of Nissan North America)

soldered connections. Most integrated circuit boards are built off-site and installed in the vehicle much like installing an engine.

AUTHOR'S NOTE: The original printed circuit boards installed in vehicles left a lot to be desired. They were fragile and susceptible to things such as cigarette smoke and tar, rough roads, and just plain abuse. They were almost impossible to repair at the technician's level and were usually replaced at a high cost to the consumer. But they did provide a base from which integrated circuits for the dash could be designed, developed, and built at a reasonable cost.

Lighting Systems

Shop Manual
pages 380–391

Most of the exterior lighting is required by federal and state regulations. The only mandated interior lighting is the instrument panel. Each of the mandated lighting systems, interior or exterior, must meet requirements for the amount of illumination generated and for positioning. Other lights may be provided for passenger comfort and convenience provided they do not distract from the mandated lights.

Lamp Specifications

The vehicle manufacturer specifies the type of lamp. There are many types used on a modern vehicle. Lamps for the marking, turn, brake, and headlight systems must meet illumination capacity specified by federal and state regulations. Other lamps like the courtesy lights are selected by type to provide the level of illumination needed and the mounting space available.

The lamps are connected to the circuits using several types of connections. Most turn and brake lamps use **bayonet** housings to lock the lamp in place (Figure 12-10). A single filament lamp will have locking tabs across from each other. Dual-filament, lamp-locking tabs are offset from each other to prevent improper installation. Headlight lamps use weather-resistant, hard-plastic shell connectors. Interior lamps may be the bayonet-type or the slide-in-type and are usually single-filament lamps (Figure 12-11). Instrument panel lamps may be bayonet-, or slide-in-type, or may be LED-type lights.

A **bayonet** connection is so named because of its push in, twist, and lock steps to make the connection.

358

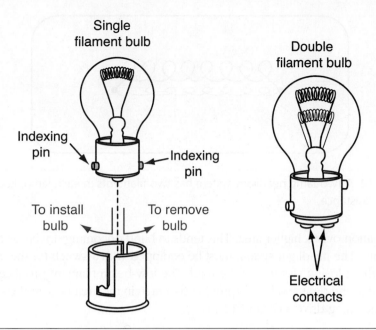

Figure 12-10 The offset pins ensure that the two contacts at the bottom of the lamp connect to the correct circuit.

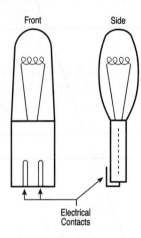

Figure 12-11 Slide-in lamp contacts are small wires folded up over the base. The female socket will have two brass strip contacts.

Exterior Lighting

The mandated exterior lights are:

- Headlights—one on each front corner of the vehicle, white, must provide high and low levels of illumination.
- Marking lights—one on each corner of the vehicle, amber in front, red in rear.
- Turn signal lights—one on each corner of the vehicle, amber in front, red or amber in rear.
- Brake or stop lights—one on each rear corner in the following driver's direct line of sight, red. A third brake light must be installed above and centered on the two lower brake lights.

Headlights

Headlight systems may use two or four lamps provided they meet the illumination and position requirements. The two-lamp system uses two filaments in each lamp (Figure 12-12). One filament is for the high beam and the second is for the low beam. The high beam is the strongest and

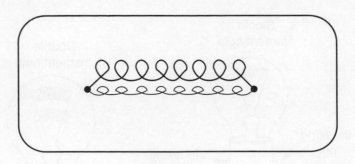

Figure 12-12 A two-lamp headlight system has two filaments in each lamp. Each filament has a different resistance.

provides illumination over a higher area. This tends to blind oncoming drivers as the vehicles approach each other. The headlight system must be equipped with a switch for the driver to change to low beam without losing visibility of the road. The low-beam filament produces less illumination and is aimed to the right and low to provide coverage in the area nearest the vehicle and prevent blinding oncoming drivers (Figure 12-13).

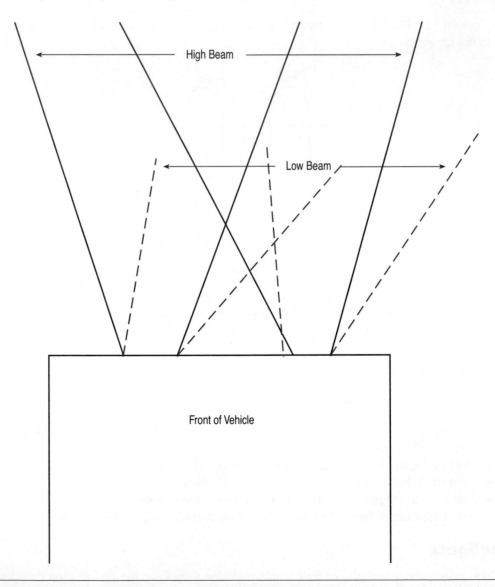

Figure 12-13 As indicated by the highlight, the low-beam illumination aimed is to the lower right, and is not as bright as the high-beam illumination.

The four-lamp system has two headlamps at each front corner of the vehicle (Figure 12-14). Two inside or outside lamps are for high beams while the other pair provides low beam. In this case, the high/low beam switch usually turns the high beam pair on and off. The two low beam lamps are lit in both modes.

The high/low beam selector or **dimmer switch** is most commonly part of the **multifunction lever** mounted to the left of the steering column (Figure 12-15). This lever also controls the turn signal and may include wiper and cruise control switches. A blue light mounted in the top

The **dimmer switch** was adopted because the headlights were much dimmer on low beam.

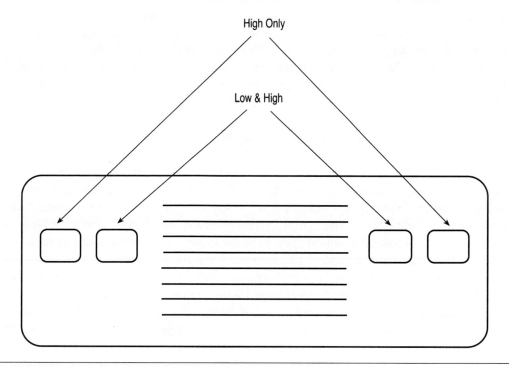

Figure 12-14 The smaller, four-lamp system allows the vehicle's grill and hood to be slightly lower and provides a wider field of illumination on high beam. This system is no longer used on the newer light vehicles.

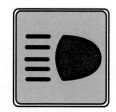

Lit when high beam selected

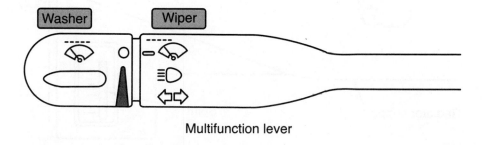

Multifunction lever

Figure 12-15 The high-beam headlight indicator is usually near the center of the instrument cluster and has a blue lens. Note the symbol for high beam on the multifunction lever (right side).

center of the instrument panel indicates that high beam has been selected. On older vehicles and some present-day heavier trucks, the high/low beam switch is mounted in the floorboard and is operated by the driver's foot. Automatic switching was tried on some of the more expensive vehicles in the past with mixed results. One of the problems then was the state of electronics. The sensor used to detect oncoming headlights was not always accurate, requiring the driver to make a panic changeover manually. Today's electronics are more reliable and automatic high/low switching has reappeared.

Headlamps must be aimed to prevent blinding other drivers and for the best illumination of the area ahead of the vehicle (Figure 12-16). In most cases, the lamps are installed into an adjustable mount. The mount is connected to the frame by springs and adjusting screws. There are usually only two screws per mount, one each for vertical and horizontal adjustment.

Lamps used for the headlight systems can be sealed beam or a single bulb installed into a reflective chamber. The sealed beam lamp has filaments or an iodine vapor bulb installed in a chamber lined by highly reflective material (Figure 12-17). The chamber is sealed by the permanent installation of the lens. Any damage to the lamp requires replacement of the entire sealed beam lamp. Sealed beam lamps may be found on the two- and four-headlamp systems.

With the manufacturer's desire and need to increase fuel mileage, the front of many vehicles had to be sloped to reduce air resistance. The federally required height-above-ground of the lights and the bulky sealed beam lamps hindered the engineers. The solution called for a redesigned headlight system. The manufacturers removed the filaments from the lens and reflector assembly. The assembly was shaped to fit the lower front contours of the vehicle while maintaining the height-above-ground requirements. A separate lamp was then installed into the rear of the assembly. The lamp is a high-intensity halogen bulb capable of producing comparable illumination as the sealed beam (Figure 12-18). Failure usually requires only the changing of the bulb, but not the lens and reflector. Most vehicles now use this design and a two-lamp headlight system. Some newer vehicles are equipped with low-voltage headlights. The lights are on any time the engine is operating. This safety feature better highlights the vehicle to other drivers and pedestrians during daylight. They are referred to as **daytime running lights.**

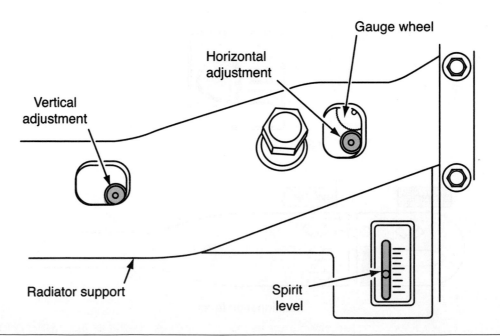

Figure 12-16 The newest headlight system has built-in angle gauges to help aim the lamps.

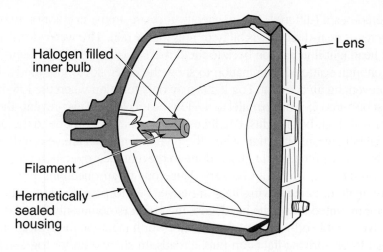

Figure 12-17 The interior of the sealed beam lamp is reflective and has the lamp or filaments mounted in the center of the cavity.

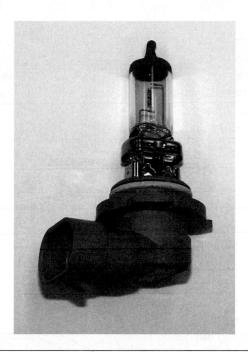

Figure 12-18 This small, high-intensity lamp actually provides more illumination than the older large bulb, but it is more a result of a better reflecting lens than the bulb itself.

Marking Lights

Marking or parking lights are used to identify the vehicle when visibility is low but headlights are not needed. Marking lamps are usually single-filament in front and dual-filament in the rear. The lights are turned on and off with a separate set of contacts within the headlight switch. The marking lamp is also used in most vehicles as the turn signal lamp. Most of the newer vehicles leave the marking lights on during headlight operations.

Fog and Driving Lights

Many vehicles are fitted with **fog** or **driving lights** or both. The two light systems are designed differently and must meet separate requirements. Fog lights are mounted below the marker lights and have scattered-beam (diffused) lens. Generally they are amber in color (may be clear) and are provided to increase near-distance visibility during fog or rain. The water droplets in fog or rain will reflect the headlight illumination back to the driver's eyes and basically blind him or her. Fog lights provide enough scattered illumination to show the outer side of the road and the center, in the hopes of preventing an accident. Fog lights may only function when the low-beam headlights are on and must be wired in series-parallel to the low-beam circuit. This prevents them from switching on singly or with high-beam lights. Fog lights must be aimed to not blind the oncoming driver.

Driving lights may only function when the high-beam lights are selected and are wired in series-parallel with that circuit. Again, this is done to prevent their operation unless the high-beam headlights are on. Driving lights can be very dangerous to an oncoming driver because of the lights' high illumination. Some driving light are advertised to provide illumination up to 1 mile. Driving lights are mounted above the headlights and are most commonly found on pickup trucks. Though they have a viable off-road use, they are not worth much on the typical highway. In many states it is illegal to use driving lights on public roads. In snow, rain, or fog conditions, they are totally useless and can present a danger to the driver and others on the road. Most driving lights are mounted by the driver using aftermarket components.

AUTHOR'S NOTE: At one time I had a 1986 Dodge Ramcharger (called an SUV today) that had fog lights on the bumper and driving lights (four) with a center-mounted, high-intensity single light mounted on a light bar above the cab. Unofficially, one or all could be turned on through a system of switches and relays. The only time they proved their worth was lighting an accident scene on a dark country road while the first responders worked the scene. After the situation was cleared, I was thanked for my assistance by the state trooper while he wrote me a warning ticket for improper lighting. The moral: Every good deed deserves another, and illegal is illegal regardless.

Turn Signals

The lamps in this system are used to communicate the driver's intentions to other persons on the road. Most turn signal systems use the marking lamps as the signaling devices. The turn signal switch is mounted behind the steering wheel on the column. It is controlled by a lever extending to the left from the column and has three positions: off, left, and right (Figure 12-19).

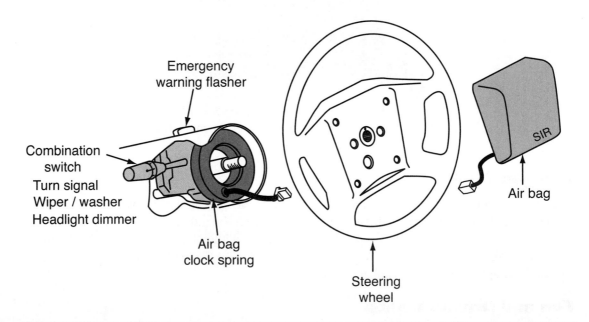

Figure 12-19 All standard modern vehicles in the United States have the turn signal lever on the left side of the steering column.

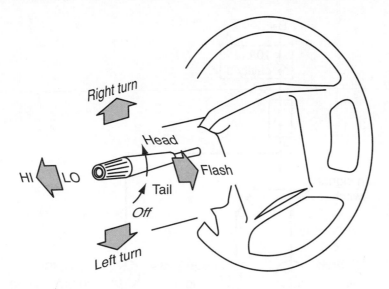

Figure 12-20 The turn signal lever can be moved up for right turns and down for left turns as shown here.

Moving the end of the lever downward supplies power from the fuse through a **flasher** to the left marking lamps at the front and rear of the vehicle (Figure 12-20). The flasher is a **bimetallic** circuit breaker. As current flows through it, the heat causes the breaker to trip and open the circuit. As it cools, the breaker resets and closes the circuits to the lamps. This process cycles the flashers on and off between .5 second and 1.0 second on the average. The result is a flashing light at each end of the vehicle on the left side. The same process happens for a right selection except for the lamps that are flashing. Adding additional lamps such as those for a trailer can overload the flasher. Installing a heavy-duty flasher is usually required before adding additional lighting.

A **bimetallic** strip has two pieces of metal laid side by side. One piece reacts to heat faster than the other one does.

The rear turn signal lens may be amber or red. If the brake lamp is housed in the same reflective and lens assembly as the signal, the lens will be red. This is usually referred to as the single lamp system because the marker, turn, and brake use one lamp with dual filaments. The lens for a two-lamp system, which uses a separate lamp for turns, has amber for signaling and red for the marker or stop. Some vehicles are equipped with three lamps, one each for marking, turning, and braking. The single requirement is that the turn lens is amber or red and the marking and brake lights are red.

The **emergency flasher** or hazard lights work much the same as a turn signal. A separate switch is used to make all four turn signal lamps flash together. The four-way circuit bypasses the turn signal circuit and directly feeds each of the four lamps. Brake applications will override the four-way on the rear lamps to alert following drivers to the situation. Brake applications will not override regular turn signals. The marker lights, main brake lights, emergency flashers, and turn signal are all wired through the turn signal switch.

Brake Lights

Brake lenses must always be red and have much higher illumination than the other lamps near them. A mechanical switch on the brake pedal rod controls the brake lights. Power is fed directly to the brake (stop) switch from the battery (Figure 12-21). When the brake switch is closed, current travels to the rear stop or brake lamps via the turn signal switch. If a turn signal is on, the flasher interrupts the current to that lamp. When the four-way circuit is operating and the brakes are applied, the brake light current enters below the flasher and provides a normal steady current to the brake lamps.

The **Center High-Mounted Stop Light (CHMSL)** is a federally mandated third brake light, usually located on the lower rear window or on the trunk lid of a vehicle.

A third brake light was mandated by federal regulations to gain quick the attention of a following driver. The third brake light is officially known as the **Center High-Mounted Stop Light (CHMSL).** It may be located on the lower edge of the rear window or mounted to the trunk lid

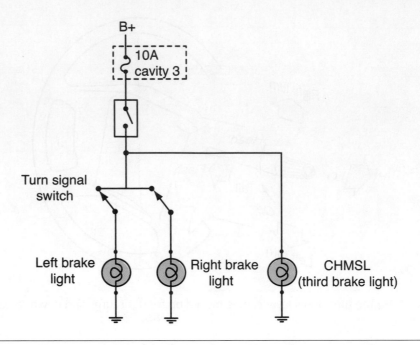

Figure 12-21 The third brake light, CHMSL, is wired parallel to the other brake lights and will work even if there is a problem in the turn signal switch.

(Figure 12-22). The third brake light is within the normal line of sight for a following driver. The lamp is wired into the brake switch output between the switch and the turn signal switch. In this manner, the light will function regardless of the turn signal switch or the other two brake lights. This light is not affected by the turn signal switch or the hazard light switch. There are some systems that have a flasher included in the CHMSL that will cause this light to flash. It is supposed to more readily alert the following driver, but in heavy, slow traffic it can become a nuisance to a following driver.

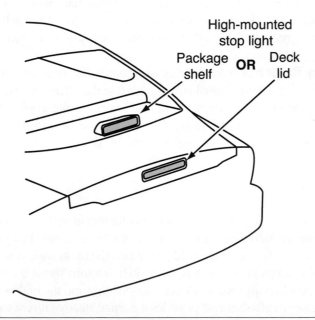

Figure 12-22 The CHMSL must be centered on the rear of the vehicle and not more than three inches below the bottom of the rear windshield.

Tag Lights

The tag light is switched on any time the marking or headlight is selected. The light is used to illuminate the rear license plate so officers can read the tag. Sometimes, two lamps are used. A working tag light is required in all states.

Other Exterior Lights

There are usually only three other exterior light systems offered. They are the engine compartment light, trunk light, and turning lights. The hood and trunk lights are provided on most vehicles and are simple circuits. Raising the hood or trunk closes a mercury or mechanical switch and the circuit. The two lighting systems are provided as a convenience item. The turning lights are more complicated and are not offered on many vehicles.

The turn or cornering lights have a clear lens lamp mounted on the front sides of the vehicle, normally just behind the turn signal assembly. When a turn signal is selected with the headlights on, the circuit closes a relay that turns on the cornering light for that side. The lamp emits a high-intensity, low-range light to aid the driver in negotiating a corner. This system is not very popular and is offered on limited vehicles.

Interior Lighting

Interior lighting is provided for entry or exit from the vehicle and illuminates the passenger compartment. The instrument gauges are **backlit** by twist locked into the rear of the panel. The lamps are positioned to light the gauges indirectly. The front side of the panel is coated with a non-glare finish. Contacts in the headlight switch turn on the lamps any time the marking or headlights are selected. The intensity of the illumination can be controlled by a rheostat mounted within the headlight switch or on a separate switch on the dash. The lights can be turned off with the rheostat switch without affecting the exterior lights. The headlight switch or the rheostat does not affect the warning lights in the instrument panel.

Included in the group known as the instrument lights are lamps to highlight various controls located on the dash. Most dash panels house the controls for heating and air conditioning, radio, blower, and on-board computer input keys. Lamps placed behind the control panels or in the button illuminate the markings and the buttons.

Backlit means the light is coming from behind or from the side of the object being viewed.

Dome and Courtesy Lights

The dome light and courtesy lights are switched on when a door is opened (Figure 12-23). Most have separate switches to control individual lights when the door is closed (Figure 12-24). The dome light is mounted near the center of the roof. Courtesy lights, if equipped, are usually mounted on the kick panels, under the instrument panels, or on the doors. Rear courtesy lights are usually mounted to the left and right side of the passengers. In addition to the courtesy lights listed, reading or map lights may be positioned to shine toward each seat. This allows a passenger to read or search for an item without hindering the driver's front or rear visibility. Overhead dome lights tend to block the driver's vision through the inside rearview mirror.

Glove Box and Key Lights

The glove box light works similar to the hood and trunk lights. Opening the glove box door closes a mechanical switch. A small lamp is lit to illuminate the interior of the glove box. This is a common problem when tracing a drain on the battery. Items placed in the box may prevent the door from opening the switch and the light slowly drains the battery overnight.

The slot for the key on the outside of the door and the ignition switch may have a small lamp, possibly an LED, for illumination. This helps the driver insert the key in the door quickly at

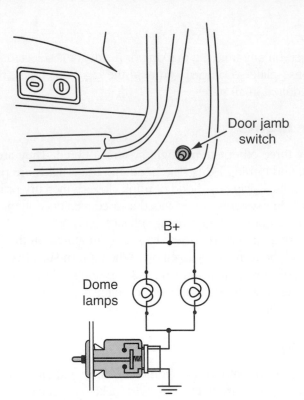

Figure 12-23 A typical location for a door switch. This switch will turn on the dome lights and the "door ajar" lamp if equipped.

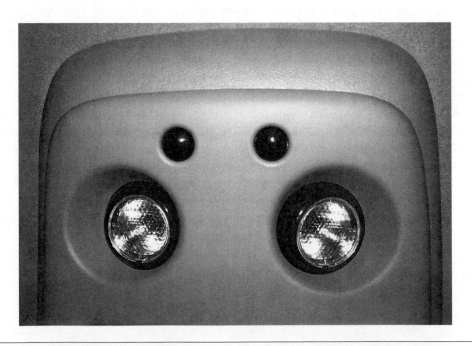

Figure 12-24 Map or reading lights can provide a narrow field of illumination for use by the passenger. Similar lights may be installed over the rear seats.

night and helps prevent marring the door's finish. The ignition switch is in an awkward position for the typical interior light. Providing a small lamp near the switch helps the driver insert the key at night.

Accessories

Shop Manual
page 391

Accessories include systems like power door locks and windows, custom seat covers and carpets, trailer hitches, and many other much-liked but not-always-necessary items. In this section, we will discuss power door locks and windows, antitheft devices, and power seats as examples of system operation. The next section will cover a major and expensive accessory: climate controls, commonly known as the heat and air-conditioning system.

Power systems such as door locks, windows, and seats were designed to help the driver control security and provide comfort without distracting from driving conditions. The door locks have recently been incorporated into antitheft systems and remote control.

Power Door Locks

The electrical/mechanical mechanism for the locks is basically a reversible polarity solenoid or motor moving the manual lock mechanism. There is a solenoid for each door controlled by parallel-wired switches in the front doors (Figure 12-25). Operating either switch locks or unlocks all doors. A switch built into the key lock on the outside of the driver's door will also perform the same function. On some vehicles, one twist of the door key will unlock the driver's door and a second twist operates the other door locks.

The solenoid is positioned to move the locking mechanism when activated. The solenoid can either push or pull the locking mechanism. The operation depends on the type of lock and the space available to install the solenoid or motor. Most vehicles have manually operated locks to back up the electric system.

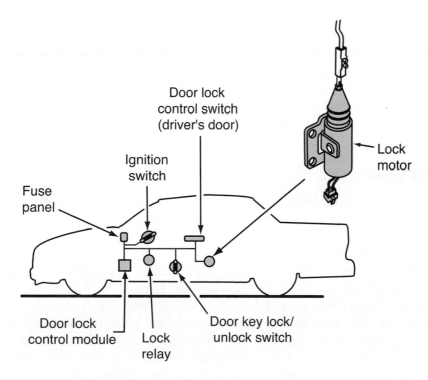

Figure 12-25 The power door locks may be controlled by the door-mounted manual switch, the remote control, or by using the key in the exterior door lock.

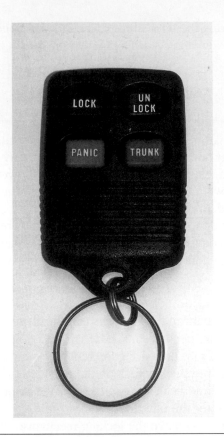

Figure 12-26 This remote control or FOB has a button to lock or unlock the door or trunk plus a "panic" button for the operator to use in case of emergency.

With the availability of dependable electronics, the door locks can be activated with a small remote control device. The body computer receives the remote signal and operates the locks. Antitheft security systems are connected to the body computer so the alarm system is turned on or off whenever the locks are activated (Figure 12-26). Many vehicles have automatic systems that lock the doors when the vehicle reaches about 20 miles per hour or a drive gear is engaged. The locks must be unlocked with the key, switch, or the remote control unit. A common fuse protects all of the lock circuits.

Antitheft Systems

Antitheft systems range from a toggle switch placed in the ignition circuit to ignition keys that must identify themselves to the PCM before the vehicle can be powered up. A toggle switch can easily be found and flipped on. This may prevent some thefts but a professional thief will be gone with the vehicle in two or three minutes.

Most antitheft systems in use today rely on noise and flashing lights to alert anyone nearby that the vehicle is being stolen. However, once the thief is inside and the engine is running, the antitheft system will usually disarm and the vehicle can be driven safety.

A later system uses an ignition key with an embedded resistor pellet. When the key is inserted into the ignition switch, a small current is sent through the key and pellet. If the resistor is present and correct, the current flows to the antitheft module to power up the vehicle.

In 1999 a new system was installed on some General Motors vehicles. Ignition keys containing small **transponders** used with specific security codes. The system's control module was placed around the steering column under the column covers. A round antenna ran from the module and encircled the ignition key slot. The antitheft module and the PCM were programmed to

A **transponder** will transmit its code by radio or radar when actuated by a specific signal from a radio or radar transmitter (interrogator).

recognize the key's security code and save it. This was done when the vehicle started for the first time at the assembly plant. It was also done at the dealership for replacement of additional keys. The key's code was cross-matched against other selected components so the key could not be used in another car. When a driver inserted the key into the ignition switch, the control module asked for identification from the key. The key transmitted its code and the antitheft control module compared it to the saved codes for the key and the selected components. If all codes were correct, the module signaled the PCM to switch on the electrical systems. Presently, multiple key codes can be stored in the module and PCM. If a key, control module, or PCM is replaced, the system has to be reprogrammed.

> **AUTHOR'S NOTE:** Aftermarket antitheft systems, including some dealer-installed units, offer some disarming problems during vehicle repairs. In some systems, if the battery is disconnected for long periods of time the security system will not disarm when repowered and the engine can not be started. There may be a specific sequence of steps required to reset the system computer. In most cases, the instructions are contained in the operating manual for the system, and that manual cannot be located. Sometimes the owner may know of the sequence, or the data must be obtained from research on the Internet and/or paper manuals.

Power Windows

Power windows use reversible electric motors to raise or lower the windows through gears or a plastic strip punched to fit over gear teeth. There is a window switch at each door with a **master switch** on the driver's side. The master switch may also have a lock switch that deactivates the other switches. The nondriver window motors are parallel to the master switch.

The master switch consists of a switch for each window motor and the window lock-out switch, if so equipped (Figure 12-27). The other switches placed on the other doors or panels will only operate that window. A common fuse or circuit breaker protects all of the window circuits.

The output gear moves the mechanical device that actually moves the window on the motor. In a straight gear system, the output gear is meshed with a large, flat sector gear. The sector is attached to the arms or regulator that push up or pull down on the glass frame. Plastic strap (ribbon) or steel stranded cable mechanisms use the motor's output gears to pull the cable or ribbon

The **master switch** houses a switch for each window and can lock out all other individual window switches.

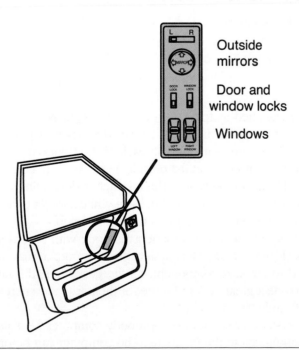

Outside mirrors

Door and window locks

Windows

Figure 12-27 Note the three different systems controlled from this master switch cluster.

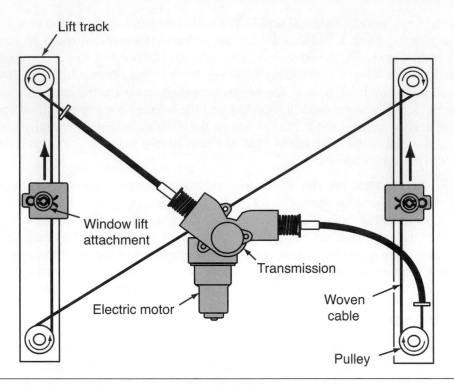

Lift track

Window lift
attachment

Transmission

Electric motor

Woven
cable

Pulley

Figure 12-28 The part shown is a replacement for a cable-type window regulator.

that is attached to the glass frame (Figure 12-28). The ribbon is punched at even intervals and fits over the output gear. Two or three pulleys and a metal guide keep the ribbon or cable in contact with the gears and pulleys to prevent slippage.

Some vehicles are equipped with automatically lowering driver's-side windows. When the switch is operated under specified conditions, the window, once started, will lower automatically to its lowest position. This allows the driver to direct attention to the environment.

 **AUTHOR'S NOTE:** Some test lights use light emitting diodes (LEDs), which can be used to test for voltage on electronic circuits. Before using, ensure it is an LED unit. It is too late after the damage is done.

Power Seats

Power seats are usually classified as four-way or six-way. Four-way seats can move electrically forward or backward and they can tilt the backrest. Six-way types have the same movements plus tilting of the seat bench. The systems may be on each of the front seats or just on the driver's.

The controls are usually mounted at the outside end of the seat bench, but may be mounted in the door or console. There will be two or three switches depending on the system. One will control the forward and rearward movement of the seat while another controls the tilt of the backrest. A third switch is used to control the motors to tilt the bench.

The actual movement of the seat components is done with reversible electric motors and gears, cables, or levers (Figure 12-29). Most systems use an output gear meshed with a toothed rail. The rail is attached to the seat component. As the gear turns, the rail and component are moved. The motors provide a great deal of torque for their size and energy usage and can cause damage to anything left under the seat.

Many power seat systems are controlled by a body computer. Sensors on the undercarriage of the seat provide position data to the computer. The computer can be programmed to remember the data and relate it to a particular operator. Selecting the proper code activates the seat and returns it to the position recorded in the computer's memory.

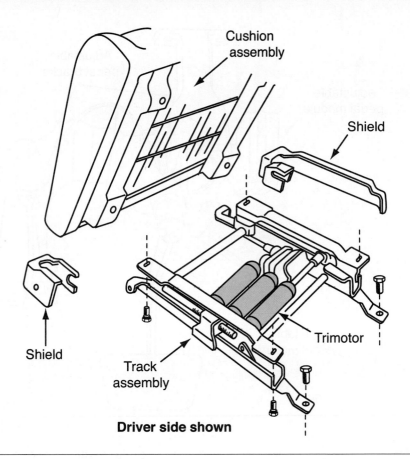

Cushion assembly

Shield

Shield

Track assembly

Trimotor

Driver side shown

Figure 12-29 A typical power seat motor is one piece with all three motors mounted in one housing.

Power Adjustable Pedals and Steering Columns

For persons who are taller or shorter than the height of a "normal" person, reaching the various pedals on the floor can at times be difficult. Many times the steering wheel position is not always comfortable for the driver. A taller person might have to move the seat back for pedal clearance, but then would be required to reach for the steering wheel. A shorter person might slide the seat forward, but then he or she would be against or very near the steering wheel. The adjustable steering wheel or actually the steering column helps alleviate some of this, but it was not until the supplemental restraint system (SRS) was mandated for the driver that other, even more serious concerns were brought to light. The concerns associated with air bag deployment were emphasized during the late 1990s when some deaths and injuries were actually caused more by the air bag than the accident. The National Highway Traffic Safety Administration recommends the driver be at least 10 inches (25 centimeters) from the steering wheel. Safer air bags were developed, but something more sweeping was needed. Ford introduced APS in some of its 1999 model vehicles.

Adjustable pedal systems (APS) can be mounted on a moveable platform or singularly mounted. The pedals can be moved up to 3 inches (eight centimeters) to accommodate the physique of most drivers (Figure 12-30). Screws or transmissions driven by electric motors move the platform or pedal along an axis that is close to the angle of the steering column. Singularly mounted units require a separate motor/transmission for each pedal. Some units are designed for either linear or arcing motion. Linear motion moves the entire pedal mechanism in a straight line while the arcing motion moves the lower end closer to the driver with minimum movement of the pedal(s) pivot mount. The APS has been accepted by customers, generally with few problems to date. The original APS offered by Ford was deemed by drivers to have the pedals too close together. Ford has since corrected that problem through a recall and a new design for its later models.

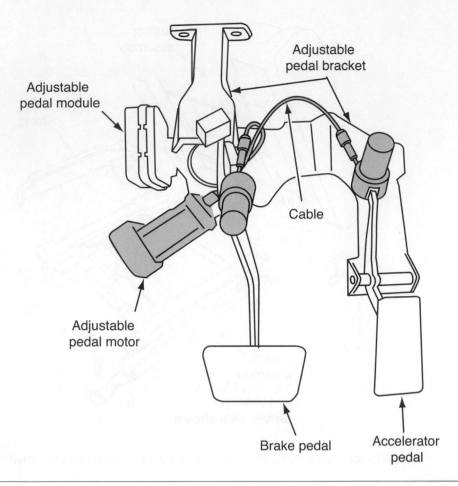

Adjustable
pedal module

Adjustable
pedal bracket

Cable

Adjustable
pedal motor

Brake pedal

Accelerator
pedal

Figure 12-30 Adjustable pedals are offered in some high-line vehicles, with more to be offered in the next several years.

AUTHOR'S NOTE: APS is not new. It was introduced in 1910 to the public on a 1908 American Simplex (any one remember that one?). It was a mechanical system where the operator had to use a wrench to adjust the pedal. General Motors obtained a patent for a manually adjusted system in 1966 and the Dodge Viper offered a manual system in 1996. Viper is one of the few vehicles offering an APS for a manual-transmission-equipped vehicle.

The steering column can be lengthened, shortened, tilted, or all three. Memory systems have been added so multiple drivers can use the same vehicle just by punching a code, and the pedals and steering column move to preset positions much like the memory seats. In fact, the controlling computer will usually position the seats, pedals, steering column, and outside rearview mirrors.

Climate Control Systems

Shop Manual
pages 391–397

Climate control system is the inclusive term for the heating and air-conditioning systems. The term is used to designate either a straight heating system or a heater/air-conditioning combined system. The heater system is used to warm the interior of the vehicle during cold weather. Naturally, the air conditioning is used to do the opposite.

Heater Systems

The heating system uses the engine's coolant to heat air that is being forced into the passenger compartment. Hoses run from the engine block to a small heater core and back to the engine

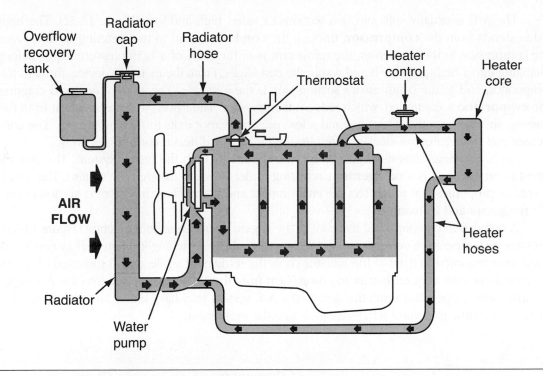

Figure 12-31 The heater hoses carry hot coolant from the engine to the heater core and back to the engine. They connect to the engine in various locations depending on engine design and placement within the engine compartment.

block (Figure 12-31). The core is located in front of the front passenger seat, either within the dash or on the engine side of the firewall. A mechanical or vacuum-operated valve controls the flow of hot coolant into the core. A variable-speed motor drives a blower and forces air around the heater core fins and through ducts into the passenger compartment. The hot coolant in the heater core heats the passing air. The core is similar to a radiator in appearance and operation.

Newer heater systems, especially those installed with an air-conditioning system, have a full-flow heater core. The coolant flows through the heater core any time the engine is running. Air is directed away from or through the core by a temperature blend door. This eliminates the valve and its controls while providing the driver more precise control over the conditioned air entering the vehicles.

The control panel mounted in the dash houses buttons or levers to direct the air, change blower speed, and adjust temperature. The operator can select outlets that direct warm air to the windshield, dash panel, or to the floor. The air lever moves blend doors within the ductwork mechanically with cables or small vacuum motors, or small electric motors. The blower switch has three or four speed settings. The temperature selector is used to adjust the amount of hot coolant allowed into the heater core, to block the flow completely, or to adjust the air flowing through the core. The heater system depends on the engine and its thermostat for operation. A poorly performing engine or a stuck thermostat may result in poor heater operation.

Air-Conditioning

Before discussing the air-conditioning (A/C) system, a quick review of heat and pressure is needed. When a liquid is heated enough, it changes to a vapor. Adding pressure to the liquid raises the boiling point and reduces vaporization. By the same standard, removing heat from a vapor causes the element to cool and liquify. Pressure and heat are the two theories that make air-conditioning systems work.

The **compressor** is a high-pressure pump built specifically for air-conditioning systems.

A **condenser** cools down a vapor and assists in changing the liquid from a vapor state to a liquid state (condensation).

The **evaporator** helps a liquid heat enough so it will change from liquid to vapor (evaporation).

A **refrigerant** is a chemical that can absorb, transport, and release a great deal of heat in a fairly small space under easy-to-create conditions.

The A/C is usually split into two sections or sides: high and low (Figure 12-32). The high side extends from the **compressor,** through the **condenser,** and to the metering device. From the compressor to the condenser, the refrigerant is in the form of a high-pressure vapor before changing into a high-pressure liquid within the condenser. From the metering device, through the evaporator, and to the compressor's suction side is the low side. The amount of liquid entering the **evaporator** is controlled, which reduces the pressure. The lower pressure and heat from the passing air vaporizes the refrigerant and a low-pressure vapor exits to the compressor. The condenser and evaporator are similar in form to the radiator and heater core, respectively.

An A/C system consists of several more components than the heater system. The cooling agent in an A/C unit is a **refrigerant** circulating under pressure through the system. The refrigerant has properties that may affect the environment and human health. The next section covers the refrigerant and its hazards.

A belt-driven compressor at the front of the engine pressurizes the refrigerant (Figure 12-33). Pressure or temperature switches protect the compressor by switching it on and off as needed. A condenser mounted in front of the radiator cools the refrigerant while a dash-mounted evaporator core allows the cool refrigerant to absorb heat from the air being directed into the passenger compartment. Depending upon the design, the A/C system may have either an expansion valve or an orifice tube to control refrigerant flow into the evaporator.

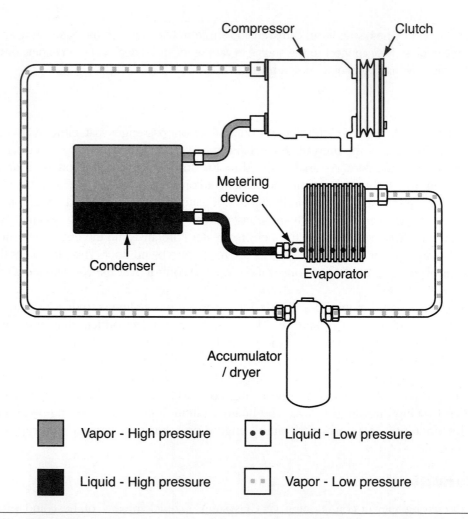

Figure 12-32 The refrigerant changes states from a liquid to vapor (evaporator) and from a vapor to a liquid in the condenser. Differences in refrigerant pressure are a prime reason for these changes.

Figure 12-33 The air-conditioning compressor may be mounted high or low on the engine block and its positioning is based on space within the engine compartment and the design of the engine.

Figure 12-34 The two units shown have similar functions, but are used with different metering devices. On the right is an accumulator/dryer and the left is a receiver/dryer.

In addition, some systems will have an **accumulator/dryer** or **receiver/dryer** to filter and clean the refrigerant (Figure 12-34). An orifice tube is used with accumulator drier systems and expansion valves are used with receiver/dryers. The lines are usually made of aluminum with a high-pressure, flexible-hose section. High-side pressures may reach 300 psi or more on a hot day. Low-side pressures are proportional to the high-side pressures and usually range from 30 psi to 50 psi. The operating pressure rises as the outside temperature increases.

When the A/C switch is closed and the engine is running, current flows through low- or high-pressure switches to the magnetic coil and clutch. The clutch locks the compressor shaft to the pulley driven by the belt. The compressor draws in low-pressure vapor refrigerant from the evaporator, pressurizes it, and pumps it to the condenser. Air flowing over the condenser fins cools and liquifies the refrigerant. The incoming high-pressure vapor at the top of the condenser forces the refrigerant down through and out of the condenser and back to the metering device. Before entering the evaporator, the refrigerant passes through an expansion valve or orifice tube. Too much refrigerant forces the evaporator pressure upward and heat cannot be transferred. If there is too little refrigerant, ice or frost forms on the outside of the evaporator and its outlet. The warm air passing over the evaporator is cooled by transferring its heat to the liquid refrigerant. The refrigerant is heated, vaporized, and loses pressure quickly. The low-pressure vapor is forced by compressor suction through an accumulator/dryer and back into the compressor. The accumulator/dryer has a cleaning filter and a moisture-absorbing material, **desiccant,** to remove any water in the refrigerant.

The expansion valve is a pressure- or temperature-sensitive metering device (Figure 12-35). An expansion valve is used in conjunction with a receiver/dryer. If the pressure or temperature within the evaporator or low side is too high, less refrigerant is allowed to enter the evaporator. The valve may be mounted on the evaporator's intake line or within the evaporator housing.

The orifice tube used with an accumulator/dryer is basically a series of small screens (Figure 12-36). The very small holes in the screens restrict the refrigerant, thereby controlling the volume of refrigerant entering the evaporator. Most orifice tubes are about 2 ½ inches to 3 inches long and roughly the diameter of a pencil. A tube is inserted into the low-side line before the evapo-

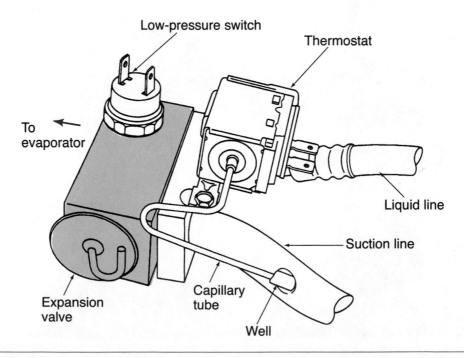

Figure 12-35 This expansion valve reacts to pressure differentials between the liquid and suction lines. The thermostat controls compression clutch operation based on the temperature of the suction line.

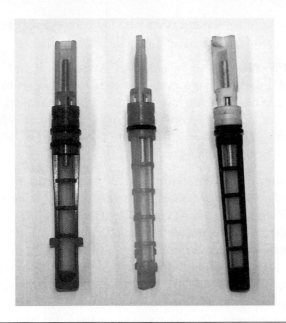

Figure 12-36 An orifice tube is a set of finely holed screens that restrict the movement of refrigerant into the evaporator. It is a fixed metering device.

rator. Most orifice tubes can be replaced by disconnecting the line and pulling the tube out with a special wrench. The new tube is pushed into place and the lines reconnected. Some Ford vehicles have the tube inserted deep into the line. The line must be replaced to replace the tube.

The high-pressure switch is mounted in the high side, either on the line or on the compressor. The low-pressure switch is on the low side and usually mounted on or near the receiver/dryer, but may be mounted on the compressor. Some units use a dual pressure switch to sense both high and low pressures (Figure 12-37). The switches are used to protect the system when the pressures do not meet specifications. It should be noted that many systems do not have high-pressure switches. Instead, they rely on the expansion valve to control high-side pressure. Almost all systems have a low-pressure switch to protect against low refrigerant charge and

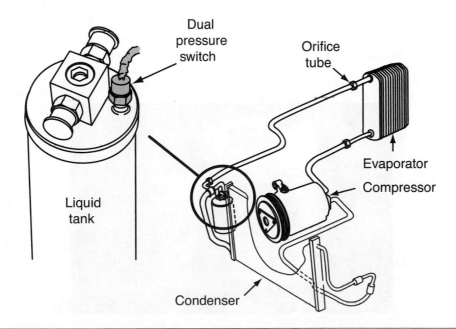

Figure 12-37 A liquid tank is common on Asian imports. The dual pressure switch is becoming more commonplace on many brands and models.

compressor damage. The compressor and other working components are lubricated in part with lubricant that is circulating through the system with the refrigerant.

AUTHOR'S NOTE: Some older A/C systems without a high-pressure cutoff switch used a pressure relief valve. The valve would open suddenly to vent refrigerant if the pressure got too high. Most worked well, but some failed. In either case the venting of refrigerant was loud, produced a cloud of steam (boiling refrigerant), and could be unnerving to the driver. Pressure relief valves are not allowed on any A/C system since the early 1990s because of ozone depletion.

Special oil is added to the system to lubricate the compressor. Most of the lubricant is circulated with the refrigerant to lube the hoses, switches, and the valves of the compressor. The older R12 systems use different oil than the newer R134a units. The R134a units use PAG or Ester oils. The new and old oils cannot be mixed (Figure 12-38).

Since the air-conditioning system is installed in vehicles with a heater, the blower and ductwork are shared (Figure 12-39). In most cases, the air-conditioning controls are on the same panel as the heater controls. The most common A/C controls are buttons labeled "MAX" and "NORMAL," or something similar. The MAX switch turns on the A/C and opens the door that draws air from inside the passenger compartment. This allows cool air from the interior to be circulated back through the system. The passenger compartment is cooled quicker using this setting. The NORMAL setting turns on the A/C, but draws in outside air. Normally MAX is used to quickly cool a vehicle upon startup and NORMAL is used after the initial cool down to keep the vehicle comfortable.

Some vehicles are equipped with automatic temperature control (ATC) systems. The operator selects the desired in-car temperature and the ATC control module will select and operate the climate control system to fulfill the operator's request. The system may operate components of the heater and the air-conditioning system at the same time or use just one of them to keep the temperature at the desired level. The module can also control blower speed operation.

The newest ATC systems allow passengers in different seats to select a temperature for their area. This individual choice provides individual comfort to each passenger instead of just the driver.

AUTHOR'S NOTE: Before we leave the air-conditioning section, let's take a look at a portion of the climate control that many technicians seem to forget. The defrost mode of the climate control uses dehumidified air to clear the windshield of moisture or fog. The air is dehumidified by the air-conditioning system. If a customer states the windshield "fogs over" in damp or cool air, turn on the defroster and check the operation of the A/C

Figure 12-38 The three main types of air-conditioning lubricants.

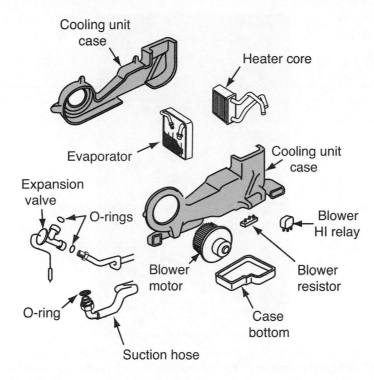

Cooling unit case

Heater core

Evaporator

Cooling unit case

Expansion valve

O-rings

Blower HI relay

Blower motor

Blower resistor

O-ring

Case bottom

Suction hose

Figure 12-39 This package contains both the heater and the air-conditioning evaporator.

Shop Manual
page 391-398

compressor. If it is not working, then the air from the defrost vents will not be dehumidified and will usually make the fogging condition worse. This is discussed more fully in the Shop Manual.

Refrigerant Control and Usage

The Environmental Protection Agency (EPA) is charged with protecting the environment of the United States. One of its responsibilities is clean air. Gas- or diesel-powered vehicles are prime causes of dirty air. The Clean Air Act of 1990, as amended, is the latest law authorizing the EPA to take action concerning air quality. One of the issues addressed is A/C refrigerant.

The **ozone layer** helps protect the earth and its inhabitants from harmful **ultraviolet** rays and some of the heat from the sun. The older air-conditioning systems in homes and automobiles used **carbon tetrachloride** or R12 as the refrigerant (Figure 12-40). If refrigerant is released into the atmosphere, it breaks apart and the chloride rises to the level of the ozone layer. There, the chloride atom acts to break down the ozone molecule. The result is direct ultraviolet radiation reaching the earth's surface.

AUTHOR'S NOTE: Parts cleaners used up to the early 1990s also contained ozone-depleting chemicals. This was particularly true of hand-held sprays that were propelled by a mixture containing a chemical similar to carbon tetrachloride. Cleaning chemicals many times were based on a chloride mixture. All parts-washer chemicals containing chloride have been banned by the EPA and common sense.

The EPA and the Clean Air Act specifically prohibit the release of any refrigerant into the atmosphere, imposing very heavy fines for violators. Technicians working on A/C systems are required to pass an EPA-approved test before they can legally handle refrigerant. The test involves the control of refrigerant, the effect of refrigerant on the ozone layer, and some of the legal requirements

The **ozone layer** is located about 9 to 18 miles (15 to 39 km) above the earth's surface. It protects the earth from the sun's ultraviolet rays.

Ultraviolet is the term given to electromagnetic radiation and is not visible to the human eye. It is dangerous to humans and can cause skin damage.

When the **carbon tetrachloride** molecule is broken by sunlight or age, the chloride atoms rise to the ozone layer and destroy the ozone atoms.

Figure 12-40 A typical container of R12 refrigerant. It also can be purchased in 30-pound and 50-pound containers. The containers are white or grayish-white.

of the Clean Air Act. The test does not include technical information on mechanical diagnosing and repairing an A/C system. The test is available from several approved sources including ASE and MACS. Some parts stores have the test booklet on hand for anyone interested in taking the test. The test is basically open-book, with twenty-five multiple-choice questions. Complete the test and mail the answer sheet with the noted fee to the address given. The results, with an identification card (if passed), are sent back. It should be noted that similar requirements are applied to fixed A/C units like the ones used in homes and businesses and refrigerators, private and commercial. Any person desiring to purchase any refrigerant must show proof that he/she passed the EPA test by presenting the identification card.

The EPA requires very specific actions to occur during **recovery, evacuating,** and **recharging** of an air-conditioning unit. All of the requirements are based on controlling refrigerant leaks into the atmosphere. The requirement allows each shop to make individual decisions on repairing a system or just adding refrigerant to a leaking system. The customer can request that a leaking system be topped off with refrigerant, but the shop may legally refuse to do so.

In addition to testing and control procedures, the EPA directed research and development of an ozone-friendly refrigerant. Research in this area has been ongoing since the mid-1970s, but the growing evidence of ozone depletion has increased the amount of funding and the number of laboratories.

A new refrigerant, R134a, was developed and is being used by most vehicles produced since 1994 (Figure 12-41). R134a consists of hydrogen-based molecules that do not affect the ozone layer. However, regulations on releasing refrigerant still apply and the EPA test on refrigerant handling must be completed before buying or using R134a. Systems using R134a normally run at slightly higher operating pressures and there is a small loss in system efficiency. There are other ozone-friendly refrigerants on the market, but R134a is the one approved by vehicle manufacturers.

In order to encourage the public to use the new refrigerant, federal, state, and local governments have imposed a progressively increasing tax on R12. Since 1990, the price of R12 has gone from 79 cents for a 14-ounce can to 18 dollars or more at the parts store. Most repair centers charge 30 dollars or more for a pound (16 ounces) of R12. Most R12 units can be retrofitted

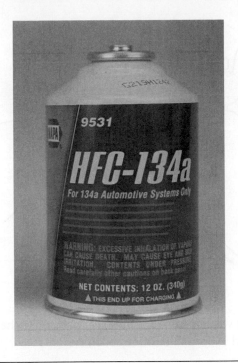

Figure 12-41 R134a containers are light blue and can be bought in 30- and 50-pound containers. The color is required by the EPA as an identification of the chemical inside.

to R134a. During retrofit, the system is emptied of R12 and most of the old lubricant, seals, and gaskets are replaced. The R134a refrigerant and the correct oil is then charged into the systems. Many shops recommend that a damaged R12 system be retrofitted during routine repairs. The savings on the refrigerant will pay for the kit to make the change. The price of the old and new lubricants is about the same.

AUTHOR'S NOTE: In 2005, the price of R12 had dropped greatly from 1999. This price drop is due to the lower number of R12-equipped vehicles now in operation. Most of the older vehicles have been retrofitted, and since 1994 almost all models come equipped with R134. However, China and other nations in Asia and Africa were given a 10-year delay to enact the Montreal Protocol on their nonexported vehicles. They were required to ban ozone-depleting chemicals starting in the year 2000. If the United States is used as an example, it will be about 2010 before all or the vast majority of their vehicles will be R134 equipped.

Restraint Systems

Restraint systems hold passengers in their seats and away from vehicle parts during an accident. This not only protects passengers from further injury, but helps the driver maintain control of the vehicle.

Seat Belts

Lap seat belts have been around for years. One end of the lap belt is anchored to a small axle assembly, anchored to the vehicle frame, and snapped locked into a second anchor at the other end of the belt. The belt, if properly used, reduces the chance of a passenger slamming forward into the steering wheel, dash, or other interior component. But, more important, it would keep the person inside the vehicle. Many people died in survivable crashes because they were thrown from

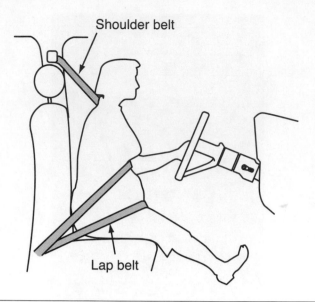

Figure 12-42 Pregnant women do not need to fear damage to the baby provided the seat belts are properly placed under and over the abdominal area.

their vehicle and then run over by the same vehicle. Later, lap belts were used with shoulder belts and the term seat belt came to mean a restraint with both lap and shoulder belts.

The shoulder belt functions similar to the lap belt, but it is anchored at the top of the door-post behind the driver's or passenger's shoulder. It should extend over, down, and across the shoulder and connect to the lap belt at the traveler's opposite hip. The seat belt now has both ends permanently connected together near the person's hip.

The belts are wrapped around a small axle at each main anchor. The axle has a ratchet or lock on one end. One type of ratchet has a small weight that is sensitive to the rapid deceleration that occurs during an accident. During sudden deceleration, the weight swings forward and locks the axle, thereby stopping the movement of the belt. In some cases, a quick jerk on the belt will lock it. Releasing the force on the belt allows the axle to unlock.

There are two major operational problems with seat belts. The critical point to seat belt operation is that vehicle occupants must hand-connect their belts in almost every belt system. If belts are not connected, they do nothing to protect drivers and passengers. A second problem involves pregnant women. A seat belt must be positioned low on the abdomen (Figure 12-42) or injury to the unborn infant might occur. The belt will still work adequately even though it is set low. Seat belts will leave uncomfortable bruises and marks on passengers or drivers during an accident, but they will heal in a few days.

Air Bags

Supplement restraint systems (SRS) are passive restraints requiring no action by the passengers or driver. They are generally referred to as air bag systems and are designed to keep anyone from hitting the front of the dash and the steering wheel. The original system included a single air bag mounted in the steering wheel. Later systems added a passenger air bag mounted in the right side of the dash. Unfortunately, air bags have caused the deaths of some motorists. A shorter person might sit closer to the steering wheel and, as a result, is too close to the deploying air bag. When a child is buckled into a child seat and placed in the front passenger position, the deploying passenger-side air bag could injure or kill the child. A public outcry directed at the National Transportation Safety Board prompted further research into a safer air bag system. Regulations and standards were adopted to allow vehicle manufacturers to reduce the force of air bag deployment.

Figure 12-43 The air bag deactivation switch was commonly used in pickup trucks to turn off the passenger air bag to protect children and infants.

In addition, a switch was installed to turn off the passenger-side bag (Figure 12-43). The new regulations also allow the owner to request that the driver's air bag be deactivated under certain, specific conditions, such as when the regular driver is of smaller build and has to sit closer to the steering wheel. The ideal solution is to reduce the force of deployment to meet individual needs. However, that option is not feasible now, but adjustable pedals help solve part of this problem.

AUTHOR'S NOTE: The federal and state governments now require that children be placed in the back seat in the proper restraint system, either a child seat or seat belts. In pickup trucks without a rear seat, the passenger side air bag can be deactivated. However, to the extent possible using a child seat, seat belts, or moving the passenger seat away from the dash is better than turning off the air bag.

The latest air bag system includes the two bags mentioned previously plus air bags mounted in the doors. The side bags are designed to reduce injuries during a crash to the side of the vehicle.

Air Bag Operation

The air bag system relies on reliable electronics to function properly. It draws power directly from the battery and is turned on by switching the ignition to run. Impact or arming switches are located behind the front bumper and inside the front doors if equipped with side bags.

A crash that is strong enough to trip (close) an arming switch signals the air bag control module to deploy the air bags. A frontal crash will fire both driver and passenger air bags.

The folded air bag is placed over a chemical that will produce a large volume of gas when ignited. When the control module ignites the chemical, it expands and rapidly inflates the air bag. The system requires a backup power source in case the battery is destroyed before the air bag can be deployed. This can cause serious injury or death if the system is not deactivated during repair work.

A capacitor is used as the reserve or backup power. The capacitor is charged each time the ignition is switched on and will store an electrical current rated at about 36 volts. If the battery is destroyed and the capacitor loses power, the capacitor will discharge. This provides enough power to operate the module and deploy the air bags. During normal operation, the capacitor slowly loses its charge after the ignition is switched off. During the drain down, the air bag can still be fired if the capacitor senses a sudden loss of power or receives a surge of electrical current.

Before beginning work inside the vehicle, the capacitor must be drained of electrical current. The acceptable method is to remove the negative cable from the battery and wait awhile. The amount of time is set by the air bag and vehicle manufacturer and is shown in the service manual.

To further protect the technician, it is suggested that the air bags be disconnected after the capacitor has discharged. There are two or more bright yellow plugs located near the bottom of the steering column under the dash. Once the yellow plugs are disconnected, the battery can be reconnected to supply power for testing of the failed components. When all of the repairs are completed and covers are installed, disconnect the negative cable again. The yellow plugs can now be connected without any danger of electric surge.

Terms to Know

Accumulator/dryer
Adjustable Pedal System (APS)
Aim
Analog
Backlit
Bayonet
Bimetallic
Carbon tetrachloride
Center High-Mounted Stop Light (CHMSL)
Climate control system
Communication bus
Compressor
Condenser
Daytime running lights
Desiccant
Digital
Dimmer switch
Driving lights
Emergency flasher
Evacuating
Evaporator
Flasher
Fog lights
Global Positioning Satellite (GPS)
Grounding switch
Heads-Up-Display (HUD)
Instrument panel
Integrated Circuit
Light-Emitting Diodes (LEDs)
Malfunction Indicator Light (MIL)
Master Switch
Microprocessor
Multifunction lever
Ozone layer
Receiver/dryer

Summary

❏ Passenger comfort devices are designed to make the driver and passenger comfortable during vehicle operations.

❏ Passenger conveniences are those items that make it easier for the operator or passenger to control vehicle components or provide some type of service to the passengers.

❏ Warning systems indicate to the driver that a dangerous or damaging condition in the vehicle may occur.

❏ Warning systems have bright red or amber lights and may have audible signals to alert the driver to some condition.

❏ Information systems provide data pertaining to vehicle operations.

❏ Instrumentation usually refers to the lights, gauges, and other devices that present information to the driver.

❏ Information may be conveyed in digital, audible, analog, text, or color displays.

❏ Most exterior lighting is required to meet federal and state regulations concerning illumination and positioning.

❏ Headlights may have either two or four lamps with high and low beams.

❏ Except for the turn signals, all of the rear-mounted lights must be red.

❏ Brake lights override the rear four-way flashers but not the regular turn signals.

❏ A third brake light is mounted high to attract the attention of following drivers.

❏ The only interior light mandated by regulations is the lighting of the instrument panel.

❏ The instrument panel is finished with a non-glare coating.

❏ Most instrument panels use circuit boards.

❏ Each power lock circuit is parallel wired to the other lock circuits.

❏ One fuse protects all of the locking circuits.

❏ Most power door lock systems are also part of the vehicle's antitheft system.

❏ Each of the power window circuits is parallel to each other and the master switch.

❏ The master switch can control all of the windows plus lock out the other windows switches.

❏ Power seats can have four- or six-way positions.

❏ The ability to tilt the seat's bench is the difference between four- and six-way seat systems.

❏ Some seat systems can "remember" a designated position.

❏ Climate control includes heating and air-conditioning systems.

❏ Most A/C units operate at high pressure.

❏ The condenser cools the refrigerant.

❏ The evaporator cools the air by removing heat from it.

❏ The compressor pumps and pressurizes the refrigerant.

❏ Expansion valves or orifices regulate the amount of refrigerant entering the evaporator.

❏ The EPA is responsible for directing research and implementing procedures to protect the environment.

❏ The Clean Air Act of 1990, as amended, charges the EPA to clean vehicle emission discharges.

❏ Chloride in R12 destroys ozone molecules.

❏ R134a is an ozone-friendly refrigerant.

❏ A person must be certified by an EPA-approved test before buying refrigerant or servicing A/C systems.

❏ Seat belts must be connected to protect the motorist.

❏ Supplemental restraint systems are known as air bags.

❏ A capacitor is used as backup power for the air bag system.

Terms to Know
(*continued*)
Recharging
Recovery
Refrigerant
Snapshot
Supplement
  Restraint
  System (SRS)
Trigger
Transponder
Ultraviolet
Warning lights

Review Questions

Short Answer Essay

1. List and label the color of the lights or lenses required on mandated exterior lighting systems. Include options, if any.

2. Describe the major differences between four-way and six-way power seats.

3. Explain how non-remote-controlled power door locks are operated from outside the vehicle.

4. List the major components on the high side of an A/C system.

5. Describe how a high-pressure liquid becomes a low-pressure vapor in an A/C system.

6. Explain the process that makes the turn signal lamps blink.

7. Describe the purpose of the CHMSL.

8. Explain the purpose of the integrated circuit used in the instrument panel.

9. List and describe the purpose of the most common dash-mounted warning lights.

10. Explain how the high and low beams are selected and any light(s) used to alert the driver of the selection.

Fill-in-the-Blanks

1. The left, rear power window switch is _____ _____ _____ with the master switch.

2. The power door locks are protected with a(n) _____ _____.

3. The ignition key slot may be illuminated with a(n) _____ _____.

4. The rear power door locks may be unlocked by turning the door key _____ _____ _____.

5. The instrument panel is _____ and finished with a _____ coating.

6. The colored band type of gauge displays information similar to a(n) _____ gauge.

7. HUD is used to project information onto the _____ in the _____ line of _____.

8. Body computers can be programmed to _____ _____ _____ for a designated _____.

9. The dimmer switch can be mounted on the floor or within the _____ _____.

10. R12 and R134a purchase and use are regulated by the _____ _____ _____ _____ as required by the _____ _____ of _____.

ASE-Style Review Questions

1. Interior lighting systems are being discussed.
   Technician A says the only mandated interior lighting is the instrument panel.
   Technician B says the courtesy and dome lights may be wired to the same door switch.
   Who is correct?
   A. A only
   B. B only
   C. Both A and B
   D. Neither A nor B

2. *Technician A* says the lamp or bulb can be replaced in the sealed beam headlight.
   Technician B says the need for better aerodynamics required the headlight lamps to be redesigned.
   Who is correct?
   A. A only
   B. B only
   C. Both A and B
   D. Neither A nor B

3. The rear lights on a vehicle are being discussed.
   Technician A says the turn signals may be amber or red.
   Technician B says three-lamp systems uses dual-filament bulbs.
   Who is correct?
   A. A only
   B. B only
   C. Both A and B
   D. Neither A nor B

4. Rear dual-filament lamps are being discussed.
   Technician A says one-filament is used for turn and marking lights.
   Technician B says one-filament is used only for brakes.
   Who is correct?
   A. A only
   B. B only
   C. Both A and B
   D. Neither A nor B

5. Climate control systems are being discussed.
   Technician A says the heater uses engine coolant to heat the passenger compartment.
   Technician B says the condenser is mounted in the dash.
   Who is correct?
   A. A only
   B. B only
   C. Both A and B
   D. Neither A nor B

6. *Technician A* says R134a molecules are chloride based.
   Technician B says R134a purchase and use are regulated by the EPA.
   Who is correct?
   A. A only
   B. B only
   C. Both A and B
   D. Neither A nor B

7. Air-conditioning systems are being discussed.
   Technician A says the refrigerant changes from a liquid to a vapor in the condenser.
   Technician B says heat from the passing airflow vaporizes the refrigerant.
   Who is correct?
   A. A only
   B. B only
   C. Both A and B
   D. Neither A nor B

8. The depletion of the ozone layer is being discussed.
   Technician A says R12 released into the atmosphere destroys ozone molecules.
   Technician B says a depleted ozone layer allows more ultraviolet radiation to reach the earth.
   Who is correct?
   A. A only
   B. B only
   C. Both A and B
   D. Neither A nor B

9. *Technician A* says information systems use data from the PCM to compute some information for display. *Technician B* says the information may be computed and displayed by a computer module linked the PCM.
   Who is correct?
   A. A only
   B. B only
   C. Both A and B
   D. Neither A nor B

10. *Technician A* says the expansion valve regulates high-side pressure on an A/C system. *Technician B* says an orifice may be used to control refrigerant flow into the evaporator.
    Who is correct?
    A. A only
    B. B only
    C. Both A and B
    D. Neither A nor B

GLOSSARY

Actuator An electrical device that converts electrical signals into mechanical action.

Activador dispositivo eléctrico que convierte señales eléctricas en actividad mecánica.

Adjustable Pedals System (APS) A system that adjusts the accelerator and brake pedals to match the driver needs.

Sistema de pedales ajustable (APS) sistema que ajusta el acelerador y el freno para adaptarlo a las necesidades del conductor.

Adjustable wrench A wrench that adjusts to fit bolts of different sizes.

Llave ajustable Llave que se puede ajustar para utilizarse con pernos de diferentes tamaños.

AERA Automotive Engine Repair Association

AERA siglas en inglés de la Asociación de Reparaciones Automotriz.

Aerobic A chemical compound that cures in the presence of oxygen.

Aeróbico compuesto químico que cura en presencia de oxígeno.

Aerodynamics The science and study of objects moving through air.

Aerodinámica La ciencia e estudio de los objetos moviendose por el aire.

Air gap The gap between two electrical devices or magnets.

Abertura de aire abertura entre dos dispositivos o magnetos eléctricos.

Air gauge A gauge to check air pressure, commonly to check tire inflation.

Medidor de aire Un medidor para revisar la presión del aire, suele usarse para revisar la inflación de los pneumáticos.

Air-operated ratchet A ratchet handle powered by air for fast removal or installation of bolts or nuts.

Trinquete accionado por aire Palanca para trinquetes accionada por aire y utilizada para remover o instalar rápidamente pernos o tuercas.

Air spring A suspension spring that uses an air bladder and piston connected to the suspension components to provide the spring action.

Muelle de aire Muelle de suspensión que utiliza un saco de aire y un pistón conectados a los componentes de la suspensión para proveer un movimiento amortiguador.

Allen wrench A wrench used to tighten or loosen Allen-head screws.

Llave Allen Llave utilizada para apretar o aflojar tornillos Allen.

Alignment The procedure to align the wheels for reduced tire wear and better steering.

Alineación procedimiento para alinear las ruedas a fin de reducir el desgaste de las llantas y lograr una mejor dirección.

Amber dash warning lights A set of warning lights in the instrument panel to alert the driver to a fault in various operating systems.

Luces ámbar de advertencia juego de luces de advertencia en el tablero de instrumentos utilizada a fin de alertar al conductor sobre cualquier fallo en distintos sistemas de operación.

American Wire Gauge (AWG) A wire standard for conductors used primarily in the United States.

Calibre americano de cables patrón de cable para conductores utilizado principalmente en los Estados Unidos.

Ammeter An electrical meter used to measure amperage or current flow.

Amperímetro Instrumento eléctrico utilizado para medir la intensidad eléctrica o el flujo de corriente.

Ampere The movement of electrical current through an electrical circuit.

Amperaje Movimiento de corriente eléctrica a través de un circuito eléctrico.

Anaerobic A chemical compound that cures in the absence of oxygen when placed under pressure.

Anaeróbico compuesto químico que cura en presencia de oxígeno cuando se lo somete a presión.

Analog A signal that varies from positive to negative over its cycle range.

Análoga señal que varía de positivo a negativo en su rango de ciclo.

Analog meter A meter that uses a pointer and scale for readings.

Instrumento analógico Instrumento provisto de un indicador y una escala para las lecturas.

Anchor A fixed unmovable point to hold devices in place usually associated with the rear brake shoes.

Anclaje punto fijo inamovible utilizado para sostener dispositivos en su lugar usualmente asociados con la zapata de freno trasera.

Antilock brake system (ABS) A system that uses a computer to determine if a wheel is locking up during braking and a control unit to pulse the brakes on the wheel that is in danger of locking up.

Sistema de frenos antibloqueo (ABS) Sistema de frenos que utiliza una computadora para determinar si se produce un bloqueo en alguna de las ruedas durante el frenado y una unidad de mando para pulsar los frenos de la rueda en la cual existe el riesgo de que se produzca un bloqueo.

ARTA The Automatic Transmission Rebuilders Association offers technical certification for automatic transmission technicians.

ARTA La asociación de renovadores de las transmisiones automáticas que provee la certificación técnica para los técnicos de las transmisiones automáticas.

ASE Institute for Automotive Service Excellence, a nationwide certification agency for automotive technicians.

ASE sigla en inglés del Instituto Nacional para la Excelencia del Servicio Automotriz, agencia de certificación nacional para técnicos automotrices.

Aspect ratio The height of the tire's sidewall in ratio to the width of the tire's tread.

Proporción dimensional altura del lateral de una llanta en relación con el ancho de la banda de rodamiento de la llanta.

Atmospheric The air from the surface to about 30 miles above earth.

Atmosférico aire de la superficie de aproximadamente 30 millas sobre el suelo.

Automatic (transmission) A type of transmission that automatically selects the best forward gear based on engine load and vehicle speed.

Automática (transmisión) tipo de transmisión que selecciona en forma automática el mejor engranaje delantero según la carga del motor y la velocidad del vehículo.

Automatic Ride Control (ARC) A suspension system in which the vehicle control and ride comfort is automatically adjusted on data received from various vehicle sensors.

Control automático de marcha (ARC) sistema de suspensión mediante el cual el control del vehículo y la comodidad de marcha se ajustan en forma automática según la información que recibe de varios sensores del vehículo.

Backlit Placing light behind or at an angle to an object being lit. This reduces glare on the object.

Trasiluminado Poniendo la luz detrás o en un ángulo al objeto que se alumbra. Esto disminuye el deslumbre en el objeto.

Ball bearing A bearing made of precision-machined balls trapped in a cage. Commonly used for wheels and axles.

Cojinete balero Un cojinete hecho de las bolas de maquinado preciso atrapados en una jaula. Suelen usarse para las rudas y los ejes.

Ball joints The suspension components used to attach the control arm to the steering knuckles and that allow wheel movement.

Juntas esféricas Componentes de la suspensión utilizados para fijar el brazo de mando a los muñones de dirección y que permiten el movimiento de la rueda.

Ball-peen hammer A type of hammer with a head with one round face and one square face.

Martillo de bola Tipo de martillo que tiene una cabeza con una cara redonda y la otra cuadrada.

Battery One of the main parts of the automotive electrical system that uses two dissimilar metals and an acid to develop electrical energy.

Batería Uno de los componentes principales del sistema eléctrico del automóvil que utiliza dos metales disímiles y un ácido para producir energía eléctrica.

Bayonet A type of locking connection used primarily on automotive lamps; the method of locking a bayonet onto a rifle.

Bayoneta Un tipo de conexión de bloque que se usa primariamente en las lámparas automotrices; el metodo de fijar una bayoneta en un rifle.

Bearing journal The highly machined part of a component where the bearing is fitted.

Muñon del cojinete La parte altamente acabado a máquina del componente en donde se coloca el cojinete.

Bearing load The amount of load a bearing can support without damage to the bearing or load. Loads may be axial or radial.

Carga del cojinete La cantidad de la carga que puede soportar un cojinete sin sostener daños al cojinete o a la carga. Las cargas pueden ser axial o radial.

Bearing races The highly machined surfaces of a bearing where the balls or rollers operate. *See* ball bearing.

Pistas de los cojinetes Las superficies altamente acabadas a máquina de un cojinete en donde operan las bolas o los rodillos. *Ver tambien* cojinete balero (ball bearing).

Bias A tire that is constructed with the plies arranged on a bias crossing each other.

Bias Llanta fabricada con las estrias arregladas para que se crucen entre sí.

Bimetallic Two adjacent metal strips that respond differently to heat.

Bimetálica dos láminas de metal adyacentes que responden en forma diferente al calor.

Biodiesel A mixture of diesel fuel and oils from biological products such as the grease remaining from cooking meats.

Biodiesel mezcla de combustible diesel y aceite de productos biológicos tales como la grasa de la carne cocida.

Block The part of the engine assembly that houses the internal engine components and provides mounting surfaces for external engine components.

Bloque La parte de la ensambladura del motor que contiene los componentes internos del motor y provee las superficies para instalar los componentes externos del motor.

Block configuration An identifying term to designate engine design such as V-block.

Configuración de bloque término identificador utilizado para hacer referencia al diseño de motor, tal como el bloque en V.

Bolt A threaded fastener used with a nut to hold automotive parts together.

Perno Asegurador filetado utilizado con una tuerca para sujetar piezas automotrices.

Bore The diameter of the engine's piston bore usually measured in inches or centimeters.

Diámetro interno diámetro del pistón del motor que en general se mide en pulgadas o centímetros.

Box-end wrench A wrench designed to fit all the way around a bolt or nut.

Llave de cubo estriado Llave diseñada para cubrir completamente un perno o una tuerca.

Brake pads The frictional parts used to contact the rotor during braking to stop the rotor and stop the car.

Cojines de freno Piezas de fricción que hacen contacto con el rotor durante el frenado y ejercen presión sobre éste para detener la marcha del vehículo.

Brake shoes The friction covered parts used in the drum brake system to stop the brake drum from rotating.

Zapatas de freno Piezas cubiertas de fricción utilizadas en el sistema de frenos de tambor con el fin de impedir la rotación del tambor de freno.

Brake vacuum booster Brakes that reduce driver braking effort using engine intake manifold vacuum acting on a diaphragm connected to the master cylinder push rod.

Frenos hidráulicos auxiliares de vacio Frenos que disminuyen los esfuerzos del conductor utilizando el vacio del difusor de entrada del motor que acciona un diafragma conectado a la varilla de empuje del cilindro maestro.

Brake warning light A red light in the dash that alerts the driver to low brake fluid or when the parking brake is on.

Luz indicadora del freno Una luz roja en el tablero de instrumentos que informa al conductor de un nivel bajo de fluido de freno o de que el freno esta en uso.

Brass hammers Tools with heads made from soft brass that are used to hammer on parts without damaging them.

Martillos de latón Herramientas con cabezas hechas de latón suave que se utilizan para martillar piezas sin deteriorarlas.

Bushing A thin-walled, pipe-type component used to support a shaft in the housing, thereby reducing shaft wobble and wear.

Buje un componente tubular de muro delgado que sirve para soportar el eje en la caja asi disminuyendo la oscilación y el desgaste del eje.

Business ethics The practice of conducting business by being fair and honest to customers, employees, and the owner.

ética de negocio La práctica de dirigir el negocio en una manera justa y honesta a los clientes, los empleados y el dueño.

Butt connector An electrical connector used to connect two pieces of wire together.

Extremo de conectador Conectador eléctrico utilizado para conectar dos alambres.

Bypass valve A valve that opens when a specified pressure is achieved. Normally used as a protection device in a hydraulic system.

Válvula de desvío válvula que se abre cuando se alcanza una presión específica. Generalmente se utiliza como mecanismo de protección en el sistema hidráulico.

Caliper The hydraulic component of disc brakes that activates the brake pads that stop the rotor during braking.

Calibre Componente hidráulico de los frenos de disco que acciona los cojines de freno que detienen el rotor durante el frenado.

Camber The inward or outward tilt of a wheel assembly mounted on the vehicle and under load.

Ángulo Camber inclinación ajustable interna o externa de las ruedas cuando el vehículo se apoya sobre ellas.

Camshaft bearing Bearings formed in the shape of a small, highly machined pipe section similar in shape to a bushing, but much stronger and with a better finish.

Cojinete del árbol de levas Los cojinetes formados en la forma de una sección de tubo altamente acabado parecido en forma a un buje pero mucho más fuerte y mejor acabado.

Capacitor An electrical storage device that discharges its current on an electrical signal. Commonly used as backup power for the air bag system.

Capacitador Un dispositivo de almacenaje eléctrico que descarga su corriente con una señal eléctrica. Suele usarse para respaldo de un sistema de bolsa de aire.

Carbon monoxide (CO) A colorless, odorless, invisible gas formed by the burning of hydrocarbon fuels like gasoline.

Monóxido de carbono (CO) gas incoloro, inodoro e invisible que se forma cuando se queman combustibles de hidrocarburos, como la gasolina.

Carbon tetrachloride A nonflammable, colorless, poisonous gas used in refrigerant and spray propellant.

Carbon tetracloruro Un gas no inflamable, sin color, tóxico que se usa en el propelente de atomización y refrigeración.

Carcass The remainder of a tire after most of the tread has been removed. A good carcass can be treaded again in a process called "recapping."

Carcasa Lo que resta de un pneumático después de que la mayoría de la banda de rodadura se haya quitado. Un alma buena puede recibir un rodamiento nuevo en un proceso que se llama "recauchutado."

Caster The forward or backward tilt of the steering knuckle in relation to the vehicle's vertical line.

Ángulo Caster inclinación ajustable delantera o trasera del muñón de la dirección en relación con la línea vertical del vehículo.

Catalyst Action An element that can cause a chemical or physical change in other elements without changing itself. Commonly used in catalytic converters to change reduce harmful elements.

Catalizador Un elemento que puede causar un cambio físico o químico en otros elementos sin cambiarse. Suelen usarse en los convertidores catalíticos para cambiar los elementos nocivos.

Catalytic converter A mechanical device placed in the vehicle's exhaust system to reduce harmful emissions of which the most common are hydrocarbons, carbon monoxide, and nitrogen oxides. *See* catalyst.

Convertidor catalítico Un dispositivo mecánico colocado en el sistema de escape del vehículo para disminuir los emisiones nocivos los más comunes que son los hidrocarburos, el monóxido de carbón, y los óxidos nítricos. *Ver tambien* catalizador (catalyst).

Cavitation The process where a liquid pump draws in air instead of liquid; can cause serious damage to the pump and other components.

Cavitación El proceso por el cual una bomba de líquidos aspira el aire en vez de un líquido; puede causar los daños serios a la bomba y otros componentes.

Center High-Mounted Stoplight A brake (stop) light placed high on the rear of the vehicle in the direct line of sight of following drivers. Intended to reduce rear end collisions by attracting following drivers' attention quicker.

Lámpara de parada montada en alto en centro Un luz del freno (de parada) colocado en alto en la parte trasera del vehículo en línea directa de vista de los conductores que siguen. Su intención es atraer la atención de los conductores más rapidamente y asi disminuir las colisiones traseras.

Center link Part of a parallelogram steering system.

Varilla central parte de la suspensión de paralelogramo.

Center of gravity The point at which an object, theoretically, can be balanced. It is the point where the shape, size, and weight of the object is equal, but not necessarily the physical center.

Centro de gravedad La punta en la cual, teóricamente, ese objeto puede ser balanceado. Es la punta en donde la forma, el

tamaño, y el peso del objeto es iguala, pero que no necesariamente es el centro físico.

Chapman strut A version of MacPherson strut installed on the rear wheels.

Puntal Chapman versión del puntal MacPherson instalado en las ruedas traseras.

Check valve A valve used to allow fluid flow in one direction but not in the other direction.

Válvula unidireccional válvula utilizada para permitir el flujo de fluido en una única dirección.

Chisel A bar of hardened steel with a cutting edge ground on one end; driven with a hammer to cut metal.

Cincel Barra de acero templado dotada de un filo cortante en uno de los extremos y que es accionada por un martillo para cortar metal.

Circuit The path followed by electrical current or hydraulic fluid. Usually includes a power source, conductors, controls, and a load.

Circuito El rumbo que transcorre un corriente eléctrico o un fluido hidráulico. Suele incluir un fuente de potencia, los conductores, los controles, y una carga.

Circuit breaker A circuit protection device that uses a bimetal arm, which moves away from a contact and opens the circuit during overload.

Interruptor para circuito Este dispositivo de protección del circuito utiliza un brazo bimetal que se aleja de un contacto y abre el circuito en caso de una sobrecarga.

Clearance The distance between two surfaces. In this instance, the clearance between the engine's valves and piston with the piston at top dead center and valves opened.

Espacio libre distancia entre dos superficies. En este caso, el espacio libre entre las válvulas del motor y el pistón cuando el pistón se encuentra en el punto muerto central y las válvulas están abiertas.

Climate Control System A term for a vehicle system that can cool or heat the interior. Replaces the term Heating and Air Conditioning System.

Sistema de control de temperatura expresión utilizada para sistemas del vehículo que dan calor o frío a su interior. Reemplaza la expresión Sistema de Calefacción y Aire Acondicionado.

Clutch switch A safety switch that is closed when the clutch pedal on a manual transmission is pressed down. Prevents the engine from cranking when the clutch pedal is up. *See* park/neutral switch.

Interruptor del embraque Un interruptor de seguridad que se cierre cuando el pedal del embraque de una transmisión manual se oprime. Previene que se arranca el motor cuando el pedal del embrague se levanta. *Ver también* interruptor de park/neutral (park/neutral switch).

Coil-near-plug An ignition coil installed near the spark plug.

Bobina cerca de bujía bobina de encendido instalada cerca de las bujías.

Coil-on-plug An ignition coil installed directly onto the spark plug.

Bobina sobre bujía bobina de encendido instalada directamente sobre las bujías.

Coil pack A set of two or more ignition coils in one unit.

Paquete de bobina Un conjunto de dos o más bobinas de ignición en una unedad.

Coil spring A suspension spring made from a round bar of spring steel wound in the shape of a coil.

Muelle helicoidal Muelle de suspensión hecho de una barra redonda de acero templado devanado en forma espiral.

Combination pliers General-purpose pliers with a slip joint for two jaw openings.

Alicates de combinación Alicates para uso general con una junta deslizante para que las tenacillas se puedan ajustar en dos posiciones.

Combination switch A component of the brake system that contains a warning light switch, metering valve, and proportioning valve.

Conmutador de combinación Componente del sistema de frenos que contiene un conmutador para la luz de aviso, una válvula de medida, y una válvula dosificadora.

Combination valve A valve housing that contains different types of valves. Automotive use of this valve is usually found in the brake system.

Válvula de combinación caja de válvulas que contiene distintos tipos de válvulas. En la industria automotriz, el uso de dicha válvula se encuentra en general en el sistema de frenos.

Combination wrench A wrench with one box end and one open end.

Llave de combinación Llave con un cubo estriado y el otro abierto.

Combustion chamber The cavity in the cylinder head where combustion takes place when the piston is at or near top dead center.

Cámara de combustión cavidad situada en el cabezal cilíndrico donde se provoca combustión cuando el pistón se encuentra en el punto muerto superior o cerca de él.

Commands Electrical signals from a computer module to an actuator.

Comandos señales eléctricas del módulo de cómputo a un activador.

Communication buss A computer network on the vehicle that carries electrical sensor signals and electrical commands.

Distribución de comunicación red de ordenadores de un vehículo que cuenta con señales de sensores y comandos.

Compound gear A gear set of three or more gears used to change torque/speed ratios or to change gear rotational direction.

Engranaje compuesto conjunto de tres o más engranajes utilizados para modificar las relaciones de torque/velocidad o para cambiar la dirección de rotación del engranaje.

Compressor A pump used to compress and move the refrigerant in an air conditioning system.

Compresor bomba utilizada para comprimir y mover el refrigerante en un sistema de aire acondicionado.

Computerized service information system System in which a personal computer is used to access service information stored on compact discs.

Sistemas computarizados de información de servicio Sistema en el cual se utiliza una computadora personal para introducir la información de servicio almacenada en discos compactos.

Condenser The part of the air conditioning where high-pressure refrigerant vapor is cooled to a high-pressure liquid.

Condensador parte del aire acondicionado donde el vapor refrigerante a alta presión se enfría hasta lograr un líquido a alta presión.

Conductor A material that will carry electrical current.

Conductor material que conduce corriente eléctrica.

Continuity A complete path for current flow in an electrical circuit.

Continuidad Trayectoria completa para el flujo de corriente en un circuito eléctrico.

Control arms The "A"-shaped suspension components that pivot and allow suspension movement.

Brazos de mando Componentes de la suspensión en forma de "A" que giran y permiten el movimiento de la suspensión.

Controller A computer module programmed to receive electrical sensor signals and issue electrical commands.

Controlador módulo programado por ordenador para recibir señales de sensores eléctricos y emitir comandos eléctricos.

Conversion factor A mathematical factor used to convert English measurements to metric measurements or the reverse.

Factor de conversión Un factor matemático que se usa para convertir las medidas inglesas a la medidas métricas o de reverso.

Coolant A liquid circulated around hot engine parts to remove the heat and prevent damage.

Refrigerante Líquido que circula por las piezas calientes del motor para remover el calor y prevenir averías.

Cords The layers of rubber and other materials used to support the tire's tread.

Cuerdas Las capas de caucho o de otras materias que se usan para soportar el la banda de rodadura del neumático.

Corporate Average Fuel Efficiency (CAFE) A requirement issued by the federal government that is an average fuel mileage for all light vehicles produced by a manufacturer.

Rendimiento corporativo promedio de combustible (CAFE) requisito que el gobierno federal impone y que consiste en el millaje de combustible promedio para todos vehículos livianos que produce un fabricante.

Cotter pin puller A tool used to hook and remove a cotter pin.

Sacapasador Herramienta utilizada para prender y remover un pasador de chaveta.

Crumple zones The sections of a vehicle designed to absorb impact forces during a crash. Used to reduce the amount of impact delivered to the passenger compartment.

Zonas de aplastamiento Las secciones de un vehículo diseñadas para absorbar las fuerzas del impacto durante una colisión. Se usan para disminuir la cantidad del impacto entregado al compartimento del pasajero.

Current The flow of electricity measured in ampere or the number of electrons passing a single point in one second. One ampere is about 168 billion billion electrons.

Corriente El flujo de la electricidad medida en amperios o la cantidad de los electrones pasando una sóla punta en un segundo. Un amperio es aproximadamente 168 billones de electrones.

Daytime Running Lights Low-intensity headlights wired to the vehicle's ignition switch used to alert other drivers of the presence of a vehicle.

Luces diurnas de cruce luces de baja intensidad conectadas al interruptor de encendido del vehículo que se utilizan para advertir a otros conductores de la presencia de un vehículo.

Dealership An automotive business that sells new vehicles and has a shop for repairs. Usually sells and services only certain brands of vehicles.

Distribuidor Un negocio automotríz que vende los vehículos nuevos y que tiene un taller para las reparaciones. Suele vender y ofrecer los servicios para ciertas marcas de vehículos.

Decimal A method used to separate a whole number and part of the number. An example would be 1.1 or one and one-tenths.

Decimal Un metodo que se usa para separar un número integro de una parte de un numero. Un ejemplo sería 1.1 o uno y un decimo.

Desiccant A chemical used to absorb moisture.

Desecante químico utilizado para absorber humedad.

Diagonally split system A system of brakes used to prevent complete brake system failure in event of a leak.

Sistema de separación diagonal sistema de frenos utilizados para evitar una falla completa en el sistema de frenos en caso de fugas.

Diagonal cutting pliers Pliers with cutting edges on the jaw for cutting cotter pins or wire.

Alicates de corte diagonal Alicates dotados de filos cortantes en las tenacillas para cortar pasadores de chaveta o alambres.

Dial caliper A precision measuring device that can be used to measure the inside, outside, or depth of an automotive part.

Calibre de cuadrante Dispositivo para medidas precisas con el cual puede medirse el interior, el exterior, o la profundidad de una pieza automotriz.

Dial indicator A precision measuring tool that registers readings on the face of a dial.

Indicador del cuadrante Herramienta para medidas precisas que registra las lecturas en el cuadrante.

Diaphragm Flat, flexible devices placed between two chambers that can be used to perform or increase the performance of certain actions. Commonly used within the brake vacuum booster to increase the driver's force applied to the brake pedal.

Diafragma Los dispositivos planos y flexibles que se colocan entre dos cámaras que pueden usarse para funciones o para aumentar las funciones de ciertas acciones. Se usan comunmente con el freno de vacío con asistencia para aumentar la fuerza que coloca el conductor al pedal de freno.

Die A tool used to cut internal threads.

Troquel Herramienta utilizada para cortar filetes internos.

Digital Term normally used to express information in numerical form.

Digital Un término que suele usarse para expresar una información en forma numérica.

Digital multimeter A high-impedance electrical measuring instrument.

Multímetro digital instrumento que mide valores elevados de impedancia eléctrica.

Digital storage oscilloscope An expensive engine analyzer that can store and display a vehicle's operating information in digit form.

Osiloscópio de almacenaje Un analizador de motor muy caro que puede almacenar y mostrar la información de operación del vehículo en forma digital.

Dimmer switch A switch used to change between high and low beam headlights.

Interruptor de resistencia regulable Un interruptor para cambiar entre los luces de carretera y los de la ciudad.

Direct current An electrical current that flows one way through the circuit.

Corriente directa corriente eléctrica que fluye en un único sentido dentro de un circuito.

Direct injection A form of fuel injection where the fuel is injected directly into the combustion chamber. Most commonly used on diesel fueled engines.

Inyección directa modalidad de inyección de combustible mediante el cual se inyecta combustible directamente en la cámara de combustión. Se utiliza generalmente en motores diesel.

Distributor A device to direct current or fluid from a central point out to different points. Commonly refers to the ignition system on a gasoline fueled vehicle or the fuel system on a diesel fueled engine.

Distribuidor dispositivo que dirige una corriente o un fluido desde un punto central hacia diferentes puntos. En general, se refiere al sistema de encendido en vehículos a gasolina o al sistema de combustible en motores diesel.

DOT Department of Transportation

DOT siglas en inglés del Departamento de Transporte.

Double flare A type of fitting on a brake line.

Avellanado (abocinamiento) doble tipo de uniones en la línea de frenos.

Double-post lift A lift that has two lift cylinders or lift mechanisms. It may be mounted in the floor or above the floor.

Ascensor de doble poste Un izador que tiene dos cilindros o mecanismos de izar. Puede montarse en el piso o arriba del piso.

Dowel pin A pin used to align adjacent components.

Pasador Dowel pasador utilizado para alinear componentes adyacentes.

Drive gear The gear that transfers engine power to the output (driven) gear. Usually mounted on a shaft driven by the engine and is different is size from the driven gear.

Engranaje impulsor engranaje que transmite potencia al engranaje (impulsado) de salida. Comúnmente está situado sobre un eje propulsado por el motor y su tamaño es distinto al del engranaje impulsado.

Driven gear The output gear that transfers power from the drive gear to the output shaft. Usually mounted on the output shaft and is different in size from the drive gear.

Engranaje impulsado engranaje de salida que transmite potencia desde el engranaje impulsor al eje de salida. Comúnmente está situado sobre el eje de salida y su tamaño es distinto al del engranaje impulsado.

Driving lights Add-on headlights used off-road for greater visibility.

Luces de conducción luces accesorias utilizadas en ruta para mayor visibilidad.

Drum The braking device that is attached to and turns with the wheel. Brake shoes inside the drum are used to slow the drum.

Tambor El dispositivo de frenado que se conecta y gira con la rueda. Las zapatas adentro del tambor se usan para frenar el tambor.

Dual overhead cam An engine that has two or more camshafts mounted in the cylinder head.

Doble árbol de levas encima de cabeza Un motor que tiene dos o más árbol de levas en la cabeza del cilindro.

Dynamic balance A method of electronically dividing a wheel assembly into four equal parts and then balancing each part against each of the other three parts.

Equilibrio dinámico Un metodo de dividir electronicamente una ensambladura de rueda en cuatro partes iguales y luego equilibrar cada parte contra las otras tres partes.

Electrolyte A mixture of sulfuric acid and water in a car battery that is extremely corrosive and dangerous.

Electrolito Mezcla sumamente corrosiva y peligrosa de ácido sulfúrico y agua en la batería de un vehículo.

Electromotive force The force that moves electrons out of their orbit and is measured in volts. May be referred to as "voltage."

Fuerza electromotor La fuerza que mueva los electrones fuera de sus orbites y que se mide en voltios. Puede referirse como "voltaje."

Electronic steering A term used to indicate that the vehicle's power steering system is pressurized by an electric motor and pump controlled by a computer. Sometimes used to indicate a four-wheel steering system.

Dirección electrónica expresión utilizada para indicar que el sistema de dirección asistida del vehículo se presuriza mediante un motor eléctrico y su bombeo es controlado por un ordenador. A veces indica el sistema de dirección de cuatro ruedas.

Emergency flasher The circuit breaker used to make the front and rear turn signals flash together. The correct name is hazard light flasher.

Luz de emergencia El disyuntor de circuito que permite funcionar las señales indicadoras delanteras y traseras a la misma vez. Su nombre correcto es lámpara destello de emergencia.

End play The amount of movement a shaft can make lengthwise within its mountings.

Juego final del cigüeñal La cantidad del movimiento longitudinal que puede hacer un eje en su montadura.

Energy The physical force needed to do work. Energy can not be created nor destroyed but can be converted between types of energy, i.e. electrical to mechanical.

Energía capacidad para producir un trabajo. La energía no puede crease o destruirse, pero puede convertirse en dos tipos distintos eléctrica o mecánica.

Evaporator A device for changing liquid into a vapor.

Evaporativo El proceso de un líquido cambiando al vapor.

Exhaust gas recirculating The process and devices used to route spent exhaust gases back through the engine to reduce combustion chamber temperature that prevents the formation of nitrogen oxides.

Recirculación del gas de escape El proceso y dispositivo que se usa para desviar el gas de escape por el motor para disminuir la temperatura de la cámara de combustión que previene la formación de los óxidos nítricos.

Exhaust pipe The device that routes the exhaust gases from the exhaust manifold on the cylinder head into the exhaust systems.

Tubo de escape El dispositivo que desvía los gases desde el escape del multiple de escape hacia el sistema de escape.

Exhaust stroke The stroke of a piston engine when the exhaust valve is opened, the intake is closed, and the upward movement of the piston forces exhaust gases into the exhaust manifold.

Carrera de escape La carrera de un pistón del motor cuando esta abierta la válvula del escape, la admisión esta abierta, y el movimiento hacia arriba del pistón admite los gases del escape dentro del multiple de escape.

Expansion valve A valve that opens and closes based on either heat or pressure. Commonly used to meter refrigerant in a climate control system.

Válvula de expansión Una válvula que se abre o se cierre según o el calor o la presión. Suele usarse para medir al refrigerante en un sistema de climatizaje controlado.

Extension A tool used with a socket and ratchet to reach deep mounted fasteners.

Extensión herramienta utilizada con un conector y un trinquete para lograr mayor sujeción en el montaje.

Fasteners Small threaded and nonthreaded parts that hold automotive components together.

Asguradores Pequeñas piezas fileteadas y no fileteadas que sujetan componentes automotrices.

Feeler gauge A measuring tool using precise thickness blades to measure the space between parts.

Calibrador de espesores Lámina metálica o cuchilla acabada con precisión de acuerdo al espesor utilizada para medir el espacio entre dos piezas.

File A hardened steel tool with rows of cutting edges used to remove metal for polishing, smoothing, or shaping.

Lima Herramienta de acero templado compuesta de una barra de filos cortantes que se utiliza para desgastar, pulir o alisar el metal.

Fixed caliper A brake caliper that is bolted stationarily and uses two or more pistons to apply the brake pads against the rotor.

Calibre fijo Un calibre de freno estacionario empernado que usa dos o más pistones para aplicar las balatas del freno en el rotor.

Flasher A repeating-action circuit breaker that opens and closes to make the turn signal lamps flash on and off.

Luz de destello Un disyuntor de circuito de acción repetitiva que abre y cierre para hacer alumbrarse y apagarse las lámparas indicadoras de viraje.

Fog lights Add-on lights to aid in visibility during fogging or rainy conditions.

Luces antiniebla luces accesorias para incrementar la visibilidad en condiciones de niebla o lluvia.

Foot-pounds A measurement of twisting force being used to accomplish a task.

Pies libras Una medida de la fuerza giratoria que se usa para cumplir una tarea.

Four-stroke The number of up and down piston movement to complete one cycle.

Cuatro tiempos cantidad de veces en que el pistón sube y baja para completar un ciclo.

Four-wheel steering Mechanical or electronic steering of the rear wheels in conjunction with the front wheels.

Dirección de cuatro ruedas dirección mecánica o electrónica que cuenta con ruedas traseras y delanteras.

Fractions The expression used to describe parts of a whole number. One-half of a whole is expressed as 1/2.

Fracciones La expresión que se usa para describir partes de un número total. La mitad de una totalidad se expresa como 1/2.

Franchise An independent business operating under the regulations and policies of a larger business. A dealership is a franchise-type operation.

Sucursal Un negocio independiente que opera bajo las regulaciones y las policias de un negocio más grande. Un distribuidor es una operación de tipo sucursal.

Friction The dragging force when two materials are moved against each other. The force used to stop the vehicle using the brake system.

Fricción fuerza de arrastre presente cuando dos materiales rozan uno contra otro. La fuerza utilizada para detener el vehículo usando el sistema de frenos.

Fuel cell A type of future vehicle using hydrogen for fuel.

Celda de combustible tipo de vehículo futuro que utiliza hidrógeno como combustible.

Fuel mileage The mileage a vehicle can be driven on a given amount of fuel usually listed as miles per gallon in the United States and Canada.

Millaje de combustible millaje al que puede conducirse un vehículo con determinada cantidad de combustible; generalmente se mide en millas por galón en los Estados Unidos y Canadá.

Fuse A circuit protection device that has a metal strip that melts and opens the circuit during an overload.

Fusible Dispositivo de protección del circuito con una banda de metal que se funde y abre el circuito en caso de una sobrecarga.

Gap The space between two components. Normally the space between two non-magnetic parts on a vehicle.

Abertura espacio entre dos componentes. Comúnmente, el espacio entre dos piezas no magnéticas de un vehículo.

Gaskets A single or multi-piece seal between two adjoining stationary components such as the oil pan and engine.

Juntas sello único o de varias piezas situado entre dos componentes estáticos adyacentes tales como el cárter de aceite y el motor.

General manager The person exercising control over the day-to-day operation of a business. May or may not be the owner.

Administrador general La persona que se encarga de controlar la operación cuotidiana del negocio. Puede ser el dueño o no.

Global Positioning Satellite (GPS) One satellite in a system of satellites that are used to determine the position of an object or a point on earth. Used extensively by aircraft, ships, and military.

Satélite de posición mundial (GPS) Un satélite en un sistema de satélites que se usan para determinar la posición de un objeto o de un punto en la tierra. Se usa extensivamente por los aviones, los buques, y el militar.

Grade markings Markings on a fastener indicating its tensile strength.

Grado marca en el sujetador que indica su resistencia a la tensión.

Grease Term commonly applied to a thick lubricant.

Grasa Un término que se aplica comunmente a un lubricante espeso.

Ground The return path for an electric current to the power source.

Conexión a tierra retorno de una corriente eléctrica a su fuente de energía.

Grounding switch A switch placed on the negative side of a circuit. Used to connect the circuit to ground, thereby completing the circuit.

Interruptor a tierra Un interruptor colocado en el lado negativo de un circuito. Se usa para conectar el circuito a la terra asi completando el circuito.

Hacksaw A type of saw made to cut metal.

Sierra para metales Tipo de sierra fabricada especialmente para cortar metal.

Hall effect switch A sensor that generates an on/off digital electrical voltage or impulse. Commonly used on ignition systems and anti-lock brake systems.

Interruptor de efecto Hall sensor que genera voltaje o impulso eléctrico digital. Comúnmente utilizado en sistemas de encendido y de frenos antibloqueante.

Hazardous waste Any waste that is directly hazardous to health or environment. Most household waste is not hazardous while the majority of industrial waste is.

Desechos tóxicos Cualquier desecho que es directamente tóxico al salud o al medio ambiente. La mayoría de los desechos caseros no son tóxicos mientras que la mayoría de los desechos industriales lo son.

Heads-up-display A method of projecting information onto the windshield in the driver's line of sight.

Presentación proyectado en pantalla Un metodo de proyectar la información en la parabrisas en la línea de vista del conductor.

Horizontally-opposed A type block configuration with the pistons laid out in a pancake fashion.

Horizontalmente opuesto tipo de configuración de bloque en que los pistones se disponen en forma de crepe.

Horsepower The power produced by an engine or motor to drive a system. Usually associated with speed on a vehicle.

Potencia potencia producida por el motor para dirigir un sistema. Generalmente asociado a la velocidad del vehículo.

Hybrid power A combination of internal combustion engine and electric motor to power a vehicle.

Potencia híbrida combinación de motores de combustión interna para dar energía a un vehículo.

Hydraulics The theory, study, and application of using a liquid to transfer or create force.

Hidráulica La teoría, el estudio, y la aplicación de usar un líquido para transferir o crear la fuerza.

Hydrocarbon (HC) Particles found in the vapor formed by the burning of hydrocarbon fuels.

Hidrocarburo (HC) partículas que se encuentran en el vapor y se forman mediante la combustión de combustibles hidrocarbonatos.

Hygroscopic The tendency of certain liquid chemicals to absorb moisture.

Higroscópico tendencia de determinados químicos líquidos a absorber humedad.

Idler arms Part of a parallelogram steering linkage.

Brazo auxiliar parte del varillaje de la suspensión de paralelogramo.

Ignition timing The process of changing spark plug firing based on speed and engine load.

Tiempo de encendido proceso que consiste en disparar el encendido de las bujías, basado en velocidad y carga del motor.

Impact wrench An air-operated wrench that impacts as well as drives a bolt or nut.

Llave de impacto Llave accionada por aire que golpea además de accionar un perno o una tuerca.

Independent suspension A suspension where each wheel can act and react to road conditions without affecting the other wheels.

Suspensión independiente Una suspensión en la cual cada rueda pueda actuar y reaccionar a las condiciones del camino sin afectar a las otras ruedas.

Index line A line on a measuring device where all measures are read.

Línea graduada Una línea en un dispositivo de medición en la cual todas las medidas se leen.

Inertia A term in the laws of motion indicating the natural tendency of an object to do what it is already doing, i.e. remaining stationary.

Inercia término de las leyes de movimiento que indica la tendencia natural de cualquier objeto a permanecer en el estado en que se encuentra, por ejemplo, a permanecer estático.

In-line Block configuration where the pistons and cylinders are arranged in a straight line.

En línea configuración de bloque donde los pistones y cilindros se disponen en una línea recta.

Instrument panel The interior component of a vehicle housing the various gauges and warning lights.

Tablero de instrumentos componente interior del vehículo donde se ubican los distintos marcadores y luces de advertencia.

Insulator A material that will not allow electric current to flow.

Aislante material que no permite la circulación de corriente eléctrica.

Intake stroke The downward movement of the piston when the exhaust valve is closed and the intake valve is closed.

Carrera de admisión El movimiento hacia abajo del piston cuando la válvula de escape esta cerrada y la vávula de admisión esta cerrada.

Integrated circuits A series of circuits implanted onto a base housing many semiconductor devices.

Circuitos integrados serie de circuitos colocados en una base donde se ubican distintos dispositivos semiconductores.

Interference The actions of one part that conflicts or interferes with the movement of another part. A type of fit between two components commonly referred to as "press fitted."

Interferencia Las acciones de una parte que es en conflicto o que interfiere con el movimiento de otra parte. Un tipo de ajuste entre dos componentes que re refiere comunmente como "ajuste de presión."

Investment The amount of time, material, human resources, and funding needed to establish and maintain a business.

Inversión La cantidad del tiempo, material, recursos humanos, y los fondos requeridos para establecer y mantener un negocio.

ISO International Organization for Standardization

ISO siglas en inglés de la Organización Internacional de Normalización.

Isolation/dump solenoid A common term applied to the pressure controller in rear-wheel, anti-lock brake systems.

Solenoide isolación/descarga Un término común dado al controlador de presion en los sistemas de frenado anti-bloqueo de las ruedas traseras.

Jack stands Strong supports placed under a car that has been lifted by a floor jack.

Soportes de gato Soportes fuertes colocados debajo de un vehículo que ha sido levantado con el gato de pie.

Jam hexagon nut A type of fastener that locks onto another fastener preventing either from becoming loose.

Tuerca hexagonal de bloqueo tipo de sujetador que asegura a otro evitando que se afloje.

Jounce The upward movement of the wheel assembly when encountering a high spot in the road surface.

Sacudida movimiento hacia arriba de las ruedas provocado al pasar sobre un punto alto en la superficie de la ruta.

Key A small, hardened piece of metal used with a gear or pulley to lock it to a shaft.

Chaveta Trozo pequeño de metal templado utilizado con un engranaje o con una polea para fijarlo a un árbol.

Keyway A groove machined into a shaft to hold a guide key.

Ranura ranura mecanizada en un eje que sujeta una llave guía.

Kinetic energy Refers to energy being applied.

Kinético energía Refiere a la energía que se aplica.

Leaf spring A type of suspension spring that uses layers of spring steel stacked on each other for spring action.

Muelle de lámina Tipo de muelle de suspensión que utiliza capas de acero templado montadas la una sobre la otra para lograr un movimiento amortiguador.

Lifter A part of the engine's valve train located between the camshaft and pushrod.

Botador parte del tren de válvulas del motor situado entre el árbol de levas y la varilla de engranaje.

Light Emitting Diode (LED) A diode that will emit light when excited by an electric current.

LED diodo que emite luz al recibir estímulo de una corriente eléctrica.

Linear A straight line or a measurement of a straight line or space.

Linear Una línea recta o una medida de una línea o un espacio recto.

Load-sensing proportioning valve A brake valve that can change the amount of braking done on the rear wheel based on the load being carried.

Válvula proporcionador sensor de cargas Una válvula de frenos que puede cambiar la cantidad del frenado por la rueda trasera basado en la carga que se le aplica.

MacPherson strut suspension A suspension arrangement that uses a strut assembly in place of a shock absorber and top control arm.

Suspensión de montantes MacPherson Distribución de la suspensión que utiliza un conjunto de montantes en vez de un amortiguador y un brazo de mando superior.

MACS The Mobile Air Conditioning Society offers technical certification for climate control technicians.

MACS La Sociedad de Acondicionamiento de Aire Móvil ofrece la certificación técnica para los técnicos de control climático.

Magnetic field The magnetic force created by electrical current moving through a conductor.

Campo magnético La fuerza magnética creado por un corriente eléctrico moviendose por un conductor.

Malfunction Indicator Light A dash mounted light that alerts the driver to a malfunction in the electronic controls for the engine and transmission.

Lámpara indicador de fallos Una lámpara montada en el tablero de instrumentos que informa al conductor que existe un fallo en los controles eléctricos del motor y la transmisión.

Manifold Absolute Pressure (MAP) A sensor that converts pressure measurement into an electric current. Used to determine engine load.

Presión Absoluta del Múltiple (MAP) sensor que convierte la presión en corriente eléctrica. Se utiliza para determinar la carga del motor.

Manual (transmission) A type of transmission requiring the driver to match the forward gear with the vehicle speed and engine load.

Manual (transmisión) tipo de transmisión donde el conductor debe ajustar el engranaje frontal con la velocidad del vehículo y la carga del motor.

Mass Size of an object regardless of the weight caused by gravity.

Masa tamaño de un objeto sin considerar el peso causado por la gravedad.

Mass Air Flow (MAF) A sensor that convert air volume and temperature into an electric current. Used to determine engine load.

Flujo de masa de aire (MAF) sensor que convierte el volumen de aire y temperatura en corriente eléctrica. Se utiliza para determinar la carga del motor.

Master cylinder A device that operates when the driver pushes on the brake pedal to force fluid under pressure to activate wheel brake units.

Cilindro maestro Dispositivo que funciona cuando el conductor pisa el pedal de freno para hacer que el fluido bajo presión accione las unidades del freno de la rueda.

Master switch A switch that can control several circuits singularly or together.

Interruptor maestro Un interruptor que puede controlar a varios circuitos individualmentes o juntos.

Mechanic creeper A flat, low-riding platform on which a technician can lay and work under a vehicle.

Tabla rodante del mecánico Una plataforma móvil y bajo en la cual un tecnico puede tenderse para trabajar debajos de un vehículo.

Mechanical advantage The increase in force between the input and output of a hydraulic circuit.

Ventaja mecánica El incremento en fuerza entre la entrada y la salida de un circuito hidráulico.

Meshed The term applied to two or more gears whose teeth work against and with each other.

Endentados El término aplicado a dos o más engrenajes cuyos dientes trabajan contra el uno y otro o juntos.

Meter A gauge that displays measurements such as a water or gas metered in the home. The process of regulating liquids or vapor.

Medidor Un medidor que muestra las medidas tal como uno de agua o gas en la casa. El proceso de regular los líquidos o vapores.

Metering valve A valve used to delay the activation of front disc brakes until rear drum brakes are activated.

Válvula de medida Válvula utilizada para demorar el accionamiento de los frenos de disco delanteros hasta que se accionen los frenos de tambor traseros.

Micrometer A measuring device normally used for measuring the outside diameter of a shaft or similar component.

Micrómetro dispositivo de medición comúnmente utilizado para medir el diámetro externo de un eje o componente similar.

Microprocessor The portion of a computer that receives data, calculates solutions, and issues commands. Commonly referred to as the computer.

Microprocesador parte del ordenador que recibe información, calcula soluciones y emite comandos. En general, hace referencia al ordenador.

Mono leaf spring A spring with only one leaf. May be mounted traversal at a right angle to the vehicle's center line.

Resorte de hoja única resorte compuesto de una sola hoja que puede montarse de modo transversal en un ángulo situado a la derecha con respecto a la línea central del vehículo.

Mud and snow tire A tire tread designed for use in mud and snow. Can produce a loud roaring sound on flat hard surfaces and may lack friction on wet smooth surfaces.

Llanta para fango y nieve llanta diseñada para su uso en el fango y la nieve. Puede producir un sonido rugiente en superficies planas y puede carecer de fricción en superficies húmedas.

Muffler A term applied to a device that reduces noise. A component of the exhaust system.

Amortiguador Un término aplicado al dispositivo que disminuya el ruido. Un componente del sistema de escape.

Multifunction lever A lever mounted on the steering column that has switches for more than one electrical system.

Palanca multifunción Una palanca montada en la columna de dirección que tiene interruptores que controlan más que un sistema eléctrico.

NASCAR A race series sanctioning organization.

NASCAR Una serie de carreras sancionada por la organización.

NATEF The National Automotive Technician Education Foundation certifies automotive training program. A branch of ASE.

NATEF La Fundación de Educación de Técnicos Automotríz certifica una programa de entrenamiento automotríz. Una parte de ASE.

NATEF key A copyrighted symbol for NATEF.

Llave NATEF El símbolo del NATEF protegido por derecho de autor.

Needle bearing A bearing made of small rollers used in drive shafts.

Cojinete de agujas Un cojinete hecho con pequeñas rodillas que se usa en los ejes propulsores.

Needle nose pliers Pliers with long, slim jaws for gripping small objects.

Alicates de puntas de aguja Alicates que tienen tenacillas largas y delgadas para sujetar objetos pequeños.

Neoprene A synthetic rubber produced by the polymerization of chloroprene that is highly resistant to oil, heat, light, and oxidation.

Neoprene Un caucho sintético producido por la polimerización del cloroprene que es altamente resistente al aceite, al calor, a la luz, y a la oxidación.

Nitrogen oxides A harmful emission chemical produced when the combustion chamber temperature exceeds 2500 degrees Fahrenheit.

Óxidos nítricos Una emisión nociva química que se produce cuando la cámara de combustión supera 2500 grados Fahrenheit.

Non-independent suspension A suspension where the suspension and wheels are mounted to a solid axle. The actions of one wheel affect the other wheel.

Suspensión no independiente Una suspensió en la cual la suspensión y las ruedas se montan en un eje sólido. Las acciones de una rueda afectan a la otra rueda.

Non-inference A term applied to two or more working, mated components that do not interfere with the movements or actions of each other. An engine that has sufficient clearance in the combustion chamber so the piston will not hit the valve.

Non interferencia Un término aplicado a dos o más componentes que no interferan con los movimientos o los acciones de cada uno. Un motor que tiene bastante juego en la cámara de combustión para que el pistón no pegará a la válvula.

Non-threaded fastener A pin or key used to connect two or more adjacent components. Not suitable to medium or high torque connections.

Sujetador sin rosca terminal o llave utilizada para conectar dos o más componentes adyacentes. No es apto para conexiones de torque medio o alto.

Nut driver A tool resembling a screwdriver with a socket-type wrench on the drive end.

Llave para tuercas herramienta similar al destornillador compuesta por una especie de llave de vaso en el extremo.

Nuts Fasteners with an internal thread that are used with bolts and studs.

Tuercas Aseguradores con un orificio en que hay labrada una hélice, hecha para que ajuste en ella la de un perno o un espárrago.

Offset screwdriver A screwdriver with blades arranged at an offset to allow driving screws in small spaces.

Destornillador angular Destornillador con las cuchillas arregladas en ángulo para poder darles vuelta a los tornillos en espacios muy limitados.

One-wire return A term used to describe a certain type of electric system. Usually refers to vehicle wiring and direct current voltage systems.

Retorno en línea expresión utilizada para describir cierto tipo de sistema eléctrico. Comúnmente hace referencia al cableado y al sistema de voltaje de corriente directa.

Ohm A measurement of resistant in an electrical circuit or device.

Ohm medida de resistencia en un circuito o dispositivo eléctrico.

One-way clutch A mechanical device in an automatic transmission used to prevent reverse rotation of a gear or gear component.

Embrague de sentido único dispositivo mecánico de la transmisión automática utilizado para evitarla rotación reversa de un engranaje o componente de engranaje.

Open A fault in an electrical circuit in which the conductor is unintentionally disconnected at some point in the circuit.

Interrupción falla en un circuito eléctrico en el cual en conductor queda desconectado en forma involuntaria en algún punto del circuito.

Open-end wrench A wrench with an opening at the end that can slip onto the bolt or nut.

Llave española Llave con una apertura en el extremo que puede introducirse en el perno o la tuerca.

O-ring A flexible seal generally used on or with a rotating component. Named because of its ring shape.

Sello tipo O sello flexible que comúnmente se utiliza en o con componentes giratorios. Se lo denomina de esta forma debido a su forma de anillo.

Outside micrometer A measuring tool designed to make very precise measurements of the outside of a part in either U.S. (English) or metric units.

Micrómetro exterior Herramienta para medidas diseñada con el fin de llevar a cabo medidas sumamente precisas del exterior de una pieza en el Sistema Imperial Británico (EEUU) o el métrico.

Overhead cam (OHC) An engine that has one camshaft mounted in each head. A type of engine valve train.

Árbol de levas sobre cabeza Un motor que tiene un árbol de levas montado en cada cabeza. Un tipo de tren de válvulas del motor.

Overhead valve (OHV) A type of engine valve train where the camshaft is in the block and the valves are in the cylinder head.

Válvula sobre cabeza Un tipo de tren de válvulas de motor cuyo árbol de levas esta en el bloque y las válvulas estan en la cabeza del cilindro.

Ozone layer A layer of the atmosphere at about 30 miles altitude which helps reduce radiation reaching the earth. Can be destroyed by vapors containing chlorides.

Capa de ozono capa de la atmósfera situada aproximadamente a 30 millas de altura que ayuda a reducir la cantidad de radiación. Esta capa puede ser destruida por vapores que contienen cloruros.

Palladium A catalyst element used in catalytic converters. A very expensive metal.

Paladio Un elemento catalítico usado en los convertidores catalíticos. Un metal muy caro.

Parallel circuit An electrical current having more than one path to ground.

Circuito paralelo Un corriente elétrico que tiene más que un rumbo a tierra.

Parallelogram A leaning, rectangular figure. A type of automotive steering system.

Paralelogramo Una figura rectangular inclinada. Un tipo de sistema de dirección automotivo.

Park/neutral switch A starter safety switch located in the shift mechanism of an automatic transmission. May be referred to as the manual lever position sensor.

Interruptor de estacionamiento/neutro interruptor de seguridad de arranque situado en el mecanismo de cambio de la transmisión automática. Puede denominarse sensor de posición de palanca manual.

Payload The amount of weight or load a vehicle can transport within its limits.

Carga máxima La cantidad de peso o de carga que puede transportar un vehículo en sus limites.

Perpetual energy An unending power supply not capable of being achieved at this time.

Energía perpetua suministro de energía incesante que no se puede alcanzar.

Pigtail A replacement part for an electrical connection consisting of the connector and short lead wires.

Flexible pieza de reemplazo de una conexión eléctrica que está compuesta por un conector y pequeños cables.

Pin A small, round length of metal that is driven into two parts to hold them together.

Pasador Pequeña pieza metálica y redonda que se inserta de manera forzada en dos piezas para sujetarlas.

Pin punch A tool used with a hammer to drive out pins from automotive parts.

Punzón de pasadores Herramienta utilizada con un martillo para remover pasadores de piezas automotrices.

Pitch gauge A tool with toothed blades used to match with threads for identification.

Calibrador de paso Herramienta con cuchillas dentadas que se aparean con filetes a fin de identificar el tamaño de los mismos.

Pitman arm Part between the steering gear and linkage on a parallelogram steering system.

Brazo Pitman parte entre la caja de dirección y el varillaje de la suspensión de paralelogramo.

Planetary gear set A set of two gears mounted within a third gear. The gears are ring, sun, and planetary.

Juego de engranajes planetarios Un juego de dos engranajes montados dentro de un tercer engranaje. Los engranajes son de anillo, sol, y planetarios.

Platinum A catalyst in a catalytic converter.

Platino Una catalista en un convertidor catalítico.

Pneumatic Driven or operated by air.

Neumático Impulsado o operado por aire.

Polarity The type of electrical charge either positive or negative.

Polaridad El tipo de carga eléctrica sea positiva o negativa.

Positive crankcase ventilation A system used to vent the pressures in the crankcase during engine operation. An emission control system. Commonly referred to as PCV.

Ventilación positiva del cárter Un sistema que se usa para ventilar las presiones dentro del cárter durante la operación del motor. Un sistem de control de emisiones. Suele referirse como PCV.

Potential energy Energy stored or contained in a stationary object that is released when object goes into motion.

Energía potencial energía acumulada o almacenada en un objeto estático que se libera cuando el objeto se pone en movimiento.

Potentialometer A sensor commonly used to change mechanical motion into electrical current. Most commonly used as an engine throttle position sensor.

Potenciómetro sensor comúnmente utilizado para convertir movimiento mecánico en corriente eléctrica. En general, se usa como sensor de posición regulador del motor.

Pound-foot The amount of force needed to move a one-pound mass one foot.

Libra pie La cantidad de fuerza requerida para mover una masa de una libra a un pie.

Power piston A piston that produces or increases power or force. Commonly refers to the boost piston in brake boosters.

Piston de potencia Un piston que produce o incrementa la potencia o la fuerza. Suele referirse como el piston de asistencia en los frenos con asistencia.

Powertrain Control Module (PCM) The first and still the primary computer on a vehicle.

Módulo de control del tren de potencia (PCM) primer y principal ordenador de un vehículo.

Power stroke The stroke of an engine that delivers power to the crankshaft.

Carrera de potencia La carrera de un motor que provee la fuerza al cigüeñal.

Pressure The force per square inch on a surface. Commonly used in hydraulic systems to determine the amount of force a pressurized liquid exerts against all of the interior walls of a sealed hydraulic system.

Presión La fuerza por pulgada cuadrada en una superficie. Comunmente usado en los sistemas hidráulicos para determinar la cantidad de fuerza con que oprime un líquido contra todos los muros interiores de un sistema hidráulico sellado.

Pressure controllers Electrical actuators used to control hydraulic pressures or flows.

Controladores de presión activadores eléctricos utilizados para controlar la presión o flujo hidráulicos.

Pressure dressing A first-aid bandage used to apply force to a wound to slow the bleeding.

Vendaje con presión Un vendaje de primera cura que se usa para aplicar la fuerza en una herida para controlar las hemorragias.

Preventive maintenance A system of maintenance services that includes routine services and preventive actions to reduce vehicle break down.

Mantenimiento preventivo Un sistema de servicios de mantenimiento que incluye los servicios rutinas y las acciones preventativas para disminuir las averías de los vehículos.

Primary piston The piston in the brake's master cylinder that is closest to the driver; the master cylinder's rear piston.

Piston primaria El piston en el cilindro maestro del freno que queda más cerca al conductor; el piston trasero del cilindro maestro.

Proportioning valve A valve used to lower the hydraulic pressure going to the rear drum brakes to ensure balanced braking.

Válvula dosificadora Válvula utilizada para disminuir la presión hidráulica que se transmite a los frenos de tambor traseros con el propósito de asegurar un frenado equilibrado.

Pushrod The rod between the brake pedal and the brake power booster and/or the master cylinder.

Varilla de empuje varilla localizada entre el pedal de freno y el reforzador de freno y/o el cilindro maestro.

R134a An ozone friendly air conditioning refrigerant required by vehicle manufactures and the EPA.

R134a refrigerante del aire acondicionado que no daña para la capa de ozono y que es requerido por los fabricantes y por la Agencia para la Protección del Medioambiente (EPA).

Rack and pinion steering gear A type of steering gear that uses a toothed rack driven by a pinion gear.

Engranaje de dirección por piñón y cremallera Tipo de engranaje de dirección que utiliza una cremallera dentada accionada por un piñón.

Radial ply tire A tire constructed with plies that run at right angles to the circumference of the tire.

Llanta con la estria radial Llanta fabricada con las estrias arregladas en ángulos rectos con respecto de la circunferencia de la llanta.

Ratchet reamer A reamer fitted with a ratchet handle.

Escariador de llave de trinquete Un escariador que tiene una palanca de trinquete.

Rear wheel antilock An antilock brake system first required on 1988 model light truck and vans. It was the first mass-produced, antilock system sold on U.S. domestic vehicle.

Antibloqueo de la rueda trasera Un sistema de frenos de antibloqueo requerido primero en los modelos de camiones ligeros y vans de 1988. Fue el primer sistema de antibloqueo producido en masa y vendido en un vehículo domestico de los EU.

Rebound The downward movement of the wheel assembly after the tire travels past a bump or when it drops into a depression.

Rebote movimiento hacia abajo de las ruedas luego de que las llantas pasan por un tope o caen en un hoyo.

Receiver/dryer A climate-control component used to filter and clean the refrigerant.

Recibidor/desecador Un componente de control climatizaje que se usa para filtrar y limpiar el refrigerante.

Recharging The process where an air conditioner is recharged or filled with refrigerant.

Recarga proceso mediante el cual se recarga el aire acondicionado con refrigerante.

Recirculating ball and nut The gear mechanism within the parallelogram steering gear normally found on heavier trucks or older light vehicles.

Bola y tuerca de recirculación mecanismo de engranajes dentro de la caja de dirección de paralelogramo que en general se encuentra en camiones pesados o vehículos livianos antiguos.

Recovering The process for capturing the refrigerant of an air conditioning system during service.

Recuperación proceso de captura del refrigerante en un sistema de aire acondicionado durante su reparación.

Rectifier A component in the alternator used to change alternating current to direct current.

Rectificador Un componente en el alternador que se usa para cambiar el corriente alterno al corriente directo.

Refrigerant The medium used to transfer heat from the passenger compartment to the outside air.

Refrigeración medio utilizado para transferir calor desde el compartimiento del pasajero al exterior.

Relay An electrical mechanical switch using a low current to control a high current.

Relevador interruptor de un mecanismo eléctrico en el que se utiliza corriente baja para controlar la corriente alta.

Resistance The opposition to current flow in an electrical circuit.

Resistencia Oposición que presenta un conductor al paso de la corriente eléctrica en un circuito eléctrico.

Resonator A sound-reducing device usually placed near the exit end of the exhaust system.

Resonador Un dispositivo para disminuir el ruido que suele ser colocado cerca de la extremidad de salida del sistema del escape.

Rhodium An element used as a catalyst in a catalytic converter.

Rodio Un elemento usado como catalizador en un convertidor catalítico.

Rivet A soft, metal pin with a head at one end that is used to hold two parts together.

Remache Chaveta blanda de metal con una cabeza a un extremo utilizada para sujetar dos piezas.

Rocker arm The part of the valve train between the top of the push rod and the valve.

Balancín parte del tren de válvulas situado entre la parte superior de la varilla de empuje y la válvula.

Rod bearing The bearing between the piston's connecting rod and the crankshaft. An insert-type bearing.

Cojinete de la biela El cojinete entre la biela del piston y el cigüeñal. Un cojinete de tipo inserción.

Roll pin A non-threaded fastener used to connect low torque components.

Pasador de rodillo sujetador sin rosca utilizado para conectar componentes de bajo torque.

Roller bearing Slender, machined rollers trapped between cages or separators. Used on shafts and wheel assemblies. *See* ball bearing.

Cojinete de rodillos Los rodillos acabado por máquina atrapados en jaulas o en separadores. Se usan en los ejes y las ensambladuras de ruedas. *Ver tambien* cojinete de bolas (ball bearing).

Room temperature vulcanizing A chemical sealant that cures at room temperature. Commonly known as RTV.

Vulcanización a temperatura ambiente Un sellante químico que cura en temperaturas de ambiente. Se conoce comunmente como RTV.

Rotary A component that functions in a rotary motion. An engine that uses rotary actions instead of pistons to produce and transmit power. Mazda's rotary engine.

Rotario Un componente que funciona en un movimiento rotativo. Un motor que usa las acciones rotarias en vez de los pistones para producir y transferir la potencia. El motor rotario de Mazda.

Rotary engine An internal combustion engine that uses one or more rotating rotors instead of linear moving pistons for intake, compression, power, and exhaust.

Motor de rotación motor de combustión interno que utiliza uno o más rotores giratorios en lugar de pistones de movimiento lineal para la admisión, compresión, potencia y escape.

Rotor The rotating component of a disc brake that is attached to the wheel hub and is stopped to stop the car.

Rotor Componente giratorio de un freno de disco que se conecta al cubo y que es retenido para detener la marcha del vehículo.

Runout The amount of wobble in a disc, wheel, or shaft.

Corrimiento La cantidad de oscilación en un disco, una rueda, o un eje.

Scan tool An electronic tool used to communicate with the vehicle's computer to retrieve DTCs and operating data.

Herramienta exploradora Una herramienta electrónica que se usa para comunicar con la computadora del vehículo para recobrar los DTC y los datos de operación.

Scavenging A method of removing exhaust gases on two-stroke engines.

Barrido método usado para quitar gases de escape en motores de dos tiempos.

Scissor lift An automotive lift that functions similar to the actions of a pair of scissors.

Izador tijera Un izador automotríz que funciona parecido al acción de un par de tijeras.

Screw A fastener that fits into a threaded hole.

Tornillo Asegurador que se inserta dentro de un agujero filetado.

Screwdriver A tool used to turn screws.

Destornillador Una herramienta que se usa para dar vuelta a los tornillos.

Sealing lock nuts Nuts that will lock in place and seal the threads against leakage of fluids or gases.

Tuerca de seguridad y sellado tuercas que se cierran y sellan las roscas para prevenir fugas de fluidos o gases.

Secondary Air Injection An older emission control system that injected fresh air into the exhaust system to promote catalyst in the converter.

Inyección de aire secundario Un sistema de control de emisiones anticuado que inyectaba aire fresca al sistema de escape para promover un catalizador en el convertidor.

Secondary circuit The electrical circuit between the ignition coil and the spark plug.

Circuito secundario El circuito eléctrico entre la bobina del encendido y la bujía.

Secondary piston The front piston in the master cylinder. *See* primary piston.

Piston secundario El piston delantero en el cilindro maestro. *Ver tambien* piston primario (primary piston).

Sector shaft The steering component between the ball and nut steering gear and the pitman arm.

Eje del sector componente de la dirección situado entre la bola y la tuerca de la caja de dirección y el brazo Pitman.

Sensor An electrical device used to convert some form of energy into electrical energy.

Sensor dispositivo eléctrico usado para convertir parte de la energía en energía eléctrica.

Series circuit An electrical circuit with only one path for current flow.

Circuito en serie Un circuito eléctrico con un sólo rumbo para el flujo del corriente.

Series-parallel circuit A series circuit that controls or regulates a parallel circuit.

Serie en paralelo Un circuito de serie que controla o regula un circuito paralelo.

Service manager The person responsible for controlling and managing the service department of an automotive business.

Gerente de servicio La person que se encarga del control y la administración del departamento de servicio de un negocio automotríz.

Service manual A paper book or computerized database with technical information and specifications for vehicle repairs.

Manual de servicio Un libro de papel o un archivo computerizado con la información técnica y las especificaciones para las reparaciones en los vehículos.

Service writer A person who meets the customer and initiates the repair order. Reports to the service manager.

Representante de servicio Una persona que recibe a la clientela y inicia el orden de reparación. Da información al gerente de servicio.

Setback The angle formed between two wheels on the same axle when the wheels are not aligned with each other.

Retroceso El ángulo formado entre dos ruedas en el mismo eje cuando las ruedas no son alineadas la una con la otra.

Shank The part of a bolt between the end of the threaded end and the bottom of the bolt head or the non-cutting portion of a drill bit.

Vástago La parte de un perno entre la extremidad de la parte enroscada y la extremidad de la cabeza del perno o la parte de una broca de taladro que no corta.

Shoe web The metal frame of a brake shoe where the friction material is bonded or riveted.

Alma de la zapata El armazón metal de un zapato (balata) de freno en donde esta pegada o remachada la material de fricción.

Sidewall The sides of a tire between the tread and bead.

Lateral laterales de la llanta situados entre la banda de rodamiento y el cordón.

Single-post lift An automotive lift with one lift cylinder, which is usually mounted in the floor.

Izador de un poste Un izador con un cilíndro de izar, que típicamente esta montado en el piso.

Slant-6 A block configuration of an in-line six cylinder engine with the cylinder tilted or slanted to the right.

Motor de 6 cilindros configuración de bloque del motor de seis cilindros en línea donde el cilindro se inclina o se dispone en diagonal hacia la derecha.

Sliding caliper A disc brake caliper that has all of the applied pistons on one side (usually inside the rotor) causing the whole caliper to slide away from the brake rotor which applies the outer pad.

Calibre deslizante Un calibre de freno disco que tiene todos los pistones aplicados en un lado (típicamente dentro del rotor) causando que todo el rotor desliza del rotor de frenos que aplica la balata exterior.

Small hole gauge A measuring tool used with an outside micrometer to measure the inside of small holes.

Calibrador de agujero pequeño Herramienta para medidas utilizada con un micrómetro exterior para medir el interior de agujeros pequeños.

Snap ring An internal or external expanding ring that fits in a groove and works to hold parts together.

Anillo de muelle Anillo interior o exterior expansible que se inserta dentro de una ranura para sujetar piezas.

Snap ring pliers Pliers that fit into internal- or external-type snap rings and allow them to be removed safely.

Alicates de anillo de muelle Alicates que se insertan dentro de anillos de muelle interiores o exteriores con el propósito de permitir una remoción segura de los anillos.

Snapshot Short electronic data records (memory) stored in a scan tool and collected during a road test for playback later. May be called a "movie."

Instantáneo Unos datos cortos eléctronicos (memoria) almacenado en un detector y colectados durante un ensayo sobre carretera para recobrarse despues. Pude llamarse una "pelicula."

Socket wrenches Wrenches that fit all the way around a bolt or nut and can be detached from a handle.

Llaves de cubo Llaves que cubren completamente un perno o una tuerca y que pueden removerse de la palanca.

Solder A tin-lead or copper-zinc metal alloy that melts easily and is used for making electrical connections.

Soldadura Una aleación metálica de estaño-plomo o cobre-zinc que se derrite fácilmente y se usa para formar las conecciones eléctricas.

Solenoid An electrical actuator that converts electrical energy into mechanical energy.

Solenoide activador eléctrico que convierte energía eléctrica en mecánica.

Speed sensor An electronic sensor that signals the PCM of wheel or vehicle speed. Used with antilock brake systems and engine/transmission controls.

Sensor de velocidad Un sensor electrónico que señala al PCM de la rueda de la velocidad del vehículo. Se usa con los sistemas de frenos antibloqueo y los controles de motor/transmisiones.

Spindle A short, machined, high-strength shaft for supporting a wheel assembly. Used on non-driving wheels.

Husillo Un eje, corto, maquinado de acero de alta tensión para soportar una ensambladura de rueda. Se usa en las ruedas que no impulsan.

Splines Internal or external teeth cut in a part used to hold it in place in another part.

Lengüetas Dientes internos o externos labrados en una pieza para que ésta se mantenga en su lugar dentro de otra pieza.

Split ball A type of small hole gauge used with a micrometer to measure small bores.

Bola hendida Un tipo de medidor de hoyos pequeños usado con un mircrómetro para medir los pequeños huecos.

Spring nut A type of fastener that locks in place by the spring action of its internal clips. It is usually installed by pushing it onto a stud.

Tuerca de resorte tipo de sujetador que se asegura mediante la acción de resorte de sus presillas internas. Comúnmente se instala empujándola dentro de un soporte.

Sprung weight The weight of the vehicle supported by the suspension system.

Peso suspendido peso que soporta el sistema de suspensión de un vehículo.

Square-cut O-ring A specially-cut O-ring commonly used as a seal for disc brake caliper and caliper piston.

Sello en O cortado cuadrado Un sello en O cortado especialmente que se usa comunmente como un sello para un calibre de frenos de disco y piston de calibre.

Start The act of the engine beginning to run under its own power. Takes place during and after cranking.

Marcha El acto del motor en funcionar independentemente. Ocurre despues del arranque.

Starter punch A punch with a taper on the end that allows the starting of pin removal.

Punzón iniciador Punzón dotado de un extremo afilado para facilitar la remoción de un pasador.

Static balance Balancing a tire using only two planes of the wheel assembly instead of four quarters. *See* dynamic balancing.

Equilibrio estático Equilibrar un neumático usado dos planos de la ensambladura de la rueda en vez de los cuatro cuadros. *Ver tambien* equilibrio dinámico (dynamic balancing).

Steering axis inclination (SAI) An angle formed by the manufacture and installation of the suspension system. It can not be adjusted without a frame machine and/or replacement of parts.

Inclinación del eje de la dirección (SAI) ángulo formado durante la fabricación e instalación del sistema de suspensión. No se puede ajustar sin ayuda de un armazón y/o reemplazo de piezas.

Stepped feeler gauge A type of feeler gauge where the end is .002 inch thinner than the remainder of the blade. May be referred to as a "go/nogo" gauge.

Galga palpadora de pasos Un tipo de medidor de hoja en el cual la extremidad mide .002 de una pulgada más delgada del resto de la hoja. Puede referirse como un medidor "va/no va."

Stroke The linear distance a piston travels from top dead center to bottom dead center usually measured in inches or centimeters.

Recorrido distancia lineal que recorre el pistón desde el punto muerto superior al inferior; generalmente se mide en pulgadas o centímetros.

Stud A fastener that has threads at both ends and is used with a threaded hole and a nut.

Espárrago Asegurador provisto de filetes a ambos extremos que se utiliza con un agujero fileteado y una tuerca.

Supercharger A belt-driven air compressor that compresses and feeds air into the engine's intake boosting engine output.

Supercargador compresor de aire accionado por correas que comprimen y llevan aire al motor, reforzando su potencia.

Supplemental restraint system The official designation for the air bag system.

Sistema restricción suplementaria La designación oficial del sistema de bolsa de aire.

Supplier A maker and/or seller of automotive parts. May be a local parts store or national manufacturer. *See* vendor.

Abastecedor El fabricador y/o vendedor de las partes automotivos. Puede ser una tienda de refacciones local o un fabricante nacional. *Ver tambien* vendedor (vendor).

Sway bay A suspension component that helps control body lean during cornering. It may be mounted at the front or rear or both.

Bastidor oscilante componente de la suspensión que ayuda a controlar la inclinación al girar. Puede colocarse en la parte frontal, en la parte trasera o en ambas partes.

Switches Electrical devices that switch current on and off and signals a computer module to perform a function.

Interruptores dispositivos eléctricos que interrumpen y restablecen la corriente y emiten señales al módulo del ordenador a fin de ejecutar una función.

Synthetic oil Non-petroleum oil product made of synthetic elements. It is used in most vehicles' components requiring pressurized or splash lubrication.

Aceite sintético aceite que no deriva del petróleo y se obtiene con elementos sintéticos. Se utiliza en la mayoría de los componentes de los vehículos que necesitan lubricación regulada o a chorros.

Tail pipe The section of the exhaust extending from the rear of the muffler to the rear of the vehicle.

Tubo de escape La sección del escape que se extiende de la parte trasera del amortiguador hacia la parte trasera del vehículo.

Tap A tool used to cut external threads.

Aterraje Herramienta utilizada para cortar filetes externos.

Tapered pin A non-threaded fastener. It is tapered to help in the proper attachment of adjacent parts and helps the pin stay locked in place.

Pin cónico sujetador sin rosca. Tiene dicha forma para facilitar la apropiada colocación de partes adyacentes y para asegurar el pin en su lugar.

Tapered roller bearing A roller bearing where the rollers are laid at an angle to the centerline of the bearing, thereby creating a bearing with the hole on one side larger than the hole on the opposite side.

Rodamiento de cilindros cónico Un cojinete de rodillos en el cual los rodillos se colocan a un ángulo al línea central del cojinete creando un cojinete con el hoyo de un lado más grande que el hoyo del lado opuesto.

Taps A set of taps are used to make or clean internal threads in a fastener.

Terrajas Un conjunto de terrajas que se usan para hacer o limpiar las roscas interiores en un asegurador.

Telescopic gauge A tool used along with a micrometer to measure the inside of holes or bores.

Calibrador telescópico Herramienta utilizada con un micrómetro para medir el interior de agujeros o calibres.

Terminal connector An electrical connector installed on the end of a wire to connect it to an electrical component.

Conectador de borne Conectador eléctrico instalado en el extremo de un alambre para conectarlo a un componente eléctrico.

Thermodynamics The theory of using radiant heat to generate mechanical energy.

Termodinámica teoría mediante la cual se utiliza la irradiación de calor para generar energía mecánica.

Thermostatic air cleaner An air cleaner that uses a temperature sensor and vacuum to direct warm air into the filter.

Limpiador de aire termostático Un limpiador de aire que usa un sensor de temperatura y un vacío para dirigir el aire cálido hacia el filtro.

Thimble The rotating end of a micrometer used to move the spindle to and from the work piece.

Manguito La extremidad rotativa de un micrómetro usado para mudar el husillo hacia o del pieza de trabajo.

Threaded fasteners Fasteners that use spirals called "threads" to hold automotive parts together.

Aseguradores fileteados Aseguradores que utilizan espirales llamados "filetes" para sujetar piezas automotrices.

Thread restoring file A special file used to repair or straighten threads with minor damage. *See* thread chasers.

Lija restaurador de roscas Una lija especial que se usa para reparar o enderezar las roscas con daños mínimos. *Ver también* peine de roscar (thread chasers).

Threads-per-inch (TPI) A unit of measurement for USC fasteners.

Rosca por pulgada (TPI) unidad de medida de los tornillos USC.

Throttle Position Sensor (TPS) A potentialometer used to measure the mechanical movement of the throttle valve and transmit that movement by electrical signal to the PCM.

Sensor de la posición de la mariposa del acelerador (TPS) potenciómetro utilizado para medir el movimiento mecánico de la válvula de la mariposa del acelerador y para transmitir el movimiento mediante una señal eléctrica al módulo de control del tren de potencia (PCM).

Tie-rod The steering part between the tie-rod end and the steering gear or to the center link.

Barra de acoplamiento parte de la dirección situada entre el extremo de la barra de acoplamiento y la caja de dirección o la varilla central.

Tire bead The part of the tire that fits and seals against the wheel.

Costura del neumático La parte de un neumático que queda y sella en la rueda.

Toe The difference between the leading edges of the front wheels and the trailing edges of the front wheel. May also be applied to the rear wheels on some vehicles.

Convergencia diferencia entre los extremos principales y de control de las ruedas delanteras. También puede aplicarse a las ruedas traseras de algunos vehículos.

Toe-in When the distance between the leading edges of the tires on the same axle is less than the trailing edges of the same tires. *See* toe-out.

Convergencia Cuando la distancia entre los bordes delanteros de los neumáticos del mismo eje es menos que los bordes traseros de los mismos neumáticos. *Ver también* divergencia (toe-out).

Toe-out When the distance between the leading edges of the tires on the same axle is more than the trailing edges of the same tires. *See* toe-in.

Divergencia Cuando la distancia entre los bordes delanteros de los neumáticos del mismo eje es más que los bordes traseros de los mismos neumáticos. *Ver también* convergencia (toe-in).

Toe-out-on-turn The difference between the leading edges of the steering tires and the trailing edges of the same tires. Is usually greater than toe. Cannot be corrected using alignment methods.

Divergencia en vuelta La diferencia entre los bordes delanteros de las ruedas de dirección y los bordes traseros de los mismos neumáticos. Suele ser mayor que convergencia/divergencia. No puede ser corregido usando los metodos de alineamiento.

Tolerances The amount of allowable error from desired specifications.

Tolerancia La cantidad de error que se permite en los especificaciones deseados.

Tone ring A toothed ring placed around a shaft or inside a rotor that causes a sensor to pulse a signal each time a tooth passes the sensor.

Anillo de tono Un anillo con dientes que se coloca alrededor de un eje o dentro de un rotor que causa que un sensor impulsa una señal cada vez que ese diente pasa por el sensor.

Torque stick A limited torque tool used with an impact wrench to tighten a fastener to a specified torque.

Pala de torsión Una herramienta de torsión limitada que se usa con una llave de impacto para apretar un asegurador a una torsión especifica.

Torque wrench A wrench designed to tighten bolts or nuts to a certain tightness or torque.

Llave dinamométrica Llave diseñada para apretar pernos o tuercas a una tensión o torsión específicas.

Torsion bar A type of suspension spring that uses the twisting of a bar attached to the frame and suspension to control suspension movement.

Barra de torsión Tipo de muelle de suspensión que utiliza el movimiento de torsión de una barra fijada al armazón y a la suspensión para controlar el movimiento de ésta.

Traction control An electronic system that controls power to a spinning wheel. Shares some components with the antilock brake systems.

Control de tracción Un sistema electrónico que controla el poder a una rueda giratoria. Comparte algunos componentes con los sistemas de frenos antibloqueantes.

Transducer An electrical device that changes one form of energy into another.

Transductor dispositivo eléctrico que alterna de una forma de energía a otra.

Transmission Control Module (TCM) The main computer for transmission operation. Works in concert with the PCM and other modules.

Módulo de control de transmisión (TCM) principal ordenador de la operación de transmisión. Trabaja en forma conjunta con el módulo de control del tren de potencia (PCM) y otros módulos.

Transponder A small, electronic device that will transmit its identification code to a receiver when commanded to so by radar or radio signal.

Transpondedor Un dispositivo electrónico pequeño que transmite su código de identificación a un recibidor bajo la comanda de un señal de radar o de radio.

Tread The part of the tire that contacts the road and provides friction for vehicle operation.

Banda de rodadura La parte del neumático que toca el camino y provee la fricción para la operación del vehículo.

Trigger A specific electronic or mechanical command that will cause a specific action to take place.

Gatillo Una comanda específica electrónica o mecánica que causará ocurrir una acción especifica.

Turbocharger An exhaust gas driven turbine and compressor used to boost engine output.

Turbocargador gas de escape emitido mediante turbina o compresor que se utiliza para reforzar la potencia del motor.

Twist drill A drill bit with cutting edges and flukes running in a twisting pattern up the length of the bit's shank.

Barrenadora de columna Una brocade taladro cuyos bordes cortantes y las colas estan puestos para que dan la vuelta alredeor de la longitud del váatago del taladro.

Two-stroke An engine takes two strokes to complete its power cycle.

Dos tiempos todo motor tiene dos tiempos para completar su ciclo.

Ultraviolet Radiation rays whose wavelengths are outside of a human's vision range.

Ultra violeta Los rayos de radiation cuyos ondas son afuera del rango de la vista del ser humano.

Unibody A vehicle design in which the body doubles as most of the vehicle's frame. Uses sub-frames to support the engine and driveline components.

Monocasco Un diseño del vehículo en el cual la carrocería tambien funciona como el armazón del vehículo. Usa los bastidores auxiliares para soportar los componentes del motor y del eje propulsor.

Uniform tire quality grading A system designating the quality of a tire's traction, tread wear, and temperature resistance.

Evaluación uniforme de la calidad de una llanta Sistema que indica la calidad de la tracción de una llanta, el desgaste de la huella, y la resistencia a la temperatura.

Unified system A system of thread classification based on the U.S. (English) system of measurement.

Sistema unificado Sistema de clasificación de filetes basado en el Sistema Imperial Británico (EEUU) de medida.

Unsprung weight Vehicle weight that is not supported by the springs.

Peso no suspendido peso del vehículo que no es soportado por los resortes.

Vacuum brake booster A brake booster using engine vacuum to assist the driver in braking.

Reforzador de frenos al vacío reforzador de frenos que utiliza un aspirador para ayudar al conductor a frenar.

Valve overlap A valve timing method used to hold the exhaust valve open long enough to allow the intake valve to partially open. Aids in boosting engine output and exhaust emissions.

Traslape de válvulas método de tiempos de válvulas utilizado para dejar la válvula de escape abierta lo suficiente como para permitir que la válvula de admisión se abra parcialmente. Ayuda a reforzar la potencia del motor y las emisiones de escape.

Valve timing The electrical or mechanical system for timing the opening and closing of the valves to match piston location within the cylinder.

Sincronización de válvulas sistema eléctrico o mecánico para sincronizar la apertura y el cierre de válvulas a fin de adaptar la localización del pistón dentro del cilindro.

Valve train A group of engine components that open and close the engine valves during the four-stroke cycle.

Tren de válvulas Conjunto de componentes del motor que abre y cierre las válvulas del motor durante el ciclo de cuatro tiempos.

Variable rate springs Vehicle springs that will support more weight the more they are compressed.

Resorte de nivel variable resortes del vehículo que soportan más peso cuando tienen mayor compresión.

V-block A block configuration with the half of the engine's cylinder set at either 60-degrees or 90-degrees to the other half, i.e. V-8.

Bloque en V configuración de bloque mediante el cual la mitad de los cilindros del motor se encuentran a 60 ó 90 grados respecto de la otra mitad; por ejemplo, bloque V-8.

Vehicle speed sensor (VSS) An electronic device that converts the rotary motion of the driveshaft into an electrical current. The PCM and TCM use this signal as one part of the engine and transmission controls.

Sensor de velocidad del vehículo (VSS) dispositivo eléctrico que convierte el movimiento giratorio del eje de transmisión en corriente eléctrica. El PCM y el TCM utilizan esta señal como parte de los controles del motor y de la transmisión.

Vendors A direct seller to the public or businesses or both. Normally not a manufacturer. *See* supplier.

Vendedores Un vendedor directo al público o a los negocios o los dos. No suele ser fabricante. *Ver también* abastecedor (supplier).

Venturi A passageway that narrows and then flares creating a low-pressure below the flare. Most common use is the venturi in a carburetor.

Difusor pasaje que se angosta y se ensancha creando mejor presión debajo del sistema de abocinamiento. El uso más común es el difusor en un carburador.

Vernier scale The 1/10,000ths markings on a micrometer.

Gama Vernier Las calibraciones de 1/10,000 en un micrómetro.

Vise grip pliers Pliers that can be locked on a part to hold it tightly like a hand-held vise.

Tenazas de sujeción Alicates que pueden enclavarse a una pieza para sujetarla con firmeza como un tornillo operado manualmente.

Volatile The flammability of an element.

Volátil La inflamabilidad de un elemento.

Voltage The amount of electrical pressure pushing current through an electrical circuit.

Tensión Cantidad de presión eléctrica que empuja la corriente a través de un circuito eléctrico.

Warning lights Lights, usually red or amber, to alert the driver to a vehicle fault.

Luces de advertencia luces, comúnmente de color rojo o ámbar, utilizadas para advertir al conductor sobre fallas del vehículo.

Washers Fasteners used with bolts, screws, studs, and nuts to protect machined surfaces and prevent nut loosening.

Arandelas Aseguradores utilizados con pernos, tornillos, espárragos, y tuercas para proteger superficies maquinadas y prevenir que las tuercas se aflojen.

Weather-tight connector An electrical connector with seals to prevent moisture or dirt from entering the connector.

A prueba de agua Un conector eléctrico con los sellos para prevenir que entra en el conector la humedad o el polvo.

Wheel cylinder The hydraulic part on the drum brake that activates the brake shoes when the driver pushes the brake pedal.

Cilindro de rueda Pieza hidráulica en el freno de tambor que acciona las zapatas de freno cuando el conductor pisa el pedal de freno.

Worm gear A spiral gear on a straight shaft that drives another gear. Usually found in recirculating ball and nut steering gears.

Engranaje sinfín Un engranaje espiral que propulsa otro engranaje. Se encuentran comunmente en los engranajes de dirección tipo bola y tuerca.

Yoke The part of the drive shaft where one side of the u-joint fits.

Brida La parte de un eje propulsor en donde queda un lado de la junta.

INDEX